T0138440

THE ECONOMICS OF AGRICULTURE

THE ECONOMICS OF AGRICULTURE

VOLUME 1

Selected Papers
of
D. Gale Johnson

Edited by John M. Antle
and
Daniel A. Sumner

THE UNIVERSITY OF CHICAGO PRESS

CHICAGO AND LONDON

D. Gale Johnson is the Eliakim Hastings Moore Distinguished Service Professor Emeritus in the Department of Economics and the College, University of Chicago.

The University of Chicago Press, Chicago 60637
The University of Chicago Press, Ltd., London
©1996 by The University of Chicago
All rights reserved. Published 1996
Printed in the United States of America
04 03 02 01 00 99 98 97 96 1 2 3 4 5
ISBN: 0-226-40172-3 (cloth)

Library of Congress Cataloging-in-Publication Data

Johnson, D. Gale (David Gale), 1916–
    The economics of agriculture / edited by John M. Antle and Daniel
A. Sumner.
        p.   cm.
    Includes bibliographical references and index.
    Contents: v. 1. Selected papers of D. Gale Johnson—v. 2. Papers
in honor of D. Gale Johnson.
    1. Agriculture—Economic aspects.   2. Johnson, D. Gale (David
Gale), 1916–   . 3. Johnson, D. Gale (David Gale), 1916–   —
Bibliography.   I. Antle, John M.   II. Sumner, Daniel A. (Daniel
Alan), 1950–   .   III. Title.
HD1408.J64   1996
338.1—dc20   95-12444
                                                                CIP

∞ The paper used in this publication meets the
minimum requirements of the American National Standard
for Information Sciences—Permanence of Paper for
Printed Library Materials, ANSI Z39.48-1984.

CONTENTS

SUMMARY CONTENTS OF VOLUME 2

The Economics of Agriculture:
Papers in Honor of D. Gale Johnson

PREFACE

As D. Gale Johnson was approaching his seventy-fifth birthday it seemed an appropriate tribute to invite a group of colleagues and former students to participate in a festschrift "workshop" in his honor. The papers presented at that event were subsequently prepared for publication and are contained in the second volume of this series, *Essays on Agricultural Economics in Honor of D. Gale Johnson*. We had originally planned to publish, as introductory chapters to each section, one or two classic research papers by Professor Johnson. That was natural, since the new research contributions in each section drew from earlier work by Johnson. A reviewer of that manuscript was enthusiastic about the plan but went on to make an obvious point that we had recognized but not appreciated fully—the Johnson papers were a highlight of the whole collection and publishing more of those articles would be of significant value in its own right. These *Selected Papers* are the result of the decision to make more D. Gale Johnson more conveniently available.

The task for the editors was to select a few of the papers from among a publication list of over three hundred entries. That job has been enjoyable but difficult. It afforded the opportunity to rediscover some of the most illuminating papers in the field of agricultural economics, but picking out only a relative handful was not trivial. The overriding criterion for inclusion was that the research discussed be of interest and use to current scholars. However, where several papers covered some of the same ground we chose one, thus avoiding some repetition but missing interesting alternative statements of ideas that may have been helpful, as it was when the original literature on a topic was initiated. We have, however, included more than one paper on topics whenever the independent contributions are sufficiently large. We have included pieces that represent all the major subject areas within agricultural economics to which Professor Johnson contributed; the included papers thus provide an indication of the breadth of his contribution.

We found it possible to limit ourselves to selecting from the academic journal articles and papers from collections of articles. Selections from his many books or monographs are not included here, in part because it is difficult to excerpt only a small section of a longer treatment and retain the value of the original. Fortunately, in most cases, as with *Forward Prices for Agriculture* and *Prospects for Soviet Agriculture in the 1980s*, shorter statements of the basic points made in the longer works are available in articles from the same period. In the case of *World Agriculture in Disarray*, the second edition has already become a classic and is widely available. Therefore it has not been necessary to reprint chapters from this important book.

The research represented below spans a half century of work in agricultural economics. The chapters are organized (somewhat arbitrarily) into sections by topic. In each section they are presented in chronological order. This arrangement facilitates appreciation of each individual paper as well as evolution of research on each topic. We are confident that the reader will find this volume as satisfying as we found the time we spent with the original publications themselves. We also expect that after reading these papers, many readers will do as we did: they will use the complete list of D. Gale Johnson publications provided below as a guide to additional reading on these topics.

ACKNOWLEDGMENTS

This volume of selected papers grew out of an idea by some of his students and colleagues to hold a research workshop to honor D. Gale Johnson on the occasion of his seventy-fifth birthday. The Chicago Mercantile Exchange and the University of Chicago deserve thanks for providing organizational support, meeting space, and financial support for the festschrift workshop held on 3 and 4 May 1991. The home institutions of the contributors to volume 2, as well as the Farm Foundation and the American Enterprise Institute, provided financial support for participants' travel to the symposium and for preparation of the manuscripts. The editors would like to thank the University of Chicago Press and a reviewer for the suggestion to prepare a full set of *Selected Papers* as a separate volume.

PART ONE

The Organization and Contribution of Labor Resources

1

Allocation of Agricultural Income

Despite extensive improvements in the available data on agricultural income over the past 15 years, it is still true that we know relatively little about the distribution of agricultural income. Most of the efforts in estimating agricultural income have been devoted to determining agriculture's share in the national income or the income of farm people from farming. These endeavors are obviously necessary first steps. However, the income data required to provide the necessary information for many types of research work and policy formulation are still lacking.

The additional data on agricultural income need to take two directions. One is to determine the functional distribution of agricultural income, i.e., the allocation of agricultural income among the factors of production. The other is the personal (family) distribution of agricultural income.[1] The first is important in analyzing the allocation of resources, while the second is significant in a welfare context.[2]

The major purpose of this paper is to develop a series of estimates of the functional distribution of agricultural income. Though major emphasis is given to net agricultural income, some attention is devoted to the allocation of gross agricultural income. In addition an attempt is made to derive estimates of the net income of commercial farms, defined here to include the 50 percent of American farms producing 85 percent or more of total agricultural output and employing the great bulk of all non-labor resources. It is recognized that the results obtained in all cases depend to a considerable extent upon the particular assumptions chosen to derive the estimates. As a consequence care is taken to spell out the assumptions very specifically.

Reprinted by permission from the *Journal of Farm Economics* 30 (November 1948): 724–49.

I. Allocation of Net Income

The allocation of income is estimated on both a net and gross income basis. Since there is little difficulty in allocating gross income once the net income is allocated, major attention is given to allocating net income.

A. The Assumptions

Estimates of the functional distribution of net agricultural income present a number of difficulties. A very large proportion of all types of agricultural resources do not receive a market return nor have a market determined price. If only one resource were not hired, the return to that resource could be obtained as a residual. With a corporate form of enterprise usually only the capital is supplied in whole or part by the firm, with both "management" and labor being hired. As a consequence a division of income along functional lines can be made with only moderate errors. In agriculture some firms own or control all resources employed and no direct market prices are available. Other firms own part of the land, all of the non-land capital and supply directly all of the labor. In only a limited number of cases would all of the land be rented and all labor hired. As a result a number of assumptions must be made in deriving the estimates.

The basic income series for the calculations is net income from agriculture.[3] This concept is an estimate of the total income produced by agriculture—agriculture's contribution to the national income. Net income from agriculture is the sum of net operator income[4]—including adjustments for changes in value of inventories—wages paid, rent to landlords not on farms,[5] and farm mortgage interest. One minor agricultural income item, interest on non-real estate debt, has not been included because the official statistics do not separate it from a large category of miscellaneous items. However, the resulting error is small.[6]

Three different combinations of assumptions have been used in estimating the division of agricultural income among the resources.

A. The return to land was estimated by "blowing up" the total net rent on rented land to include all land. The extension was made by assuming that the net rent on rented land would bear the same relationship to the total net return on all land as the value of rented land did to the value of all land for each year. The estimates of the relationship of the value of rented land to all land were based on census data, with interpolations being made for intercensus years.[7] The returns to capital were calculated by multiplying the value of non-real estate inventories on January 1 of each year by an estimated rate of interest. The rates of interest used were 6 percent for 1910 to 1934, 5.5 percent for 1935 to 1937, and 5 percent from 1938 to date. The returns to labor were determined as a residual, being the difference between net agricultural income and the computed returns for land and capital.

B. The return to land was determined by multiplying the annual average rate of interest on farm mortgages by the estimated value of farm real estate on January 1 of the current year. The returns to capital were determined as in Method A.

C. To derive an independent estimate of labor returns as contrasted to treating labor as a residual claimant, the assumption was made that all family workers, including the operator, received the same return as hired workers. The income series for hired workers used was the total wage bill paid by farmers. When this labor income series is added to returns to land and capital as calculated under assumptions A or B, the total of the three shares never exactly equals total agricultural income. As a result the labor returns calculated as noted in this paragraph are used only for comparative purposes.

The three combinations of assumptions include most of the procedures used in farm management or farm income studies. A further distinction is sometimes made by separating a management return as a residual after returns to land, capital and unpaid family labor have been estimated. We could, for example, by combining estimated returns to land and capital from Methods A or B and to labor from Method C obtain a residual which might be termed a management return.[8]

B. Method A
The estimated returns to land, capital and all labor as derived from Method A are given in Table I.

C. Method B
The results obtained by applying Method B are shown in Table II.

D. Comparison of Methods A and B
The proportions of the total returns going to land and labor are much more variable under Method B than Method A. This difference results from the assumptions as to the appropriate method of valuing the current contribution of land to total income. Method B assumes that current land values multiplied by the appropriate interest rate indicates the income received by land. However, land values presumably depend upon anticipated returns over an extended period of time and a capitalization ratio that is undoubtedly quite inflexible. In addition, the mortgage rate of interest varies little from year to year, showing little responsiveness to changing agricultural price conditions. Every element entering into the calculation of land returns as in Method B is very inflexible compared to farm income.

Table I. Estimated Distribution of Agricultural Income to Land Capital and Labor According to Method A, 1910–1946

Year	Return to land[1]	Return to capital[2]	Return to labor[3]	Total[4]	Return to land	Return to capital	Return to labor
	(Millions of dollars)				(Per cent of total)		
1910	1,659	430	3,159	5,248	31.6	8.2	60.2
1911	1,716	451	2,586	4,753	37.0	8.6	54.4
1912	1,765	440	3,022	5,227	33.8	8.4	57.8
1913	1,767	481	3,052	5,310	33.4	9.1	57.5
1914	1,822	515	3,137	5,474	33.3	9.4	57.3
1915	2,018	532	2,882	5,432	37.1	9.8	53.1
1916	2,556	546	3,185	6,287	40.6	8.7	50.7
1917	3,711	603	5,631	9,945	37.3	6.1	56.6
1918	3,877	761	6,782	11,420	33.9	6.7	59.4
1919	4,117	838	6,854	11,809	34.9	7.1	58.0
1920	2,557	841	6,713	10,111	25.4	8.3	66.3
1921	1,897	695	2,679	5,271	36.0	13.2	50.8
1922	2,215	542	3,671	6,428	34.5	8.4	57.1
1923	2,506	520	4,265	7,291	34.6	6.9	58.5
1924	2,839	531	3,930	7,300	38.9	7.3	53.8
1925	2,632	530	5,368	8,530	30.9	6.2	62.9
1926	2,437	558	5,271	8,266	29.4	6.8	63.8
1927	2,758	560	4,708	8,026	34.3	7.0	58.7
1928	2,663	595	5,117	8,375	31.8	7.1	61.1
1929	2,590	635	5,191	8,416	30.8	7.5	61.7
1930	1,990	630	3,948	6,568	30.4	9.5	60.1
1931	1,221	503	2,896	4,620	26.5	10.9	62.6
1932	858	399	1,960	3,217	26.7	12.4	60.9
1933	1,207	333	2,384	3,924	30.8	8.5	60.8
1934	1,579	341	2,619	4,539	34.8	7.5	57.7
1935	1,854	319	3,976	6,149	30.2	5.2	64.6
1936	1,995	428	4,084	6,507	30.6	6.6	62.8
1937	1,924	436	4,882	7,242	26.6	6.0	67.4
1938	1,696	416	3,998	6,110	27.8	6.8	65.4
1939	1,854	420	4,081	6,355	29.2	6.6	64.2
1940	1,904	423	4,083	6,410	29.7	6.6	63.7
1941	2,693	442	5,893	9,028	29.8	4.9	65.3
1942	3,877	576	8,547	13,000	29.9	4.4	65.7
1943	4,571	773	10,770	16,114	28.4	4.8	66.8
1944	4,892	823	9,896	15,611	31.3	5.3	63.4
1945	4,981	855	9,981	15,817	31.5	5.4	63.1
1946	5,261	894	12,734	18,889	27.9	4.7	67.4

[1]See Appendix Table I.
[2]See Appendix Table III.
[3]Calculated as a residual.
[4]BAE, "Net Farm Income and Parity Report, 1943," (Washington, USDA, 1944) p. 10 and BAE, *The Farm Income Situation*, June–July, 1947, pp. 22. Includes government payments.

Table II. Estimated Distribution of Agricultural Income to Land, Capital and Labor According to Method B, 1910–1946

Year	Return to land[1]	Return to capital[2]	Return to labor[3]	Total[4]	Return to land	Return to capital	Return to labor
	(Millions of dollars)				(Per cent of total)		
1910	2,088	430	2,730	5,248	39.8	8.2	42.0
1911	2,163	451	2,139	4,753	45.5	8.6	45.9
1912	2,276	440	2,511	5,227	43.5	8.4	48.1
1913	2,346	481	2,483	5,310	44.1	9.1	46.8
1914	2,415	515	2,544	5,474	44.1	9.4	46.5
1915	2,415	532	2,485	5,432	44.4	9.8	45.8
1916	2,621	546	3,120	6,287	41.6	8.7	49.5
1917	2,775	603	6,567	9,945	27.8	6.1	66.1
1918	3,049	761	7,610	11,420	26.6	6.7	66.7
1919	3,327	838	7,644	11,809	28.2	7.1	64.7
1920	4,045	841	5,225	10,111	40.0	8.3	51.7
1921	3,812	695	764	5,271	72.3	13.2	14.5
1922	3,403	542	2,483	6,428	52.9	9.5	37.6
1923	3,373	520	3,398	7,291	46.3	6.9	46.8
1924	3,179	531	3,590	7,300	43.5	7.3	49.2
1925	3,116	530	4,884	8,530	36.5	6.2	57.3
1926	3,041	558	4,667	8,266	36.8	6.8	56.4
1927	2,906	560	4,560	8,026	36.2	7.0	56.8
1928	2,897	595	4,883	8,375	34.6	7.1	58.3
1929	2,873	635	4,908	8,416	34.1	7.5	58.4
1930	2,873	630	3,065	6,568	43.7	9.5	46.8
1931	2,640	503	1,477	4,620	57.1	10.9	32.0
1932	2,234	399	584	3,217	69.4	12.4	18.2
1933	1,843	333	1,748	3,924	47.0	8.5	44.5
1934	1,852	341	2,346	4,539	40.8	7.5	51.7
1935	1,807	319	4,023	6,149	29.3	5.1	65.6
1936	1,729	428	4,350	6,507	26.6	6.6	66.8
1937	1,703	436	5,103	7,242	23.5	6.0	70.5
1938	1,633	416	4,061	6,110	26.7	6.8	66.5
1939	1,561	420	4,374	6,355	24.6	6.6	68.8
1940	1,548	423	4,439	6,410	24.1	6.6	69.3
1941	1,507	442	7,079	9,028	16.7	4.9	78.4
1942	1,590	576	10,834	13,000	12.3	4.4	83.3
1943	1,703	773	13,638	16,114	10.6	4.8	84.6
1944	1,914	823	12,874	15,611	12.3	5.3	82.4
1945	2,227	855	12,735	15,817	14.1	5.4	80.5
1946	2,501	894	15,494	18,889	13.3	4.7	82.0

[1] See Appendix Table II.
[2] See Appendix Table III.
[3] Calculated as a residual.
[4] BAE, "Net Farm Income and Parity Report, 1943." (Washington, USDA, 1944) p. 10, and BAE, *The Farm Income Situations*, June–July, 1947, pp. 22. Includes government payments.

Consequently the land returns indicated by this method are only indirectly related to current conditions impinging upon factors influencing the demand for land for use, rather than for ownership. Differences in the estimates of land returns by the two methods may be taken to indicate the gains and losses involved in farm ownership for any particular year. The farm owner will presumably be at an advantage whenever rental payments are greater than the interest cost, and vice versa.

Rental contracts either directly reflect changes in prices and/or yields, as in crop and livestock share leases, or are subject to fairly continuous bargaining.[9] As a consequence, rental rates are likely to more accurately reflect the value of alternative uses of land for any particular year than a return based on a percentage of land value or price.[10]

Theoretically one would expect approximately equal proportionate short run changes in the returns to labor, land, and capital. Only very modest changes are made in the employment of any one of the three resources from year to year. Part of the reason for the small change in employment is due to the inelasticity of the supply of capital, land, and labor; part is due to technical reasons which make more than modest changes in the total employment of capital and land difficult. Were it not for the existence of capital rationing and uncertainty agriculture could have an elastic supply curve for capital, but in actual fact the supply curve is inelastic. In addition, technical reasons make capital transfers out of agriculture difficult; the capital instruments already in agriculture cannot be readily transferred to other uses. New capital that can be added in the short period can constitute only a small proportion of the existing supply. Additions or subtractions to the supply of land can obviously be made only over a considerable period of time. Because of the interrelationships between the demand for and supply of labor, labor employment remains practically constant from year to year. Since the proportions of the various resources utilized in agriculture do not change appreciably from year to year, the changes in the absolute level of returns to each should be approximately equivalent. This result is obtained for land and labor with Method A, and if we could have a more satisfactory procedure for attributing returns to capital, e.g., a market price for the use of capital instruments such as machinery, the theoretically anticipated result might be obtained for all three resources. As a consequence, Method A has been accepted as superior to Method B in estimating returns to land and labor.

Over the average of the 37 years Method B attributes a larger percentage of agricultural income to land than Method A. This difference results from the fact that, particularly during the early years, net land rentals did not provide a rate of return equivalent to the farm mortgage rate of interest. In other words, farm owners were paying more for farm land than the going rental rates, capitalized at the mortgage rate of interest, would indicate as an appropriate valuation during the period. Two possible explanations may be given for the discrepancy. One was the desire for land ownership for what it was presumed to provide in prestige, social standing and security. The other was the anticipated gain from appreciation in land values, particularly from 1910 to 1919.

Table III provides a comparison of the estimated returns for land (in dollars) derived by the two methods and the percentage net rent was of current land valuation.

Three periods of stability in the rate of return from rents should be noted. One was from 1910–14 when the average return was approximately 4.7 percent as compared with a rate of interest of slightly less than 6.1 percent.[11] The second was during 1924 to 1929 with an average net rental return of approximately 5.5 and an average interest rate of slightly less than 6.2 percent. The third period was from 1935 to 1940 when an average rental return of 5.5 and an average rate of interest of 5 percent prevailed.[12]

The very high rates of return during the two war periods and the low rates during the two major depressions are worthy of note. These reflect how far current market valuations of the use of land may depart from the valuation of the land in terms of ownership. This in turn reflects the extent to which long run considerations override the short run in the case of land prices.

The above estimates of returns to land may be used to make certain judgments about the long run feasibility of current levels of land values. Assuming that land will continue to receive 30 percent of the total net agricultural income and that the capitalization rate is 5 percent, net agricultural income will have to average $10 billions to support a total value of land and buildings of $60 billions.[13] If the capitalization rate is 6 percent, net agricultural income would have to be $12 billions. These incomes should be compared with $9.0 billions in 1941; $13.0 billions in 1942; $15.8 billions in 1945 and $18.9 billions in 1946.

The Department of Agriculture has estimated that under full employment conditions in a normal post war year with retail prices at the 1943 levels net agricultural income would be about $12–12.5 billions. Under conditions of intermediate employment net agricultural income was estimated at about $8.5 billions.[14]

If the general price level remains at or above the 1943 level, it certainly is not outside the realm of possibility that net agricultural income will average in excess $10 billions. It is, of course, impossible to say at this time whether the level of land values of late 1946 are generally too high. However, it does seem clear that much of the alarm over rising land values has been unwarranted. Perhaps such alarm and the associated propaganda

Table III. Estimated Total Net Land Rental, Interest Charge on Current Land Valuation and Rate of Return on Rented Land, 1910–1946

Year	Total net land rental[1]	Interest charge on land[2]	Difference	Rate of return on rented land[3]
		(Millions of dollars)		(Percent)
1910	1,659	2,088	−429	4.7
1911	1,716	2,163	−447	4.8
1912	1,765	2,276	−511	4.7
1913	1,767	2,346	−579	4.6
1914	1,822	2,415	−593	4.6
1915	2,018	2,415	−397	5.1
1916	2,556	2,621	−65	6.0
1917	3,711	2,775	936	8.2
1918	3,847	3,049	798	7.7
1919	4,117	3,327	790	7.5
1920	2,557	4,045	−1,488	3.9
1921	1,897	3,812	−1,915	3.1
1922	2,215	3,403	−1,188	4.1
1923	2,506	3,373	−867	4.8
1924	2,839	3,179	−340	5.6
1925	2,632	3,116	−484	5.3
1926	2,437	3,041	−604	5.0
1927	2,758	2,906	−148	5.8
1928	2,663	2,897	−234	5.6
1929	2,590	2,873	−283	5.4
1930	1,990	2,873	−883	4.2
1931	1,221	2,640	−1,419	2.8
1932	858	2,234	−1,376	2.3
1933	1,207	1,843	−636	3.9
1934	1,579	1,852	−273	4.9
1935	1,854	1,807	47	5.6
1936	1,995	1,729	266	5.9
1937	1,924	1,703	221	5.5
1938	1,696	1,633	63	4.9
1939	1,854	1,561	293	5.5
1940	1,904	1,548	356	5.7
1941	2,693	1,507	1,186	7.7
1942	3,877	1,590	2,287	10.9
1943	4,571	1,703	2,868	12.1
1944	4,892	1,914	2,978	11.5
1945	4,981	2,227	2,745	10.7
1946	5,261	2,501	2,760	10.1

[1]See Table I.
[2]See Table II.
[3]Calculated by dividing estimated net land rental by the estimated value of land and buildings as of January 1 of the current year. Value of land and buildings taken from BAE, "Net Farm Income and Parity Report, 1943" (Washington: USDA, 1944), p. 29 (mimeo.) and BAE, *Farm Income Situation*, June–July, 1947, p. 28.

Table IV. Estimated Percentage Distribution of Agricultural Incoe, Comparisons of Methods A and B, 1910–1946

Period	Method A			Method B		
	Land	Capital	Labor	Land	Capital	Labor
1910–14	33.8	8.7	57.5	43.4	8.7	47.9
1915–19	36.7	7.7	55.6	33.7	7.7	58.6
1920–24	33.9	8.8	57.3	51.0	9.0	40.0
1925–29	31.4	6.9	61.7	35.6	6.9	57.5
1930–34	29.8	9.8	60.4	51.6	9.8	38.6
1935–39	28.9	6.2	64.9	26.1	6.2	67.7
1940–44	29.8	5.1	64.9	15.2	5.2	79.6
1945–46	29.7	5.0	65.3	13.7	5.0	81.3
1910–1946	31.9	7.5	60.6	35.4	7.5	57.1

were important in preventing greater increases than actually occurred. In any case, it seems clear that farmers who bought land as late as the spring of 1945 have made rather excellent purchases.[15]

A condensation of Tables I and II, giving the distribution of total income by five year periods, is given in Table IV. This permits an easier comparison of the two methods, though much of the variability is covered up by the use of averages. For the whole period, labor received 61 percent of total income according to Method A and 57 percent by Method B. Method B results in rather variable estimates of the shares going to land and labor, with labor's share being particularly low in 1920–24 and 1930–34, and particularly high in 1940–44.

According to the results of Method A, labor has been receiving about two-thirds of the total agricultural income in recent years. During 1910–14 labor received about 58 percent of the total income, while from 1925–29 the percentage attributed to labor was approximately 62 percent.

Over the period the percentage share going to land has declined moderately. The percentage attributed to land was about 34 percent from 1910–14; approximately 31 percent during 1925–29 and 29 percent in the years 1935 to 1940. However, most of the lower percentage is explained by the two years 1937 and 1938.

Contrary to what one might expect, the percentage of the total income going to capital apparently has a downward trend. During the first few years of the period the return to capital (mostly machinery and livestock) was about 9 percent of the total. During the last 10 years of the period, the return hardly averaged 6 percent of the total. What this decline in proportion received indicates is difficult to say. The decline is in part the result of the lower interest rate used in the later years (6 percent from 1910 to 1934, 5.5 percent for 1935–36 and 5 percent since), but is also due to the absence of any significant increase in the value of capital assets employed by farmers.[16]

E. Method C

Method C assumes that we may appropriately use market returns to estimate labor returns in a manner similar to that used for land in Method A. The extension of the return for rented land to all land may be accepted with only moderate qualification, but the extension of a return to hired labor to all farm labor involves greater difficulties.

The first problem involved in such an estimate is the choice of the appropriate farm wage rate. The Department of Agriculture currently publishes four separate farm wage rates and a composite rate. Of the four rates the one theoretically comparable with the estimates of net labor return is the monthly wage without board. Since the net farm income figures include the value of home consumption and house rent, any wage rate which included perquisites would presumably be too low. Likewise a monthly rate is perhaps more suitable than a daily rate.

The first difficulty which the writer confronted was that the employment data on hired workers by regions used in calculating a national average wage would be different from the relative regional employments of all workers. In other words, the wage rate was weighted by regional hired employment weights, and since a larger proportion of all workers are hired workers in the regions with the highest wage rates, the actual weighting of the wage rates results in a higher wage rate than if all farm workers were used in calculating the weights. Reweighting the wage rates by total employment reduced the national rate by about 6 percent in recent years.[17]

The second difficulty was the lack of comparability between the estimates of labor income derived by using the monthly wage rate without board and that derived by using the total wage bill as used in the income and expense estimates of the Bureau of Agricultural Economics. When the total wage bill is divided by the average number of hired workers an estimate of hypothetical annual income per worker comparable to that obtained by multiplying a monthly rate by 12 should result. Actually there are marked absolute differences, and relative movements in time are not at all comparable.[18] The decision was made to use estimates of hired labor income calculated from the wage bill.[19]

The third difficulty was the comparability of the estimates of the number of unpaid family workers and the number of hired workers employed in agriculture. Prior to 1935 estimates of the number of people employed in agriculture have been based largely on the Census of Occupations. Since it is more obvious what constitutes a hired farm worker than an unpaid family worker, the estimated number of hired workers may be more accurate than that of unpaid family workers. In addition, many workers doing the equivalent of as little as one or two days work in the week were included as unpaid family workers. This may result in the inclusion of many more "part-time" unpaid workers, such as school children, housewives, and farm operators with non-farm jobs than in the case of hired workers. Further, there was a significant discrepancy between the estimates of the number of unpaid family workers in the Agriculture and Occupation Censuses in 1940. The Census of Agriculture reported 7,941,000 family workers and the Population Census only 6,307,000, some 26 percent less. Much of the difference in estimates was due to the exclusion of a large number of female workers in the Population Census because of a failure to explicitly instruct enumerators with reference to work at chores. A later cooperative study by the Bureau of Census and Bureau of Agricultural Economics tends to substantiate the estimates derived from the Census of Agriculture, if two days or more of work per week is considered sufficient to classify as a full-time worker.[20]

It is perhaps safe to assume that except for periods similar to 1932–1934, the number of hired workers represent essentially full-time workers. However, one can be much less confident in making the same assumption about unpaid family workers. Even if family workers actually worked time equivalent to that of hired workers, the large numbers of young people, women and workers over 65 in the family worker category must reduce the labor effectiveness of the group.[21]

Though one may suspect the comparability of the two employment series, the present writer is not in any position to attempt the systematic analysis which is required. As a result, the BAE employment estimates have been accepted for family workers as well as for hired workers. Table V shows the labor returns determined by assuming family workers receive the same income as hired workers, using the total wage bill as the basis for the estimate. The table also compares these results with those obtained from Method A.

The years in which the differences are positive are years when farm operators and their families received less return per person than hired workers. This condition prevailed in 21 out of the 37 years. Only since 1935 have operator families rather consistently, had a labor return more favorable than hired workers. However, even during this period the difference was of a very small magnitude, except for the war years.[22]

II. Income of Commercial Farmers

Our previous calculations have shown that operators and unpaid family workers have received a labor return no larger than that of hired workers.[23] This particular result forces attention on the effect of including 3,000,000 very small and unproductive farms in our estimates of agricultural income.

Table V. Comparisons of Estimated Labor Returns Derived from Assumptions of Methods A and C, 1910–1946

Year	Total labor return		Difference	Income per worker[3]	
	Method A[1]	Method C[2]		Method A	Method C
	(Millions of dollars)			(Dollars)	
1910	3,159	3,194	−45	259	263
1911	2,586	3,191	−605	215	265
1912	3,022	3,298	−276	246	274
1913	3,052	3,357	−325	251	279
1914	3,137	3,312	−175	261	276
1915	2,882	3,331	−449	241	278
1916	3,185	3,665	−480	265	305
1917	5,631	4,527	1,104	478	384
1918	6,782	5,334	1,448	599	470
1919	6,854	6,042	812	617	544
1920	6,713	7,010	−297	591	617
1921	2,679	4,565	−1,886	235	400
1922	3,671	4,406	−735	322	385
1923	4,265	4,793	−528	377	421
1924	3,930	4,840	−910	345	426
1925	5,368	4,957	411	470	433
1926	5,271	5,052	219	457	438
1927	4,708	4,881	−173	418	434
1928	5,117	4,846	271	450	429
1929	5,191	4,854	337	460	430
1930	3,948	4,447	−499	353	398
1931	2,896	3,515	−619	260	315
1932	1,960	2,590	−630	177	234
1933	2,384	2,315	69	216	210
1934	2,619	2,778	−159	241	256
1935	3,976	3,395	581	356	305
1936	4,084	3,800	284	370	344
1937	4,882	4,302	580	448	395
1938	3,998	4,121	−123	371	382
1939	4,081	4,063	18	380	378
1940	4,083	4,213	−130	386	390
1941	5,893	5,056	847	577	473
1942	8,547	6,654	1,893	822	616
1943	10,770	8,570	2,200	1,049	803
1944	9,896	9,846	50	986	940
1945	9,981	10,681	700	1,014	1,085
1946	12,734	11,824	910	1,272	1,181

[1]See Table I.
[2]See Appendix Table IV.
[3]Determined by dividing columns 2 and 3 by estimated number of farm workers.

Table VI. Estimated Total Income per Farm, Distribution of Income Among Capital and Land and Labor per Farm, and Labor Return to Family and Hired Workers per Worker, for Commercial Farms, 1939–1946

Year	Total income per farm[1]	Capital and land income[2]	Labor income[3]	Labor income per family worker[4]	Labor income per hired worker[5]
			(in dollars)		
1939	1,850	655	1,195	615	395
1940	1,865	675	1,190	602	400
1941	2,625	910	1,715	930	515
1942	3,780	1,300	2,480	1,370	690
1943	4,685	1,555	3,130	1,725	875
1944	4,540	1,660	2,880	1,540	1,005
1945	4,600	1,695	2,905	1,645	1,095
1946	5,490	1,790	3,700	2,065	1,195

[1]For the period 3,130,000 commercial farms were assumed to receive 91 percent of total agricultural income. Estimates for 1939 calculated from L. J. Ducoff and M. J. Hagood, "Differentials in Productivity and in Farm Income of Agricultural Workers by Size of Enterprise and by Regions," U.S.D.A., August, 1944, Mimeo, p. 21 and unpublished data supplied by Mr. Ducoff.

[2]Capital and land income assumed to be the same proportion of farm income as derived from Method A for all farms for each year. This may result in a slight overestimate of capital and land income since these farms used about 85 per cent of total capital in 1939 while receiving 91 per cent of the income.

[3]Obtained by subtracting land and capital income from total income.

[4]Total number of family workers was assumed to be 4,665,000 for 1939. This was approximately 57 per cent of all family workers in 1939. In subsequent years the rate of change in family workers on commercial farms was assumed to be one-third of the change in all family workers. Estimate for 1939 was taken from L. G. Ducoff, *Wages of Agricultural Labor in the United States*, USDA, Tech. Bull. No. 895, July, 1945, p. 10.

[5]Total number of hired workers was assumed to be 2,235,000 in 1939. This was approximately 86 per cent of all hired workers in that year. In subsequent years the number of hired workers was assumed to change at one-half the rate of the number of all hired workers. It was assumed that commercial farmers paid 90 per cent of the total wage bill in 1939 and 1940, 91 per cent in 1941 and 1942, 93 per cent in 1943, and 94 percent in 1944, and 95 percent in 1945 and 1946.

The comparative labor returns of hired workers and family workers do not indicate that those operators hiring labor receive a labor income no larger than paid to hired workers. The low earnings of family workers compared with hired workers is obtained because of the inclusion of a large number of subsistence, part-time, and resident farms in the total.[24]

An attempt has been made to carry through the particular assumptions of Method A for the commercial sector of American agriculture. The results are presented without any pretensions of having achieved a high degree of accuracy, but in the belief that some little light is thrown on a significant issue usually ignored in agricultural income data. Commercial farms have been defined to include all farms selling, trading, or using more than $600 worth of products in 1939. There is no particular merit in the lower limit chosen except that it is commonly used and results in the inclusion of about half of all farms.

Table VII. Total Income per Farm and Labor Income per Worker, All Farms and
Commercial farms, 1939–1946
(in dollars)

Year	Total income per farm		Labor income per worker	
	All[1,4]	Commercial[2]	All[1,4]	Commercial[3]
1939	1,020	1,850	380	545
1940	1,050	1,865	386	545
1941	1,485	2,625	577	795
1942	2,160	3,780	822	1,150
1943	2,725	4,685	1,049	1,455
1944	2,660	4,540	986	1,375
1945	2,700	4,600	1,014	1,425
1946	3,225	5,490	1,272	1,795

[1]1939–1943 calculated from BAE, "Net Farm Income and Parity Summary: 1943," pp. 8 and 12. 1944–1946 from *The Farm Income Situation*, June–July, 1947, pp. 22 and 24.

[2]From Table VI, Method A.

[3]Calculated from Table VI.

[4]These columns do not represent the income of the non-commercial farms but all farms including commercial farms. Income of non-commercial farmers are much less. In 1946, for example, the labor income on non-commercial farms was of the magnitude of $300 to $350.

On the commercial farms, operators and other family workers apparently received a labor return almost twice that of hired labor. (See Table VI.) The war period, of course, has resulted in a very marked improvement in the position of all farm workers.

A comparison may be made between the incomes on commercial farms and all farms. Table VII shows the total income produced per farm and the labor income per worker. There is a greater difference in total income per farm than in labor income because the commercial farms, while comprising 50 to 55 percent of all farms, use approximately 65 percent of all labor. Labor returns tend to be approximately a third higher on the commercial farms than on all farms.

Much might be said for a revision of our agricultural income data to provide a separate series for commercial farms defined in some reasonable manner. In terms of public policy these farms present very different problems than those of the various categories of smaller farms. Current income data are of little help in determining the income position of the farms that produce the great bulk of the agricultural output. Furthermore, the wholly inadequate income position of many other farm groups tends to be covered up during periods similar to the present by the significant income strides made by the commercial farmers.

Table VIII. Allocation of Gross Agricultural Income[1]
(in percent)

	1910–14	1915–19	1920–24	1925–29	1930–34	1935–39	1940–44	1945–46
Products and services purchased[2]	12.2	11.6	16.1	15.6	19.6	16.3	13.6	12.9
Taxes	3.2	2.9	5.2	5.1	6.8	4.7	2.9	2.5
Depreciation[3]	10.3	9.1	12.6	10.2	12.8	10.5	8.9	9.3
Net income	74.3	76.3	66.1	69.1	60.8	68.5	74.6	75.3
Land	24.4	27.7	21.9	21.4	17.6	18.9	22.2	22.4
Capital	6.5	5.9	5.9	4.8	6.0	4.1	3.7	3.8
Labor	43.5	42.7	39.5	42.8	37.3	43.6	48.7	49.1

[1]Gross agricultural income is cash income from marketings adjusted for changes in inventory, value of home consumed products, rental value of homes and government payments minus the value of livestock and feed purchased.

[2]Includes all products and services purchased from non-farmers for use in current production. Excludes livestock and feed purchased.

[3]The depreciation item may or may not be equal to actual expenditures on buildings, machinery and motor vehicles. During the following periods net investment occurred: 1910–14, 1915–19, 1925–29, 1935–39, and 1940–44, while in the other two periods actual disinvestment occurred. Depreciation does not include the cost of maintaining soil fertility.

III. Allocation of Gross Agricultural Income

The allocation of gross agricultural income made below is essentially mechanical, accepting as valid the estimated division of gross income into production costs and net income as calculated by the Bureau of Agricultural Economics. Production costs are here interpreted to cover three types of costs: (1) Products or services purchased from non-agricultural industries for use in current production, (2) property taxes, and (3) depreciation or maintenance. We have excluded livestock and feed purchased as a production cost on the assumption that such purchases represent transfers solely within American agriculture. Two modest errors are involved in making this assumption. First, some feed and feeder livestock are imported. Second, even on domestic feed and livestock purchases a part of the purchase price represents payments to non-agricultural industries. Such payments would include transportation, commissions, and services of feed processing industries.

The estimated percentage distribution of gross agricultural income is given in Table VIII for five year periods from 1910 through 1944 and for 1945–46. Net income reached an average low of 60.8 percent of gross income during 1930–34 and an average high of 76.3 percent in 1915–19.[25] The high was closely approached by experience during 1940–44. During the inter-war period taxes accounted for a considerably larger proportion of gross income than before World War I. Products and services purchased also increased, but not to the same extent as taxes. Depreciation apparently did not have a trend during the period, being approximately the same in

1910–14, 1925–29, and 1935–39, and showing similar relationships in each of the two war periods and in each of the two depression periods.

The net income items, land, capital, and labor, show about the same relative relationships to gross income as they did to net income. Land has apparently received a smaller proportion of the gross product in recent years, while capital has shown a fairly marked decrease. If 1940–44 is excluded, labor has apparently about held its own. One might hazard the guess that if national income remains high and employment opportunities exist, the proportion of the gross income going to labor should exceed the proportion existing from 1935–39 (43.6 percent), but will probably not be as large as during 1940–44 (48.7 percent).

IV. Summary

How closely the estimates of functional distribution of income approximate the actual distribution depends upon the accuracy of the basic data and the validity of the assumptions made. As a first approximation the estimated distribution derived by estimating land returns from rents paid, capital returns from current valuation of inventories, and labor returns as a residual seems to be of reasonable accuracy. The major limitation lies in the treatment of capital. There is no reliable series of data on short term interest rates or amounts paid by farmers, and not all capital items are included, though the errors involved are unlikely to appreciably affect the estimated returns to land or labor.

However, given the limitations of the present estimates, it is possible to argue that attempts to obtain a functional distribution of income in agriculture is one of the necessary forward steps in the use of income data in policy formulation and economic analysis. Present types of agricultural income data are quite unsatisfactory for determining the nature or extent of resource and welfare problems in agriculture.

This is not to argue that the estimates made above, or improved modifications of them, are adequate. Much more needs to be done. An ambitious program of income estimates might well include the following in addition to the series now available:

    I. Annual estimates of income for the following classification of farms
        A. Family type farms
            1. High production
            2. Moderate production
            3. Low production
            4. Part-time
            5. Resident
        B. Non-family types

II. Annual estimates of functional distribution of income for each of the above categories of farms

III. Annual estimates of non-agricultural income, by sources (labor and capital), for each category of farms

IV. Annual estimates of personal (family) distribution of income from all agricultural and non-agricultural sources

V. Estimates of the differences in money incomes required to obtain equivalent standards of living in different agricultural regions and in different sizes of cities by regions.

The above is an ambitious program, yet such a program seems necessary if income data are to be utilized in economic analysis and policy formulation in any except the very crudest fashion. Income data should be useful in answering two broad questions: (1) Are agricultural resources being utilized with maximum efficiency? (2) Are farm families receiving incomes sufficient to provide a socially accepted minimum scale of living? Neither of the two questions can be answered from income data alone, but adequate income data are required for answering both.

An important part of the problem raised by the first question is the general equivalence of returns to resources in agriculture and in other sectors of the economy. Whether such general equivalence exists can be determined only by having information on the monetary returns, including their distribution, and a fairly accurate indication of the evaluation of the money incomes in real terms.

Such an analysis does not, of course, provide conclusive evidence on whether or not maximum efficiency in resource use has been achieved. Further analysis is required to determine if the income paid or attributed to a resource is equal to the value of the marginal product of the resources. In other words, we need to know not only if the income received by comparable resources are equivalent in different employments, but whether or not the incomes are equal to the value of the marginal products in the respective employments. If the resource incomes or prices are not equal to the value of the respective marginal products, resource efficiency can be improved by rearranging resources to bring about such equality.[26]

Appendix Table I. Net Rent to all Landlords, Estimated Proportion of Total Land Rented, and Estimated Total Net Rent on all Farm Land, 1910–1945

Year	Net rental to all landlords[1] (millions of dollars)	Proportion of all land rented[2]	Estimated total net rent (millions of dollars)
1910	612	36.9	1,659
1911	640	37.3	1,716
1912	667	37.8	1,765
1913	669	38.2	1,767
1914	696	38.7	1,822
1915	789	39.1	2,018
1916	1,012	39.6	2,556
1917	1,488	40.1	3,711
1918	1,570	40.5	3,877
1919	1,684	40.9	4,117
1920	1,046	40.9	2,557
1921	776	40.9	1,897
1922	906	40.9	2,215
1923	1,025	40.9	2,506
1924	1,161	40.9	2,839
1925	1,079	41.0	2,632
1926	999	41.0	2,437
1927	1,131	41.0	2,758
1928	1,092	41.0	2,663
1929	1,062	41.0	2,590
1930	810	40.7	1,990
1931	492	40.3	1,221
1932	343	40.0	858
1933	479	39.7	1,207
1934	622	39.4	1,529
1935	727	39.2	1,854
1936	788	39.5	1,995
1937	760	39.5	1,924
1938	672	39.6	1,696
1939	734	39.6	1,854
1940	754	39.6	1,904
1941	1,045	38.8	2,693
1942	1,485	38.3	3,877
1943	1,728	37.8	4,571
1944	1,864	38.1	4,892
1945	1,858	37.3	4,981
1946	1,952	37.1	5,261
1947	2,300	37.0	6,216

[1]*Agricultural Statistics, 1943*, p. 412, *Agricultural Statistics, 1946*, p. 572 and *The Farm Income Situation,* June–July, 1947, p. 22, 1946 estimated by writer. Includes government payments accruing to landlords.

[2]Based on Census data for 1910–1934. See *Sixteenth Census of the United States, 1940, Agriculture,* Vol. III, *General Report*, p. 154 and 159; 1935–1946 supplied by Mr. M. M. Regan of the Bureau of Agricultural Economics.

Appendix Table II. Value of Farm Land and Buildings, Average Farm Mortgage Interest Rate, and Estimated Interest Cost

Year	Value of farm land and buildings (millions of dollars)[1]	Farm mortgage interest rate (per cent)[2]	Estimated interest cost (millions of dollars)
1910	34,801	6.0	2,088
1911	36,050	6.0	2,163
1912	37,306	6.1	2,276
1913	38,463	6.1	2,346
1914	39,586	6.1	2,415
1915	39,597	6.1	2,415
1916	42,271	6.2	2,621
1917	45,495	6.1	2,775
1918	49,987	6.1	3,049
1919	54,539	6.1	3,327
1920	66,316	6.1	4,045
1921	61,476	6.2	3,812
1922	54,017	6.3	3,403
1923	52,710	6.4	3,373
1924	50,486	6.3	3,179
1925	49,648	6.3	3,116
1926	49,052	6.2	3,041
1927	47,634	6.1	2,906
1928	47,495	6.1	2,897
1929	47,880	6.0	2,873
1930	47,880	6.0	2,873
1931	43,993	6.0	2,640
1932	37,236	6.0	2,234
1933	30,724	6.0	1,843
1934	31,933	5.8	1,852
1935	32,859	5.5	1,807
1936	33,910	5.1	1,729
1937	34,757	4.9	1,703
1938	34,747	4.7	1,633
1939	33,931	4.6	1,561
1940	33,642	4.6	1,548
1941	33,497	4.5	1,507
1942	35,331	4.5	1,590
1943	37,855	4.5	1,703
1944	42,532	4.5	1,914
1945	46,389	4.8	2,227
1946	52,114	4.8	2,501
1947	58,604	4.6	2,696

[1]As of January first for each year, BAE, "Net Farm Income and Parity Report: 1943," (Washington: United States Department of Agriculture, July, 1944), p. 29 (mimeo.) and BAE, *Farm Income Situation*, June, 1945, p. 28.

[2]1910–1928—*Agricultural Statistics, 1941*, p. 596; 1929 to date—*Agricultural Statistics, 1944*, p. 456.

Appendix Table III. Value of Non-Real Estate Farm Property, Estimated Interest Rate, and Estimated Interest Charge on Non-Real Estate Capital, 1910–1946

Year	Value of Non-real estate farm property[1] (millions of dollars)	Estimated interest rate (per cent)	Interest charge (millions of dollars)
1910	7,160	6.0	430
1911	7,519	6.0	451
1912	7,331	6.0	440
1913	8,015	6.0	481
1914	8,589	6.0	515
1915	8,872	6.0	532
1916	9,103	6.0	546
1917	10,047	6.0	603
1918	12,676	6.0	761
1919	13,966	6.0	838
1920	14,011	6.0	841
1921	11,585	6.0	695
1922	9,035	6.0	542
1923	8,672	6.0	520
1924	8,857	6.0	531
1925	8,841	6.0	530
1926	9,296	6.0	538
1927	9,338	6.0	560
1928	9,912	6.0	595
1929	10,586	6.0	635
1930	10,498	6.0	630
1931	8,382	6.0	503
1932	6,647	6.0	399
1933	5,554	6.0	333
1934	5,685	6.0	341
1935	5,879	5.5	319
1936	7,782	5.5	428
1937	7,951	5.5	436
1938	8,328	5.0	416
1939	8,407	5.0	420
1940	8,468	5.0	423
1941	8,844	5.0	442
1942	11,523	5.0	576
1943	15,470	5.0	777
1944	16,465	5.0	823
1945	17,094	5.0	855
1946	17,887	5.0	894
1947	21,245	5.0	1,062

[1]BAE, "Net Farm Income and Parity Report: 1943," p. 29 and *The Farm Income Situation*, June–July, 1947, p. 28.

Appendix Table IV. Number of Hired Workers, Total Expenses for Hired Labor, Average Expense per Worker, Total Farm Employment and Estimated Total Labor Return According to Method C, 1910–1946

Year	Number of hired workers[1] (thousands)	Expenses for hired labor[2] ($ million)	Average per worker per year ($)	Average farm employment[1] (thousands)	Estimated total labor ($ million)
1910	2,877	757	263	12,146	3,194
1911	2,870	760	265	12,042	3,191
1912	2,889	792	274	12,038	3,298
1913	2,905	807	279	12,033	3,357
1914	2,919	805	276	12,000	3,312
1915	2,934	815	278	11,981	3,331
1916	2,966	904	305	12,016	3,665
1917	2,933	1,127	384	11,789	4,527
1918	2,841	1,335	470	11,348	5,334
1919	2,784	1,515	544	11,106	6,042
1920	2,883	1,780	617	11,362	7,010
1921	2,901	1,159	400	11,412	4,565
1922	2,915	1,122	385	11,443	4,406
1923	2,894	1,219	421	11,385	4,792
1924	2,874	1,224	426	11,362	4,840
1925	2,871	1,243	433	11,448	4,957
1926	3,027	1,326	438	11,534	5,052
1927	2,950	1,280	434	11,246	4,881
1928	2,956	1,268	429	11,296	4,846
1929	2,984	1,285	430	11,289	4,854
1930	2,850	1,134	398	11,173	4,447
1931	2,690	847	315	11,159	3,515
1932	2,498	884	234	11,069	2,590
1933	2,433	512	210	11,023	2,315
1934	2,346	601	256	10,852	2,778
1935	2,489	740	305	11,131	3,395
1936	2,561	880	344	11,047	3,800
1937	2,631	1,039	395	10,892	4,302
1938	2,620	1,000	382	10,789	4,121
1939	2,595	982	378	10,740	4,063
1940	2,566	1,020	398	10,585	4,213
1941	2,532	1,238	488	10,361	5,056
1942	2,542	1,626	640	10,397	6,654
1943	2,406	2,009	835	10,263	8,570
1944	2,227	2,184	981	10,037	9,846
1945	2,118	2,299	1,085	9,844	10,681
1946	2,148	2,536	1,181	10,012	11,824
1947	2,227	2,776	1,246	10,157	12,656

[1]BAE, "Farm Wages, Farm Employment and Related Data," (Washington: USDA, 1943), p. 155, and BAE, *Farm Labor*, September 12, 1947, p. 6.

[2]BAE, "Net Farm Income and Parity Report: 1943," p. 26 and *Farm Income Situation*, June–July, p. 26.

Notes

I wish to acknowledge the criticisms and comments of T. W. Schultz and William H. Nicholls.

1. Since the first draft of this paper was written (early in 1945) much work has been done in the U.S. Department of Agriculture to obtain data to accurately estimate the personal distribution of income in agriculture.

2. Cf. T. W. Schultz, "Income Accounting to Guide Production and Welfare Policies," a paper read before the Western Farm Economics Association, June 28, 1945 and published by the Association, and D. Gale Johnson, "Contribution of Price Policy to Income and Resource Problems in Agriculture," This JOURNAL, XXVI (Nov., 1944), pp. 637–8, 651–2, and 658–9.

3. Net income from agriculture is seldom used in income comparisons, though it has much to commend it from an analytical standpoint to the somewhat similar concept of net income to persons on farms from agriculture. The magnitude of the latter concept is affected by such considerations as shifts in ownership of land or changes in residence of hired workers.

4. All income estimates given in this paper include government payments.

5. The Bureau of Agricultural Economics currently excludes rent paid to landlords living on farms as a production expense to farm operators, though rent paid to landlords not living on farms is considered a production expense. The reason for excluding rent paid to landlords living on farms has been stated as follows:

"Rent to landlords living on farms is not included in expenses because, while it is an expense item to one farmer, it represents an item of income to another." (Net Farm Income and Parity Report: 1943, p. 25).

Consistency would require a similar exclusion of wages paid to laborers living on farms or at least wages paid to farm operators, as a production expense as such wages represent income to other farm people, in many cases to other farm operators or members of their families. It would seem more reasonable to include rent, regardless of the recipient, as a production expense in determining farm operator's income. If this were done, farm operator income would be the net income of all farm operators from the farms over which they have managerial control.

6. A. S. Tostlebe, et al., "The Impact of the War on the Financial Structure of Agriculture," U.S. Dept. of Agriculture, Misc. Pub. No. 567, 1945, pp. 25, 26, presents estimates of interest payments on non-real estate debt for 1915, 1918 and 1940 to 1943 (pp. 25–6). Inclusion of interest payments on non-real estate debt would have increased total agricultural income by about 5 per cent in 1915, 3 percent in 1918 and 4, 3, 2 and 2 percent respectively in 1940, 1941, 1942, and 1943.

An offsetting error, which was not discovered until after the estimates in this paper were made, is due to the double counting of interest paid on mortgages by non-farm landlords. In recent years this item has been about $60,000,000, or one percent or less of total agricultural income. In all estimates made by the Bureau of Agricultural Economics prior to June 1945, net rent to non-farm landlords included farm mortgage interest payments and the estimates of interest payments on farm mortgages included all farm real estate. As a result, in determining net agricultural income the mortgage interest paid by non-farm landlords was counted twice. In estimates published in June, 1945, net rent paid to non-farm landlords was calculated exclusive of interest payments thus removing the double counting. This resulted in an increase in estimates of operator's income and a reduction in estimates of total agricultural income. This revision is available only for 1940 to date.

It might be noted that compilers of national income statistics are uncertain as to the appropriate status of short-term interest. Short-term interest may be considered as a cost in the same sense as purchases of raw materials; as a factor payment, or as a transfer payment.

7. Data for 1935–1946 were kindly supplied by Mr. M. M. Regan of the Bureau of Agricultural Economics.

8. This step was not taken for two reasons—one, there does not now seem to be an appropriate agricultural wage rate available, and two, the writer is not fully confident that the estimates of employment of family labor are wholly comparable with those of hired labor. See later discussion for further comments.

9. For the nation as a whole, rental contracts fixed in cash have covered farms which include approximately 20 percent of the value of all land and buildings on tenant operated farms for the census years 1919 to 1939. In 1920 share-cash rental contracts represented about 15 percent and in 1939 about 25 percent of the value of land and buildings on all rented farms. In the share-cash rental contracts the share proportion of the total rent is generally considerably more important than the cash part. At the outside cash rent would not make up more than 25 to 30 percent of all rent payments. Part owners apparently rent about the same proportion of the rented part of the farm in cash as do full tenants, though data are available only for 1939.

10. Some doubt may be raised as to the extension of rent on rented land to obtain a return on all land because rented land may not be an adequate representation of all land. The major difficulty arises, the writer believes, because of the differences in regional distribution of rented land. In New England and Middle Atlantic states much less land is rented than in other sections, such as the South and North Central states. The Pacific and Mountain States also have less rented land than the national average. As a consequence, two types of errors are possible. First, the average level of land returns calculated by estimating from rents may be too high or low. Second, the movement of land returns in time can be affected by considerations peculiar to the cotton, hog-corn, or wheat producing areas. Except for unusual circumstances, the second error is not likely to be of importance and the size of the first error is undoubtedly small. The census regions in which tenancy is significant have approximately 75 percent of the value of land and buildings and customarily produce the same proportion of gross and net farm income. Even in the Western States tenancy is undoubtedly of sufficient importance to permit rents to be of sufficient importance to serve as a guide. However, it is probably true that land returns should be estimated on a regional basis and then combined to obtain a national estimate. The method used in this paper resulted in weighting regions in proportion to land rented rather than in proportion to all farm land. A more refined analysis should approach the problem from at least a regional standpoint.

The large drop in the proportion of net agricultural income attributed land in 1920 illustrates the desirability of using regional rather than national data. Rent payments fell very sharply from 1919 to 1920 because of the very large declines in cotton and wheat prices in late 1920 as contrasted to more modest changes in milk, fruits, and feed grains and livestock. The rent series is heavily influenced by cotton and wheat prices and thus a distortion resulted.

11. See Appendix Table II for farm mortgage interest rates.

12. These results are approximately the same as those reported by M. M. Regan, H. R. Johnson and F. A. Clarenbach, *The Farm Real Estate Situation, 1944–1945*. Circ. No. 743, Washington: U.S. Department of Agriculture, 1945, pp. 37–38.

13. As of January, 1947 the value of farm real estate was estimated to be 58,604 million dollars. See *Farm Income Situation*, June–July, 1947, p. 28.

14. *What Peace Can Mean to American Farmers, Post-War Agriculture and Employment*, USDA, Misc. Pub. No. 562, p. 25. Writer adjusted published estimates of net operator income to net agricultural income.

15. It should be noted that if farm incomes remain at the 1946 level through 1948, long run agricultural income will not need to equal the figures indicated in the text. The present high levels of income over a three year period would make possible a net downward revaluation of the land by 10 to 20 percent from the 1946 values. This point applies, of course, only to existing owners.

16. It is frequently overlooked that there has been almost no net investment in agriculture in the last 35 years. The investments made in tractors, trucks, and motor vehicles have been more than offset by the decline in investment in horses and mules. If 40 percent of the investment in automobiles is attributed to the farm, investment in motor vehicles increased from an average of $46,000,000 in 1910–14 to $1,202,000,000 in 1940, an increase of $1,156,000,000. The investment in horses and mules declined from an average of $2,811,000,000 to $1,256,000,000, a decrease of $1,546,000,000. The dollar investment in other machinery was almost identical—$1,430,000,000 and $1,431,000,000 in the two periods.

17. In calculations presented later, such reweighting was *not* undertaken.

18. The discrepancies cannot be considered as insignificant. The following tabulation indicates the differences for three very stable periods. Monthly rates have been multiplied by 12.

	Calculated from wage bill	Monthly wage without board	Monthly wage with board	Composite monthly rate	Cash wages paid
	(In dollars)				
1910–14	271	350	265	297	190
1925–29	433	608	479	530	323
1935–39	361	410	312	349	281
	Index—1910–14 = 100				
1925–29	159	181	174	178	170
1935–39	133	118	117	118	148

19. The reasons for the differences in the two series are discussed rather fully by E. W. Grove, *Income Parity for Agriculture, Part II—Expenses of Agriculture Pro-*

duction, Section 1, The Cost of Hired Farm Labor (Washington: U.S.D.A. 1939), pp. 13–15. The writer personally believes that the wage bill estimates are too low and the wage rate estimates by crop correspondents too high, with the true wage returns lying somewhat nearer the wage bill estimates.

20. Louis J. Ducoff and Gertrude Bancroft, "Experiment in the Measurement of Unpaid Family Labor in Agriculture," *Jl. of the Am. Stat. Assoc.*, XL (June, 1945), pp. 205–213.

21. The age distribution of hired workers differs considerably from the family worker distribution, with a much smaller proportion of hired workers at either extreme of the distribution.

22. See below, Section II, for an interpretation of these results.

23. This statement assumes that actual hired workers had the same length of employment each year as operators and unpaid family laborers. Hired workers were undoubtedly at work a much smaller proportion of the time. Consequently the statement as it stands is true of actual workers only if each group of workers had the same amount of employment.

24. In 1939 farms that had a total value of products sold, traded or used in the home of less than $599 accounted for 42.5 percent of the total number of family workers and only 13.9 percent of the hired workers. Farms with a total value of $4,000 to $9,000 had 4.7 percent of the family workers and 18.0 percent of the hired workers, while farms with a total value of production excess of $10,000 employed 17.6 percent of all hired labor and involved only 0.9 percent of the family labor. (See L. J. Ducoff, *Wages of Agricultural Labor in the United States*, U.S.D.A., Technical Bulletin No. 895, July, p. 10.) It has been estimated that more than 20 percent of all farms may be classed as part-time or residential farms. Another 20 percent, not falling in either of the above two categories, has a total value of product of less than $600. (See T. W. Schultz, *Agriculture in an Unstable Economy*, New York: McGraw-Hill Book Co. 1945, p. 199). Because of the methods used in defining employment in agriculture, the essentially non-commercial sector of American agriculture is given a disproportionate weight in the employment aggregates.

25. The year 1932 represented the minimum proportion of net income to gross income of 57.1 percent, while the year 1943 showed the highest proportion of gross income attributable to net income, namely 78.1 percent. In 1918 net income was 77.9 percent of gross income, the second highest proportion, while 1921 showed the second lowest proportion, 58.8 percent.

26. While the article was in press data became available permitting estimates for 1947. Such estimates were inserted in the appendix tables, but no changes were made in the text or text tables. Some of the estimates are given here. Total net income from agricultural and government payments was $21,143,000,000. According to Method A (Table I) the percentage allocation of income was 29.4 percent to land, 5.0 percent to capital, and 65.6 percent to labor. Income allocation according to Method B (Table II) indicates 12.8 percent to land, 5.0 percent to capital, and 82.2 percent to labor. The rate of return on rented land (Table III) was 10.6 percent. Total labor returns as a residual as calculated by Method A was $13,865,000,000, while labor returns derived from wages paid hired labor was $12,656,000,000. This gives a positive difference of $1,209,000,000 or a difference

of $119 per worker (Table V). Income estimates for commercial farmers are as follows: (Table VI) income per farm, $6,150; capital and land income per farm, $2,115; labor income per farm, $4,035; labor income per family worker, $2,210; and labor income per hired worker, $1,305. Labor income per worker on all farms was $1,362 (calculated according to Method A) and $1,930 on the commercial farms.

2

Resource Allocation under Share Contracts[1]

I

How are resources allocated when a share of the total product is paid for
the use of a particular physical asset, such as farm land or a retail store?
This is an important practical question because three-fourths of all rented
agricultural land is leased under share contracts and they are becoming
increasingly prevalent in the retail field. It is also an interesting theoretical
issue that has not been adequately treated in economic literature.

A share-rent contract usually requires the tenant to pay the landlord a
specified proportion of the farm's produce or of the gross sales of a retail
store. The crop-share lease, which calls for payment of a certain proportion of the crop to the landlord, is the commonest share lease in agriculture. In the cotton areas the true sharecropper pays a share rent not only
for the use of the land but for the seed and fertilizer and the use of capital
equipment, including work horses or mules.

If resources are to be allocated in an optimum manner, certain marginal
conditions must be satisfied. In the factor markets, all factors of comparable nature tend to receive the same marginal return. Within the firm, resources must be so employed that the value of the marginal product is equal
to the marginal cost of the factor, and the marginal cost of the factor must
equal its price. The stipulations of the crop-share lease create circumstances in which both the tenant and the landlord, when each views his
interest separately, consciously attempt to violate the marginal conditions
required for maximum output. Under a crop-share lease, if the landlord's
share of the crops is half, the tenant will apply his resources in the production of crops until the marginal cost of crop output is equal to half the

Reprinted by permission from the *Journal of Political Economy* 58, no. 2 (April 1950):
111–23. © 1950 by The University of Chicago. All rights reserved.

value of the marginal output. The same tenant, however, will conduct his livestock operations, where important costs are borne by the landlord and the receipts are not shared with him, in the usual manner. The landlord will not invest in land assets unless the value of the marginal product is twice the marginal cost.

It is at once obvious that, if both the tenant and landlord act as though the interest of one is distinct from the interest of the other, the net product of the farm will be less than it could be. The tenant will farm extensively, and the landlord will refuse to make improvements that would pay under a more rational method of pricing land. Before John Stuart Mill, the well-known English economists condemned share leases on these grounds. Adam Smith argued that the tenant would be extremely reluctant to employ his own capital on the farm, inasmuch as the landlord would receive a large part of the resultant product.[2]

Arthur Young, in his *Travels*, described the metayer system in France and roundly condemned it: "There is not one word in favor of the practice, and a thousand arguments that might be used against it."[3] He ascribed the low rent of land in France as compared to England to the metayer system, because it led to inefficient farming.[4]

Richard Jones, in his *An Essay on the Distribution of Wealth and on the Sources of Taxation* denounced the metayer system on two related grounds. First, "The divided interest which exists in the produce of cultivation mars almost every attempt at improvement."[5] His second objection was more subtle. The argument was as follows: The metayer system results in a very meager productivity of agricultural labor because numerous obstacles to improvement inhere in this system. If agricultural productivity stays at a low level, there will be no change in the relative numbers of agricultural and nonagricultural population. The strength of a nation is derived from the size and wealth of the nonagricultural population, while the nonagricultural population can be numerous and productive only if farmers produce much more than is required to feed themselves.[6] This represents a rather modern conception of the process of economic progress and is relevant to the present discussion because of the role given to "the nature of the conditions under which land is occupied."

McCulloch also discussed the share renting of land, though he apparently had not studied the problem in detail, and his analysis was similar to Smith's. He wrote: "The practice of letting lands by proportional rents . . . is very general on the continent; and wherever it has been adopted, it has put a stop to all improvements, and has reduced the cultivators to the most abject poverty."[7]

Mill took a much more favorable view of the metayer system. He was well acquainted with the literature referred to above and recognized the important point made by Smith.[8] Mill, however, argued that most of the defects pointed out, particularly by Young, were due to imperfections of the metayer system as practiced, not to the operation of a perfect system.[9] Mill apparently felt that insecurity of tenure was the major defect of meta-yage in France and that, if this were overcome, the metayer system would function well. Mill came to this conclusion from two bits of evidence. First, he attributed the poor state of the Irish cottier to insecurity of tenure and the setting of rents by competition.[10] Second, relying upon descriptions of metayage in Italy, where security of tenure existed, he concluded that agricultural improvement would take place if the tenant were secure. Mill also relied upon evidence presented by Sismondi, a landowner and a metayer landlord, who favored the system. In Tuscany, where Sismondi owned land, metayers had security, not by law but by custom. Sismondi described the life of metayers in romantic terms, quite contrary to Young's description.[11]

There are several errors in Mill's argument. This is particularly true of his contempt for competitive rents[12] and his belief that all that was required to perfect the metayer system is security of tenure. Under certain circumstances security of tenure would improve the lot of the tenant; but, as I shall show below, the share system can function effectively, under American conditions at least, only when tenure is of short duration. However, it should be noted that Mill took a more reasonable position than some of his predecessors, by arguing that the low incomes of metayers should not be attributed to the metayer system. Mill argued that it was the "operation of the population principle" that largely determined the economic status of farm people. "A multiplication of people beyond the number that can be properly supported on the land or taken off by manufactures" was possible under any land-tenure relationship.[13] Mill supported small holdings with fixed tenure on the ground that this system was most likely to engender population restraint and agricultural improvement.

Marshall brought to the discussion of share contracts not only a superior analytical framework but his great insight into the operation and effects of institutions. Marshall added to the precision and generality of the argument, while at the same time he considered the types of adjustments that landlords could make to avoid certain of the disadvantages of share contracts.[14]

Recently Schickele and, more particularly, Heady have rather fully developed the possible deviations in resource allocation that can result from the share contract.[15] Heady concluded that the crop-share contract can be perfected under the following condition: "The cost of variable factors (where one such factor as land is fixed) must be divided between the landlord and tenant in the proportions that hold for the division of the product."[16] He does not indicate whether he would include the sharing of the labor factor by the tenant and the landlord. If he did, the share going to the tenant would be very small, though this need not reduce his income. However, if the landlord paid the tenant for half his labor, the nature of the lease would be drastically altered.

II

There are two important problems relating to the share contract that have not been considered adequately by the writers whose work was described in the previous section. These writers seem to have considered the analytical problem solely in terms of the tenant allocating his resources on a given farm. As a result, economists have neglected the interesting theoretical issue of how the tenant determines the amount of land to rent.

One might conclude from the previous work on share contracts that they inevitably lead to misuse of agricultural resources, low crop yields, and meager land improvements. Though the crop-share contract has probably induced some misuse of agricultural resources, the nature of the deviations from optimum are quite subtle and are not immediately obvious from a cursory examination of American farms operating under different types of tenure arrangements. Consequently, the second problem that we shall consider is the type of adjustments that landlords and tenants have made in their mutual relations to make crop-share tenancy function reasonably well.

III

A crop-share tenant, like any tenant, must decide how large a farm he will lease. If the lease were a cash lease, economists would say that the amount of land to rent was determined by the point at which the marginal return from the last acre of land rented equaled the rent. Presumably, the crop-share tenant makes his decisions in the same way. If he does, he continues to rent land until the value of the marginal product of the last acre rented is zero. This is surprising, but it can be proved readily.

Let us assume that the tenant combines only two resources—labor (n) and land (z)—and that no restriction is placed upon the tenant by the landlord other than the payment of the stipulated share of the product. The wage rate (w), the price of the product (P), and the share proportion ($\frac{1}{3}$) are constant. We therefore obtain the following equations:

$$X = f(n, z) \text{—(the production function);} \tag{1}$$

$$Y = XP - nw - \tfrac{1}{3}Xp \text{—(the profit or income equation);} \tag{2}$$

$$\frac{\partial Y}{\partial n} = P\frac{\partial X}{\partial n} - w - \tfrac{1}{3}P\frac{\partial X}{\partial n} = 0; \tag{3}$$

$$W = (P - \tfrac{1}{3} P) \frac{\partial X}{\partial n} = \tfrac{2}{3} P \frac{\partial X}{\partial n};$$ (4)

$$\frac{\partial Y}{\partial z} = P \frac{\partial X}{\partial z} - \tfrac{1}{3} P \frac{\partial X}{\partial z} = 0;$$ (5)

$$(P - \tfrac{1}{3} P) \frac{\partial X}{\partial z} = 0;$$ (6)

$$P \frac{\partial X}{\partial z} = \tfrac{1}{3} P \frac{\partial X}{\partial z}; \frac{\partial X}{\partial z} = 0.$$ (7)

Equations (3) and (4) prove what was discussed above and by the other writers mentioned earlier—that labor will be employed until the wage rate is equal to the value of the marginal product to the tenant, which is two-thirds (in this example) of the actual value of the marginal product. These equations can be generalized to include all other factors except land.

Equations (5), (6), and (7) show that the tenant will continue to rent additional land until the marginal physical product is zero. This result may at first seem strange, but it can be demonstrated without the above equations.

When a tenant adds an additional acre of land, its marginal cost is one-third the value of its marginal product. The renter will equate his marginal return from the land to the marginal cost of the land. This equality will exist only when the value of the marginal product is zero; for it is a simple truism that a third of a variable can be equal to the value of the variable only when the variable has the value of zero. Consequently, the tenant can reach a position of maximum profit only when an additional unit of land adds to neither his costs nor his receipts.

The case that we are considering represents an interesting example of a falling supply curve under conditions of apparent competition. Rent per acre is a linear function of the average product of land, and, since the average product is falling, the supply price of land must also be a falling curve. The marginal cost of land, which is not equal to the average rent paid per acre, is equal to one-third the value of the marginal product of land, which is also a decreasing function. The case is depicted graphically in Figure 1.

Linear functions are shown in Figure 1 for simplicity; the example can be generalized for any type of average-product function. The average-product curve (CS) represents the average product per acre to be obtained by renting additional land, which is measured along OS. The marginal-product curve (CT) is derived directly from the total-product curve. The average-rent curve (BS) is one-third the average product, since the share rent is one-third the total product. The marginal-supply price (BT) is one-

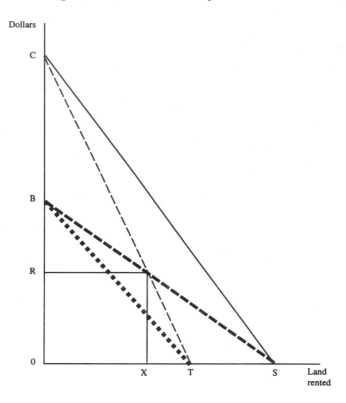

Figure 1

third the marginal product, since the additional rent per acre is one-third the additional product resulting from the use of additional acres. The marginal-supply price must be falling if the average-product curve is decreasing.

The marginal return to the tenant from employing land is equal to the average rent per acre at a price of *OR* and a quantity of *OX*. However, the tenant is not influenced by the average-rent curve but by the curve indicating the marginal cost of land as shown by the curve *BT,* marginal-supply price of land. Thus in a competitive situation it seems that a falling supply price might emerge.

The share contract results in a curious violation of our usual conception of economic rationality. Under the conditions assumed in our example, the marginal return from land to the farm operator (tenant) is zero. This means that the marginal return to labor is at a maximum and that, if labor is paid the value of its marginal product, the total product will be exhausted by payment to labor. Yet the tenant is willing to pay the landlord a third of the total product for the use of the land, which, at the margin, returns nothing to him. This apparently inconsistent result is possible because labor actually receives only two-thirds the value of its marginal product. The return to labor, after payment of the rent, is equal to its wage and presumably the value of its marginal product in alternative uses. This fact, of course, explains why labor is willing to accept less than the value of its marginal product in its present employment.[17]

The implication of the crop-share contract to the landlord is that the land is being combined with small (relative) amounts of other factors. If these factors get a return equal to earnings in alternative employments, the landlord cannot be receiving as much rent from a specific piece of land as he could under a fixed-rent contract. Under such a contract more labor and other factors would be applied to each unit of land, since labor and other factors would be used until their prices equaled the value of the respective marginal products. Consequently, there would be a larger total product and a greater share (residual) available for payment to land.[18]

IV

Will landlords permit a tenant to use land until the marginal product is zero? By definition, the total amount of land that bears a price is in limited supply relative to existing demands. All such land is presumably capable of producing a marginal product greater than zero in some use. Though there may be some empirical evidence that share-rented farms are exploited less intensively than cash-rented farms in the same area and of the same general type, there are no empirical data which indicate that the marginal product of land has fallen to zero where crop-share contracts prevail.

If the share proportion going to the landlord were variable, it would be easy to see how the share-rent market could establish renting conditions that would achieve a relatively efficient allocation of resources. The landlord always has the alternative of renting for a sum of cash independent of current output. This presumably represents the minimum aggregate amount of rent that he will accept for the farm. Hence, if the tenant follows an extensive program of farming, the landlord will ask for a larger share of a smaller average output. If the share proportion were a function of the amount of the other factors applied per unit of land, the tenant would find that he was confronted with a very different marginal-cost curve for land (though it still might be falling) than if the share were independent of the amount of other factors applied to the land.

The equilibrium conditions can be indicated from the following set of equations. A new rent equation (8) has been added, and this equation indicates that the share proportion decreases as the labor inputs per unit of land are increased and vice versa.

$$R = g\frac{n}{z} XP \text{ (}R \text{ is total rent);} \tag{8}$$

$$X = f(n, z)\text{—(the production function);} \tag{9}$$

$$Y = Xp - nw - g\frac{n}{z} Xp\text{—(the profit or income equation);} \tag{10}$$

$$\frac{\partial R}{\partial n} = g'\left(\frac{1}{z}\right) XP + g\left(\frac{n}{z}\right) P \frac{\partial X}{\partial n} = \frac{\partial R}{\partial n} PX + rP \frac{\partial X}{\partial n}\left[r = g\left(\frac{n}{z}\right)\right]; \tag{11}$$

$$\frac{\partial Y}{\partial n} = P \frac{\partial X}{\partial n} - w - g'\left(\frac{n}{z}\right) XP - Pg \times \left(\frac{n}{z}\right)\frac{\partial X}{\partial n}; \tag{12}$$

$$P \frac{\partial X}{\partial n} = w + PX \frac{\partial r}{\partial n} + rP \frac{\partial X}{\partial n} = w + \frac{\partial R}{\partial n}; \tag{13}$$

$$\frac{\partial R}{\partial z} = g'\left(-\frac{n}{z^2}\right) XP + g\left(\frac{n}{z}\right) P \frac{\partial X}{\partial z} = \frac{\partial r}{\partial z} PX + rP \frac{\partial X}{\partial z}; \tag{14}$$

$$\frac{\partial Y}{\partial z} = P \frac{\partial Y}{\partial z} - PXg'\left(\frac{-n}{z^2}\right) - g\left(-\frac{n}{z}\right) P \frac{\partial X}{\partial z}; \tag{15}$$

$$P \frac{\partial X}{\partial z} = PX \frac{\partial r}{\partial z} + rP \frac{\partial X}{\partial z} = \frac{\partial R}{\partial z}. \tag{16}$$

Equation (13) indicates that the tenant will use labor until the value of the marginal product is equal to the wage rate plus the marginal change in rent resulting from the marginal input of labor. The quantity $\partial R/\partial n$ may be either negative or positive, since $\partial r/\partial n$ is negative and $\partial X/\partial n$ is positive. If $\partial R/\partial n$ is positive, the tenant will use less labor on a given piece of land than he would if he rented it for cash at the same average rent (R/z).

Equation (16) shows that the tenant will use land until the value of the marginal product equals the marginal cost of land, which is reflected in the change in total rent resulting from renting an additional acre of land. Thus it is apparent that the tenant will not continue to rent land until its marginal product is zero.

Is it likely that a tenant operating under these conditions will allocate his resources in exactly the same way that he would if paying a cash rent independent of the actual output? The answer is apparently in the negative. There is only one average rent per acre for which the resource allocations will be the same under a variable-share proportion and a fixed rent per acre.[19] And there is no reason to believe that this particular rent would emerge under competitive conditions.

There is no evidence that the rental share varies with the intensity of cultivation. Evidently landlords have not availed themselves of the possibility of varying the rental share (in the manner described above) as a means of increasing the rent.

V

I have stated earlier that the crop-share lease has not resulted in the gross misallocation of land that would have occurred if the tenant had actually combined so few nonland resources with the land as he would have if he had not had to consider the landlord's interest. The evidence we have on the actual rents paid under crop-share and cash leases substantiates this thesis. Though admittedly inadequate, the available evidence indicates that the crop-share contract yields at least as much, if not more, rent per acre than does the cash lease on comparable farms.[20] Since the crop-share landlord bears a large proportion of the uncertainty confronting the farm firm, he must receive a larger rent to compensate him.

How has the landlord protected his interest and achieved a reasonable level of rent? Three techniques are available to the landlord for enforcing the desired intensity of cultivation. The first is to enter into a lease contract that specifies in detail what the tenant is required to do. A second is to share in the payment of expenses to the same extent as in the sharing of the output. The third is to grant only a short-term lease, which makes possible a periodic review of the performance of the tenant.

The landlord can enter into a contract that specifies in considerable detail what the tenant must do: the amount of labor to be used, the methods of cultivation, the application of fertilizer, and so on. Though this is apparently done in Europe,[21] it is uncommon in the United States. However, there are certain conditions in this country under which the landlord has a reasonable assurance that the land will be farmed with the appropriate intensity. One set of conditions is illustrated by wheat farming in the Great Plains. Technological conditions are such that a given level of intensity of cultivation is best adapted to the region. Either lower or higher levels of intensity provide appreciably smaller marginal returns to the variable factors. Since the farmers are uncertain as to the exact effect of modest variations in inputs, most farmers follow a fairly traditional input level.[22]

The sharecropper situation in the South represents another case in point. Here the landlord assures himself of the appropriate intensity of cultivation by supplying seed, fertilizer, and most or all of the equipment. By keeping the size of the individual unit small enough, the landlord can force the sharecropper to apply more labor per acre than the sharecropper would if he were free to choose the size of his farm. If the unit is small and the sharecropper is restricted in outside earnings, he must apply his labor and that of his family until he has achieved at least some minimum level of income. He might continue to work until the marginal satisfaction derived from his share is equal to the marginal disutility of work; but, if the land area is kept small enough, the marginal utility of income will be very high. In this way the sharecropper will be forced to farm as intensively, if not more so, as he would if he rented for a cash payment.

In the two cases described, the problem of achieving the appropriate degree of intensity is fairly easily solved. However, where the tenant supplies all the (nonland) capital and labor and these resources may be combined with land in continuously varying proportions over a wide range, the bargaining process must be somewhat more subtle, as well as more complex. In the Corn Belt the crop-share contract is widely used; yet the tenant can apply his capital and labor to a greater or less degree in his livestock enterprise and slight his crop enterprises.

By sharing certain expenses, the landlord can induce a more rational allocation of resources. In this case the tenant and the landlord will make the decision jointly and can arrive at the best allocation of resources. However, within the framework of the crop-share lease, the possibility of sharing expenses is limited in scope. The most complete sharing is in the livestock-share lease, but this involves a fairly radical departure from the crop-share basis.

It seems unlikely that either of the alternatives discussed is sufficient to prevent the tenant from "exploiting" the landlord. A much more powerful restraint upon the tenant must prevail. In my opinion this restraint is the short-term lease. This type of lease is alleged to be a serious shortcoming of American tenure institutions; but, without it, there seems little likelihood that the crop-share lease would lead to a reasonably efficient use of land. The crop-share contract has fallen into disrepute throughout western Europe, and a major reason is that the tenant generally has permanent tenure. If the tenant cannot be dispossessed, it is difficult for the landlord to enforce a given intensity of cultivation.

With a short-term lease renters are obviously aware that landlords have the alternative of renting their land for a cash rent independent of current output. Consequently, the tenant must plan to produce an average output per acre that will provide a rental payment, if yields are average, equal to the possible cash rent plus any additional payment required to compensate the landlord for the uncertainty that he bears.

If the renter does not give certain assurances that the degree of intensity of cultivation will be such as to provide the appropriate rent, he will be unable to find a farm. Once he has found a farm, he may fear that his lease will not be renewed unless sufficient rent is actually paid. Since the cost of moving is substantial and his previous production performances are likely

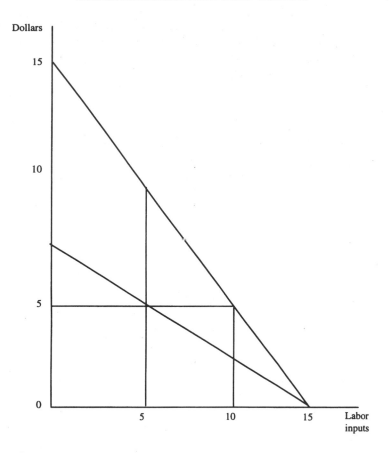

Figure 2

to affect his ability to locate a new farm, he will have an incentive to farm in a fashion that will provide the necessary rental payment.

The tenant's position may be illustrated by Figure 2. In this diagram it is assumed that he is renting a specific farm and is applying his labor to it. He owns 10 units of labor, and its price in alternative employments is $5.00 per unit. If the tenant maximized his income for the given year, he would utilize 5 units of labor on the farm and sell the rest of the labor. His income would then be $56.50 for the year. If he applied all his labor on the farm, he would receive $50. This would be true whether he rented for a fixed cash payment equal to Ricardo's residual or for a 50 per cent share.[23]

The tenant's action in this case will depend upon what he thinks "he can get by with." Looking at the alternative, as a whole, of farming this particular farm, he loses nothing by applying all 10 units of his labor to the farm. The loss on the last 5 are just counterbalanced by the "rent" on the first 5.[24] The tenant might try to use only, say, 9 units of labor on the farm and 1 unit elsewhere, on the assumption either that this would not be detected by the landlord or that it would not be resented if it were. The latter assumption hardly seems justified, and the landlord is undoubtedly sensitive to the first circumstance.

The loss to the landlord is very considerable if the tenant applies only 5 units of labor to the land. The landlord will receive only $31.50 instead of $50.00—a loss of roughly one-third the rent.

The example chosen above assumes that, if the tenant applied the same quantity of capital and labor to the farm as he would under a cash lease, the landlord's share would exactly equal the rent, defined as the value of the marginal product of the land. Since the crop share is uniform for large areas, including land of diverse productivity,[25] the correspondence between the landlord's share and "true" rent is not likely to be achieved so readily. One possible way of attaining it is to vary the cash rent for hay and pasture land and for buildings. Some cash rent is paid on most crop-share farms, though not on all. If the land productivity of a particular farm is relatively high compared to other farms with the same share rent, the cash payment may reflect this difference. If the share rent on land of the lowest productivity is sufficient to equal the possible cash rent, the exaction of higher cash rent for land used for nonshare purposes will not create additional distortions in resource use. This will be true even if the cash payment for pasture and hay land and buildings is above the value of their marginal product. One of the conditions which the landlord generally imposes upon the tenant is that a certain proportion of the farm be devoted to hay and pasture. Once this decision is made, the cash payment for this land becomes a fixed cost and affects the tenant's decision only with regard to renting the farm or not. The extra payment would not affect current input decisions and would not create any additional impediments to rational resource use.

If the share proportion were too high on land of low productivity, it would be hard to visualize how the crop-share lease could function. One would expect that, if the share were too great, either the land would lie idle or it would be farmed very extensively. For example, some hitherto cultivated land of poor quality was actually left idle during the thirties in the North Central and Great Plains states because tenants could not be obtained to farm it on the prevailing share arrangement.

The process by which the landlord and tenant enter into a lease is not well understood. The price system does not function in the normal sense, for land is not necessarily rented to the tenant offering the highest rent payment. However, there is only a difference of modest degree in the role of price rationing in the share-rental market and in the cash-rental market. When a man sells a bushel of wheat, he has no interest in the use to which the wheat is put and is consequently willing to sell to the highest bidder. However, when a man sells the use of land, he has a real interest in how the land will be used.[26] Consequently, the choice of tenant is never made without considering what the impact of the tenancy will be upon the value of the asset. The crop-share tenant is presumably chosen in terms of explicit or implicit notions concerning the level of output that he will produce on the farm. Any agreement that is reached is enforced by the short-term condition of the tenure rather than by a detailed lease contract.

VI

The material in this paper leads to the conclusion that the short-term lease serves a useful purpose in creating conditions within which the crop-share lease results in a reasonably efficient utilization of land. Most students of farm tenure strongly disapprove of short-term leases but have failed to recognize its advantages.

But it must be admitted that the conclusion rests on a premise that has not been adequately verified. The premise is that crop-share tenants do not combine nonland resources with land in nonoptimum proportions to a greater degree than do cash tenants. Yet the evidence on the equivalence of rent under the two lease contracts is consistent with only one other hypothesis: namely, that crop-share tenants have lower incomes than cash tenants.[27] The latter hypothesis assumes that crop-share tenants are exploited by the landlord, while the conditions of the crop-share lease lend credence to the opposite view.

Admittedly, the short-term lease has numerous disadvantages.[28] One of its main disadvantages is that it reduces the expected value to the tenant of the marginal product of semidurable inputs. This can be remedied in part by providing for compensation for unexhausted improvements, which can be suggested more readily than it can be accomplished. But the most serious disadvantage, from both an economic and a social viewpoint, is the high mobility of tenants inherent in the short-term-lease system. The cost of moving is important, but perhaps more important as an economic cost is the loss in productive efficiency due to unfamiliarity with the physical characteristics of the land during the first year of tenancy. The relatively low social status of the tenant can be attributed to his actual or assumed temporary residence in the community.[29]

VII

Almost sixty years ago Alfred Marshall wrote: "There is much to be gained from a study of the many various plans on which the share contract is based."[30] This same statement can be made today.[31] The interests of farmers, politicians, and agricultural and land economists have all been so diverted by the value presumption of farm ownership that all too little attention has been given to analysis of the effects of rental contracts upon resource use. There are certain exceptions to this generalization, including the articles by Heady and Schickele referred to above. But even in these cases the analysis is not based upon data that indicate how the rental market actually functions and how share leases affect the use of resources.

From the conclusions of this paper it is unreasonable to presume that the resource use on farms with crop-share leases, for example, depart as far from the optimum as is suggested by the analysis of what the tenant would do *if he were left free to do as he chooses*. He is, of course, not free to do exactly as he pleases. In fact, under certain circumstances the share contract would not result in any loss in resource efficiency. Yet we do not have the empirical information which would permit us to say whether the actual functioning of share leases results in little, much, or no misuse of resources. I hope, however, that the issues have been so formulated in this paper as to permit an empirical verification of certain of the more important effects of share leases upon resource allocation.

Notes

1. The research on which this article is based has been done on a project financed by a grant for Agricultural Economics Research at the University of Chicago, made by the Rockefeller Foundation. I am indebted to my colleagues, Professors O. H. Brownlee, E. J. Hamilton, and T. W. Schultz, for suggestions and criticisms.

2. "It could never, however, be to the interest of this last species of cultivators [the metayer] to lay out, in the further improvement of the land, any part of the little stock which they might save from their own share of the produce, because the lord, who laid out nothing, was to get one-half of whatever it produced. The tithe, which is but a tenth of the produce, is found to be a very great hindrance to improvement. A tax, therefore, which amounted to one-half must have been an effectual bar to it. It might be the interest of a metayer to make the land produce as much as could be brought out by means of the stock furnished by the proprietor; but it could never be his interest to mix any part of his own with it" (Adam Smith, *Wealth of Nations* ["Modern Library" ed.], p. 367).

Not only did Smith object to share renting, but he proposed that taxes be used to induce landlords to use other leasing arrangements: "Some landlords, instead of a rent in money, require a rent in kind, in corn, cattle, poultry, wine, oil, etc. Others again require a rent in service. Such rents are always more hurtful to the tenant than beneficial to the landlord. They either take more or keep more out of the pocket of the farmer, than they put into that of the latter. In every country where they take place, the tenants are poor and beggarly, pretty much according to the degree in which they take place. By valuing . . . such rents rather high, and conse-quently taxing them somewhat higher than common money rents, a practice which is hurtful to the whole community might perhaps be sufficiently discouraged" (*ibid.*, p. 783).

3. Arthur Young, *Travels during the Years 1787, 1788, & 1789 [in the] Kingdom of France* (Dublin, 1798), II, 241.

4. *Ibid.*, p. 239.

5. (London, 1831), p. 102. It should be noted that the writer has been able to find no discussion of the share contract by either Ricardo or Malthus, even though both were intensely interested in agricultural rent.

6. *Ibid.*, pp. 106–8. Jones's summary of this argument was: "Not only the wealth of a nation, but the composition of society, the extent and respective influence of the different classes of which it consists, are powerfully affected by the efficiency of agriculture. The extent of the classes maintained in non-agricultural employment throughout the world must be determined by the quantity of food which the culti-vators produce beyond what is necessary for their own maintenance. The agricul-turalists of England for instance produce food sufficient to maintain themselves, and double their own numbers. Now the existence of this large non-agricultural population, the wealth and influence of its employers, and of persons who traffic in the produce of its industry, affect in a very striking manner the actual elements of political power among the English, their practical constitution, and their national character and habits. To the absence of such a body of non-agriculturalists and of the wealth and influence which accompany their existence, we may trace many of the political phenomena to be observed among our continental neighbours. If the agriculture of those neighbours should ever become so efficient, as to enable them to maintain a non-agricultural population, at all proportionate to our own, they may perhaps approximate to a social and political organization similar to that seen here. At all events, they will have the means of doing so" (*ibid.*, pp. 106–7).

7. J. R. McCulloch, *Principles of Political Economy* (Edinburgh, 1843), p. 471. His analysis was stated as follows: "The widest experience shows that tenants will never make any considerable improvements unless they have a firm conviction that they will be allowed to reap the whole advantage arising from them. It is in vain to contend that, as the tenant knows beforehand the proportion of the increased pro-duce going to the landlord, if the remainder be a due return to his capital, he will lay it out. No one tenant out of a hundred would so act. There are always very considerable hazards to be run by those who embark capital in agricultural im-provements; and if to these were added the obligation to pay a half, a third, or a fourth of the *gross* produce arising from an improvement, to the landlord, none would ever be attempted by a tenant, or none that required any considerable outlay, or where the prospect of return is not very immediate" (p. 470).

8. John Stuart Mill, *Principles of Political Economy* (Ashley ed., 1920), p. 305.

9. *Ibid.*, pp. 307–8.

10. *Ibid.*, pp. 308, 315–17, and Book II, chap. ix, "Of Cottiers."

11. See Simonde de Sismondi, *Principes d'économie politique* (Paris, 1827), Book III, chap. v, and also *Études sur l'économie politique* (Paris, 1837), I, 278–330.

12. Mill argued for a "tenant right" in the land for two reasons. First, security

of tenure induces the tenant to make improvements if the land rent is fixed by law or custom. If the rent is not so fixed, any improvements may be used as an excuse by the landlord to increase rents. Second, security of tenure and interest in the land will lead to a limitation upon population. Where farms cannot be subdivided or the rent increased, each family recognizes that its fate is in its own hands and limits its family size accordingly. Where such is not the case, an increase in population will force up rents and subdivision of the land, reducing tenants to a state of poverty (*op. cit.*, p. 321).

On the other side, it must be noted that Mill failed to recognize the effect of the "tenant right" upon the landlord and upon the income of new tenants. This is true even though Mill cites evidence showing that tenants sold the occupancy of farms for "ten to sixteen, up to twenty and even forty years' purchase of the rent" (*ibid.*, p. 320). This was not paid for unexhausted improvements but for the right of occupancy. Mill does not show why there is any difference between the landlord's and the outgoing tenant's collecting the "competitive rent."

13. *Ibid.*, p. 304.

14. "For, when the cultivator has to give to his landlord half of the returns to each dose of capital and labour that he applies to the land, it will not be to his interest to apply any doses the total return to which is less than twice enough to reward him. If, then, he is free to cultivate as he chooses, he will cultivate far less intensively than on the English plan (fixed cash rent); he will apply only so much capital and labour as will give returns more than twice enough to repay himself, so that his landlord will get a smaller share even of those returns than he would on the plan of a fixed payment" (Alfred Marshall, *Principles of Economics* [8th ed.; London, 1920], p. 644; see also pp. 642–45).

15. Rainer Schickele, "Effect of Tenure Systems on Agricultural Efficiency," *Journal of Farm Economics*, XXIII (February, 1941), 185–207; and Earl Heady, "Economics of Farm Leasing Systems," *Journal of Farm Economics*, XXIX (August, 1947), 659–78.

16. Heady, *op. cit.*, p. 673.

17. What is argued here means nothing more than that the tenant considers the leasing of the farm as a single opportunity to be compared with not farming or with farming under a cash lease. The supra-marginal returns on early labor inputs counterbalance the submarginal returns on later labor inputs.

18. The above reasoning indicates that landlords cannot solve the problem by setting the rental high enough for their share of a small output to equal the cash rent they could get. If all landlords banded together and the agricultural production function were of a certain nature, this solution might be possible. Output would be restricted, and farm prices would be very high. However, it would be to the interest of each landlord individually to rent his land for a fixed cash payment, and thus the scheme would break down.

19. I am indebted to Mr. Kenneth Arrow for this conclusion.

20. I have estimated net rents on crop-share-rented farms in Iowa from 1925 through 1946. From 1925 through 1934 net rents on share-rented farms averaged perhaps a dollar per acre less than on cash-rented farms. From 1935 through 1939 the net rents were roughly the same. From 1940 through 1946 net rents were at

least four dollars an acre more on share-rented than on cash-rented farms. These data are not presented as conclusive evidence, partly because of the roughness of the statistics and partly because it cannot be proved that the lands rented under the two lease types are comparable (for data on cash rents see H. R. Johnson, *The Farm Real Estate Situation, 1946–47* [U.S. Department of Agriculture Circ. No. 780], p. 29).

21. A description of existing metayer contracts in Italy by Giacomo Giorgi is in point here. Among the conditions which the tenant accepts is the following: "to cultivate rationally the farm in accordance with the granter's orders." The metayer cannot do the following things without permission of the landlord: (1) "cultivate grounds of his own or grounds taken with a lease contract"; and (2) "harbour, also temporarily, the outsider of the farming family, to shelter cattle of others and to reduce the capacity of labour of the farming family, allowing the absence of one or more of his members" (*The Metayage in the Province of Perugia* [privately printed, 1947]).

22. Even in this case, the tenant has some freedom of action. He may own capital equipment which he rents to other farmers and may neglect his own farm as a consequence.

23. The example has been chosen in this way to illustrate a riskless situation. In actual cases the share rent seems to be in excess of cash rents largely because of the risks borne by the landlord.

24. Schickele argues—incorrectly, I believe—that whenever a farmer uses more inputs than indicated by the equality of the marginal return and the cost to the tenant of the input, "he is paying a hidden privilege rent to the landlord for being permitted to stay on the farm" (*op. cit.*, p. 193). Actually if the tenant does not apply more inputs than is indicated by this criterion, the landlord is being exploited.

25. Cf. Heady, *op. cit.*, p. 674.

26. Alfred Marshall pointed out a similar peculiarity of labor: "It matters nothing to the seller of bricks whether they are to be used in building a palace or a sewer; but it matters a great deal to the seller of labour . . ." (*op. cit.*, p. 566).

27. A comparison of the relative efficiency with which resources are used on farms leased under crop share and cash rents can be made without too much difficulty. It involves only two things: first, an estimate of the income of comparable tenants under the two leases (the most important element of comparability is the capital position of the tenants) and, second, an estimate of the net rent paid per acre of comparable land. A rough test can be made by comparing the net output of farms of approximately the same size.

28. It is difficult to understand why the short-term lease has persisted for cash-rent leases. One reason may be that American landlords are always ready to sell their farms if a capital gain can be realized. A long-term lease would make it difficult to sell to an owner-operator. Another reason may be the fear that tenants, because many of them want to be owners, will exploit a rented farm as a means of accumulating capital. A short-term lease is a protection to the landlord. Finally, it is extremely difficult to adjust rents under a long-term lease. And if rents are to be established competitively, failure to agree over the rent bargain must be left as a condition for terminating the lease.

29. See Joseph Ackerman and Marshall Harris (eds.), *Family Farm Policy* (Chicago: University of Chicago Press, 1947), chaps. xvi and xvii.

30. *Principles of Economics* (2d ed.), p. 643.

31. In making this comment, I have relied mainly upon the survey of research in this field given by Leonard Salter, *A Critical Review of Research in Land Economics* (Minneapolis: University of Minnesota Press, 1948), chap. vii.

3

Functioning of the Labor Market

During the last three decades there has been a continuous interchange of workers between agriculture and the rest of the economy. This interchange has resulted on balance over the three decades in a net movement of members of the farm labor force into the nonfarm labor force.[1] In this paper I have tried to analyze the functioning of the labor market, particularly the nonfarm labor market, as it has affected the net transfer of labor from agricultural to nonagricultural pursuits. The analysis is in a normative context—there is an interest in determining how well or efficiently the market has functioned. It is assumed that the market should function to provide for an efficient utilization of the nation's labor resources. For present purposes the following (incomplete) criterion of efficiency has been accepted: Labor of equivalent capacities should earn the same real (marginal) return in all employments.[2]

Our inquiry emphasizes two main issues: (1) What is the relationship between real returns to labor in agricultural and nonagricultural employments for comparable labor? (2) Does the functioning of the nonfarm labor market adversely affect farm migrants in their search for jobs? These, of course, are not all of the issues that could be considered.

I. Real Returns to Farm and Nonfarm Labor

The use of our restricted criteria—equal real returns to equivalent labor—requires three different types of determinations. Each is a major undertaking in itself; consequently, in this paper some will have to be considered only cursorily. First, it is necessary to determine the differences in the purchasing power of income as it may be affected by region, occupation, and community size. Second, it is necessary to determine the equivalence of

Reprinted by permission from the *Journal of Farm Economics* 33 (February 1951): 75–87.

capacities or skills of the groups being compared. Finally, it is necessary to determine the money value of the labor return. The latter is an extremely complicated undertaking in agriculture and none too simple in many non-agricultural cases. In agriculture, roughly three fourths of all labor is performed by the farm operator or his family. For these workers there is no explicit money wage to be used as a guide, but the returns must be calculated by techniques of somewhat dubious validity. Even these techniques, which might give reasonable results under ideal conditions, are made even less reliable because we do not have satisfactory measurement of the labor employment in agriculture and of the nonfarm labor incomes of individuals whose major work activity is in agriculture. (These two problems are related, of course.)

Some research has been done on the measurement of the purchasing power of income in farm and nonfarm areas.[3] In this paper I shall assume that a dollar earned in a farm area will purchase 25 percent more than a dollar earned in a nonfarm urban community. The difference would be less than this if the comparison is with a rural nonfarm community.

A study of the comparability of the farm and nonfarm residents as workers[4] that analyzed data on the work experience of migrants indicated two conclusions: (1) Recent migrants from farm areas obtained jobs that paid roughly 90 percent as much as the median received by the nonfarm population after adjustments had been made for age and sex, and (2) there was no important difference to be attributed to the region of origin in the occupational experience of migrants, if only whites are included. I do not take the 90 percent figure very seriously. Its meaning to me is that there is probably not much difference between the capacities and skills of farm and nonfarm people, taken as a group. Some of the difference, assuming it to be statistically significant, is related to the relatively short period of adjustment permitted the farm migrants. The remainder might well be attributed to differences in education.

The return to labor in agriculture has been estimated in my article, "Allocation of Agricultural Income."[5] Several of the reservations implicit in the methods are discussed in that paper and need not be repeated here.

Whether these estimates are a fairly reliable indicator of the total annual labor return of individuals employed in agriculture depends largely upon the estimate of labor employment used. If the estimate used is a good measure of the man years of work done in agriculture, there is no need to add in any returns from nonfarm work. If the estimate used is a good estimate of the number of individuals whose major activity was in agriculture, there is probably little need for adjustment.

The current estimates of farm employment published by the BAE are undoubtedly too high to serve as an estimate of either full-time employment or of the number of individuals with their major work activity in

Table I. Average Labor Returns per Worker Year from Farming by Regions 1940 and 1945[a] (in dollars)

	1940	1945
New England	510	1,090
Middle Atlantic	565	1,100
East North Central	505	1,455
West North Central	500	1,745
South Atlantic	260	710
East South Central	260	540
West South Central	315	760
Mountain	505	1,560
Pacific	530	2,200
U.S.	385	1,135

[a]Estimates made by writer from BAE and Census data. Labor return is a residual after deducting from net farm income the estimated rent on land plus a return on non-land capital. Farm employment in 1940 was assumed to be 10.5 million and in 1945 about 9.8 million. These were roughly the BAE estimates available at the time the study was made; they are about 10 percent less than current BAE estimates and about 10 percent higher than estimates made by the Bureau of the Census for workers 14 years and over.

agriculture. The estimates contained in the MRLF are probably somewhat low.[6] It is possible to obtain an independent estimate of farm employment by patching together bits and pieces, starting from the estimates of total farm population. If this is done, one gets an estimate of about 10 million individuals 14 years and over employed in agriculture in 1940 and about 8.5 million in 1948. To these should be added perhaps 500,000 children under 14. The official BAE estimates are 11.671 million and 11.08 million for 1940 and 1948 respectively.

If we use the lower estimates and ignore children under 14 years, we get an average labor return of $410 in 1940 and $1,770 in 1948, the peak year of labor income during the forties. The income of employed industrial workers was $1,275 in 1940 and $2,710 in 1948.[7] The farm labor returns after the adjustment for cost of living differences are $560 and $2,210.

But these estimates cover up the very wide diversity of agricultural labor returns associated with location. Table I gives estimates of the average labor returns per worker from farming by regions for 1940 and 1945.[8] Net income data are not available by states or regions since 1945, but on the basis of total cash receipts the South Atlantic lost ground relative to the nation exclusive of the South; the East South Central retained its relative position and the West South Central gained slightly between 1945 and 1948. Consequently, it seems reasonable to assume that 1948 data would reveal much the same relative picture.

If the data in Table I are modified to include only labor performed by whites, the figures for labor returns in the South would be increased by 10 to 15 percent. This leaves them considerably below any non-south region.

The data in Table I would seem to reflect a rough equivalence between farm and nonfarm labor returns in three and perhaps four regions in 1945. The average wage income of employed industrial workers was $2,255.[9]

II. Functioning of the Nonfarm Labor Market

Migration and the increase in the number of nonfarm jobs held by farm people have not been important enough to achieve equality of real returns between farm and nonfarm occupations requiring equivalent capacities for three regions with about half the farm population. When there is a high proportion of unemployment in the nonfarm sector of the economy, the failure of migration is inexplicable. But we have now had a decade of high level employment and the relative position of low income farm regions has not been improved appreciably.

A. Regional Nonfarm Wage Rates

It is possible that farm people in low income farm areas may not find alternative jobs outside of agriculture because either the nonfarm jobs in the immediate localities are relatively low paying jobs, or controls prevail that limit the access of farm workers to nonfarm jobs. These seem to be the two most important influences that might operate from the demand side.

The importance of nonfarm wage levels in the low farm income areas follows from a significant feature of domestic migration—the preponderance of short distant moves. Analysis of the data on farm migrants from the South, the largest area of relatively low farm labor incomes, who moved to nonfarm residences indicate that for the period 1935–40, 84 percent of the white males and 94 percent of the nonwhite males stayed in the South. Considering all farm migrants (from all regions) to urban areas, 60 percent remained in the same state and 20 percent moved to contiguous states. A slightly larger percentage of the farm migrants to rural nonfarm communities moved shorter distances—64 percent within a state and 18 percent to contiguous states.[10]

If there were the same general disparity in labor returns in the nonagricultural sectors of the economy by areas as there is in agriculture, the failure of migration (job mobility) to solve the problem of low incomes in agriculture would be understandable. But does such a similar pattern of nonfarm labor returns exist?

It is true that per capita incomes in the South are considerably lower than the national average—60 percent as much in 1935 and 70 percent as much in 1946. But the lower per capita income payments seem to be explained almost completely by the following considerations: (1) The greater relative importance of agriculture in the South than in the rest of the country; (2) The relatively lower incomes in agriculture than in the rest of the

Table II. Median Wage or Salary Incomes of Urban Residents Who Earned $1 or More and Had No Other Money Income by Region, Sex, and Color, 1939

	All Workers[a]		Worked 12 Months	
Area and Color	Male	Female	Male	Female
White Workers				
Northeast	$1,237	$728	$1,485	$920
North Central	1,282	717	1,510	883
South	1,122	686	1,359	826
West	1,334	827	1,611	1,040
U.S.	1,247	725	1,488	907
Nonwhite Workers				
Northeast	835	476	1,020	568
North Central	804	406	981	526
South	500	242	631	296
West	768	496	952	639
U.S.	602	297	739	366

[a]16th Census of the United States, 1940, *Population, the Labor Force, Wage or Salary Income in 1939*, Table 5. Excludes individuals on public emergency work.
[b]*Ibid.*, Table 5a.

Table III. Median Incomes of White and Nonwhite Urban Families by Regions, 1946[a]

	Median incomes	
Region	White	Nonwhite
Northeastern States	$3,367	$2,235
North Central States	3,244	2,294
West	3,206	[b]
South	3,014	1,549
United States	3,246	1,929

[a]Source: Bureau of the Census, *Current Population Reports, Consumer Income*, Series P-60, No. 1, Rev., Table 4.
[b]Data not given in source. Median for urban and rural-nonfarm nonwhite families was $2,659. Median for urban families probably about $2,750 to $2,800.

country; (3) The greater relative importance of nonwhites in the South; and (4) The different distribution of the nonfarm population among communities of various sizes. If these factors are isolated and their effects determined or eliminated, it appears that white families in communities of the same size have roughly the same incomes in the South as in any other region of the nation. This is not true of nonwhites, however, since the incomes of nonwhite families are significantly lower in the South. Tables II and III present some of the relevant data; other data indicating the same conclusions are available.[11]

It should be remembered that the urban classification does not result in a homogeneous distribution of community sizes among regions. Consequently, it is to be expected that if the regions could be further subdivided into groups according to community size, the southern white incomes would roughly approximate those in the rest of the country.[12]

Thus far we have been dealing with large areas. It is possible that in any given and predominantly rural areas with low farm incomes, nonfarm incomes could be low and be offset by higher nonfarm incomes in the same region, thus raising the regional average to roughly the national level. This could occur if the sub-areas within the region with low farm incomes have only a small proportion of the nonfarm population of the region.

A tentative test of this possibility has been undertaken, but the results are essentially negative. There does not seem to be any relation, either in the South or in the North Central States between the level of nonfarm income and farm income if counties are used as the basis of comparison.[13]

The wage structure in nonfarm employment may not be a good index of the wages in the jobs available to farm migrants. Employment in some jobs may be closed to a considerable degree by monopoly—either by firms or unions; in other cases, legal minimum wages may reduce the level of employment in affected industries. The wage structure of the jobs that farm migrants could obtain might reflect wages considerably below that paid for nonfarm workers with similar capacities, which would reduce the amount of migration out of agriculture.

B. Effects of Unions on Wage Rates

In 1945, total union membership in the United States was roughly 14 million or about one third of the nonfarm labor force. The fact of organization is not adequate evidence that unions have significant control over job opportunities, either by affecting wage rates or by control of the specific individuals working in the organized firms.[14] Wage rates—and thus the total of job opportunities—can be controlled, in theory at least, in two ways. First, by bargaining, backed by the threat of various punitive measures, that establishes a wage rate above what the firm would have to pay to attract the number of workers that it is willing to employ at that rate. Second, by direct control over the number of individuals permitted to be union members—through initiation fees, apprentice rules, and period of required training—and forcing employers to hire only union members.[15] If there is competition among employers, this will force wage rates to higher levels (if numbers have actually been controlled); if employers organize, the results are probably indeterminate, since the employers fear that higher wage rates will not attract more workers.

The use of either technique by unions may result in the creation of a "counter force." Employers resist wage rate changes that might well be in

Table IV. Average Level of Annual Wages in Retailing and Manufacturing in Counties Ranked by Level of Gross Farm Income in 1939

A. North Central States

	Iowa		Illinois		Nebraska		Kansas	
	R[b]	M[b]	R	M	R	M	R	M
Group I[a]	714	810	712	871	752	773	724	1,046
Group II	699	970	796	976	700	921	646	913
Group III	728	926	792	1,043	627	797	668	949
Group IV	675	816	714	836	651	742	702	961

	Minnesota		Wisconsin		Indiana		Ohio	
	R	M	R	M	R	M	R	M
Group I	749	867	807	991	798	897	834	1,061
Group II	837	1,133	959	970	734	860	741	944
Group III	789	976	697	822	719	752	758	914
Group IV	826	811			696	776	706	952

B. Southern States

	Georgia					
Percent	0–30		30–50		50 or more	
nonwhite	R	M	R	M	R	M
Group I	550	462	610	496	537	489
Group II	563	440	539	472	584	445
Group III	647	—	618	536	499	451
Group IV	516	509	489	504	596	411

	Kentucky			
Percent	0–5		5 or more	
nonwhite	R	M	R	M
Group I	633	705	789	733
Group II	629	678	723	596
Group III	557	829	624	603
Group IV	617	555	641	668

	Tennessee			
Percent	0–10		10 or more	
nonwhite	R	M	R	M
Group I	585	684	558	682
Group II	595	643	601	570
Group III	535	613	542	535
Group IV	551	484		

Table IV. *continued*

B. Southern States

	Virginia					
Percent	0–10		10–40		40 or more	
nonwhite	R	M	R	M	R	M
Group I	675	697	724	688	710	482
Group II	738	754	633	512	703	519
Group III	734	—	613	—	611	449
Group IV	819	686			625	381

	North Carolina					
Percent	0–10		10–40		40 or more	
nonwhite	R	M	R	M	R	M
Group I	723	740	650	498	665	613
Group II	586	516	649	573	679	465
Group III	622	654	733	672	680	559
Group IV	658	650	631	600	553	498

Source of data: U.S. Bureau of Census, *County Data Book, 1940.*

[a]Counties ranked by level of gross farm income per farm and then divided into roughly equal groups. Group I includes highest income counties and Group IV (or III) the lowest.

[b]R = retailing; M = manufacturing.

Table V. Relation between Changes in Average Straight-Time Hourly Earnings and Changes in Degree of Organization, 1933 to 1946

	Degree of Organization[a]				
	I	II	III	IV	V
Number of industries	4	10	9	21	6
Absolute increase (cents)[b]	54.0	56.2	61.6	64.3	50.4
Percentage increase[b]	183	142	144	136	95
Change in employment[c]	71	86	182	174	38

Source: Taken directly from or calculated from data in Ross and Goldner, *op. cit.*, pp. 258–59, 270–71, 274–75.

[a]Degree of organization: Groups I to IV, less than 40 percent organized in 1933; Group V more than 40 percent organized in 1933. Group I, less than 40 percent organized in 1946; Group II, 40–60 percent; Group III, 60–80 percent, and Group IV, 80–100 percent.

[b]Unweighted average.

[c]Percentage increase for the Groups. Coverage not quite identical with data on wage increases.

their own interest in the absence of unions. Employers also may be more prone to organize and follow a set pattern of action.

The effects of unions on wage rates is subject to dispute; research in this field consists of bits and pieces rather than systematically integrated theoretical and empirical work. The approach that has been most popular is to compare changes in wage rates in unionized industries from a time prior to unionization to some second date with the changes in wage rates in a non-unionized industry. If this could be done for a large number of pairs of industries in which each member of the pair were comparable in all other important respects, some useful insights might be achieved. But such comparisons are usually not possible, either because of lack of data or because the similar industries seem to fall into either the unionized or non-unionized categories. Studies have also been made of firms in the same industries that are unionized and not unionized, but failure to eliminate such factors as size of plant or location by region or city size make it difficult to interpret the results.

A pioneer study of the effect of unionism on wages is contained in Professor Douglas' famous *Real Wages in the United States, 1890 to 1926*. Douglas compared indexes of hourly earnings in "union" manufacturing industries and in "payroll" manufacturing industries.[16] Using 1890 as a base of 100, the union industries' full time weekly earnings increased to 261 in 1926, the payroll industries to 275. If 1914 is used as a base, the increase was to 216 for union industries and to 205 for the payroll industries.[17] During the same period, unskilled labor increased from 100 in 1890 to 293 in 1926, and from 100 in 1914 to 209 in 1926.[18] Douglas concludes as follows: "It seems clear, in any case, that the increase in real wages during the period studied has been caused primarily by the increase in productivity rather than by unionization."[19]

Douglas' results have been criticized by Ross because Douglas relied upon relative changes in wage rates rather than absolute changes. The basis for his criticism seemed to be the following: The non-unionized industries had lower earnings in the base period and if both groups of industries received equal absolute changes in wage rates, the non-unionized industries would show the larger percentage increases. If equal absolute changes in earnings are indicative of equal improvement, the percentage changes in earnings would give a fallacious indication.

The choice of the measure to be used in indicating the effects of unions upon wage rates cannot be made in so arbitrary a fashion. The choice rests on presumptions about how the wage rate pattern would have changed during the period in the absence of unions. Ross' analysis must rest on the assumption that without unions absolute changes in wage rates would be equal, with the relative or percentage differentials narrowing. Most classical analyses of the labor market assume that without unions percentage differentials would remain constant, with the absolute differentials widening. *A priori* it is impossible to say what changes would have occurred in the absence of unions. It is necessary to develop a general equilibrium model of the labor market, and from whatever information is available about the parameters of the model, estimate the changes that would have occurred in the wage rate pattern. Equal absolute or equal percentage changes are only two of a number of possibilities.

Even accepting Ross' criterion of absolute increases as a measure of the effectiveness of unions, there is no evidence in the material he presents that indicates that the degree of organization in 1946 was a factor influencing the change in wage rates. The material presented includes the absolute changes in hourly earnings for industries classified by degree of unionization. The relatively unorganized group received increases averaging 54.0 cents per hour; the most highly organized industries had an average increase of 64.3 cents per hour.[20] But analysis of variance indicates that the group means do not differ significantly.[21]

If instead of absolute increases, relative increases are used, it is found that the differences in relative increases by degree of unionization from 1933 to 1946 are very small. The most unionized groups have the smallest increases, but the differences are probably not significant. The period from 1933 through 1946 should provide a fairly good test, since most unionization in the United States occurred following 1933.[22]

Joseph Garbarino found that the degree of unionization was not significant in explaining (relative) wage changes between 1923 and 1940. Most of the changes seemed to be explained by changes in output per man-hour, degree of enterprise concentration and employment.[23]

Lester has compared the wage rates for union and non-union plants based on data from a group of wage studies made by the BLS for 1943, 1944, 1945, and 1946. The comparisons were for a city or local labor market or a region. In general the wage rates in non-union plants were as high as or higher than in union plants in about 30 to 35 percent of the cases.[24] But many of the non-union plants were much smaller than the union plants and smaller plants generally pay somewhat lower wages than large plants. Consequently it would be necessary to standardize for the size of plants before an accurate comparison could be made. In any case, these data do not indicate that unions have achieved appreciably higher wage rates than those paid to non-union workers.

The above data are not conclusive evidence that unions do not, in general, have much effect on wage rates. Nor do they preclude the possibility that unions involving upwards of three million or so workers, particularly the building trades, railroads, and printing, have not had appreciable effects on wages. But the contrary hypothesis that unions have in most cases "distorted" the wage structure in a way inimical to farm migration does not seem to be supported by available data.

If unions have not had much effect on wage rates, obviously the employment opportunities available to new entrants into the nonfarm labor force have not been restricted by high wage rates in the unionized industries. This does not prove that unions cannot affect employment opportunities in other ways. Two such avenues seem to be open to them: raising labor costs by reducing output per unit of time and by affecting lay-off and hiring policy. A few comments on the latter policy are in order. As long as we have recurring periods of unemployment, union rules affecting seniority on layoffs and re-hiring may work to the disadvantage of recent farm migrants. Such migrants will be among the first to be laid off (if they have had a job) and the last to be re-hired or hired. During a recovery period, new entrants to the non-farm labor force will be forced to wait until all workers who have been laid off (and wish to return to work at their old jobs) are taken on before the migrants can be employed. However, if we are able to maintain employment at high levels almost all of the time, the seniority rules imposed by unions should not have much impact upon the availability of job opportunities.

Job opportunities available to farm people may be influenced by governmental action, particularly by minimum wage legislation. It is unlikely that minimum wage legislation in the United States has had much effect. Rising price levels have tended to negate any influence that such legislation might have had. But minimum wage legislation that exempts certain industries (agriculture) and fails to recognize community size and occupational differences could reduce the alternatives open to many farm migrants, particularly nonwhites and those whites who could obtain only relatively low paying nonfarm jobs.[25]

III. Why Has Migration Failed?

If our two main propositions concerning regional or area wage rate differences and the effect of unions on wage rates are valid, the failure of migration has not been explained by our analysis. Two possibilities remain. One, there may be other aspects of the functioning of the nonfarm labor market that pose barriers to migration. We have mentioned one earlier, namely, the uncertainty of unemployment due to business fluctuations. If other aspects are important, I have been unable to determine what they may be. Second, the main barriers to migration may be indigenous to the farm people and the conditions in which they live.

At present, I would argue that understanding the failure of migration to achieve equality of returns to labor in agriculture and non-agriculture will come largely through analysis of influences indigenous to farm people and their immediate environment. Most farm people in low income areas may be ignorant of the economic opportunities existing elsewhere; most may have insufficient capital to permit a move; most may have so limited a set of experiences that they fear the transition to nonfarm life; many may feel strong family or community ties; many may have reached an age that inhibits seeking new experiences; others may reject the values and modes of living that they associate with nonfarm living. These are some of the possible explanations, but as yet no one has designed a detailed and acceptable research program that would provide us with new insight. The development of such a research program is one of the important challenges in the study of rural life.

Notes

The research on which this article is based has been done on a project financed by a grant for Agricultural Economic Research at the University of Chicago made by the Rockefeller Foundation. This paper is a revision of one presented at the Research Seminar on Economic Efficiency at the University of Chicago in July, 1950, and was read at a joint meeting of the American Farm Economic Association and the American Economic Association in Chicago, December 30, 1950.

1. The net movement of farm population to nonfarm areas from 1920 through 1949 was approximately 17.5 million, of which perhaps half was in the labor force. The total out-movement from farms during these three decades exceeded 50 million. Size of movements based on BAE estimates of farm population and migration.

2. This proposition is a useful criterion of efficiency if one of three conditions actually exists. First, there is nowhere in the economy a significant difference between the price of labor and the value of its marginal product. This condition would prevail, more or less exactly, if there were no monopoly or monopsony. Second, if monopoly or monopsony exists, it results in a difference between the price of labor and the value of its marginal product where the price of a given quality of labor is high relative to the price of the same quality of labor employed subject to competitive conditions. A transfer of labor from low paying jobs to higher paying jobs would increase total real output. The efficient position as defined by our condition would not, of course, represent an absolute maximum. Third, the incidence of monopoly or monopsony is randomly distributed as it affects jobs with different levels of (real) pay for labor of comparable capacities. Some transfers from lower to higher paying situations would result in a decline in the value of marginal product of some labor, but on balance a net increase would occur.

It would appear that our criterion would fail to indicate change in the right direction only if the value of the marginal product of labor exceeds its wage in those cases that the individual worker would gain by transferring to a higher paying job, in which the value of the marginal product did not exceed the wage. Even here, if the wage in the latter exceeded the value of the marginal product in the former, an improvement could result.

It should be apparent that our criterion for judging the functioning of the labor market forces us to consider only one part of a total situation—namely, do all (most) laborers—within a given structure of demand by firms—achieve the best use of their own labor resource?

3. See the article by Nathan Koffsky and the comments by Margaret G. Reid, E. W. Grove, and D. Gale Johnson in *Studies in Income and Wealth*, Vol. XI (New York, National Bureau of Economic Research, 1949). Koffsky estimates that for low income families, the cost of a farm budget at urban prices (1941 data largely) would have been 30 percent greater than at farm prices. This is, of course, an overestimate of the cost of achieving an equivalent level of consumption as measured by satisfaction. It is also likely that in the higher income brackets the purchasing

power of income on farms falls relatively to urban purchasing power. Food expenditure is a smaller proportion of total expenditure, and home produced food declines as a proportion of total food. It is the food category that is responsible for most of the difference in purchasing power of income. In recent years, the difference in purchasing power would be narrowed by the increase in the ratio of food valued at farm prices and food valued at urban retail prices.

4. "Comparability of Labor Capacities of Farm and Non-Farm Labor," (dittoed). The measure of capacity was the income earning ability of farm migrants in nonfarm occupations compared to the actual income earned by the nonfarm population as wage and salary workers. The study is subject to numerous reservations which are indicated in the paper.

5. [Reprinted as chap. 1 of this volume.]

6. See "A Critical Appraisal of Farm Employment Estimates" by M. C. Nottenberg and this writer, to be published in the *Journal of the American Statistical Association.*

7. *Agricultural Outlook Charts*, 1950, p. 3.

8. Available evidence indicates that the relative income positions of the three southern regions were no better in 1929 than in 1940 or 1945. See BAE, USDA, *Income Parity for Agriculture*, Part VI, Section I (October, 1945).

9. In any comparison of farm and nonfarm wage incomes, it must be remarked that the age and sex distribution of the farm and the nonfarm working force differs. Considering only the workers over 14 and using urban wages as weights, the 1940 farm population had an earning capacity about four percent below the nonfarm.

The regional pattern of nonfarm wages and incomes is considered in Section II. Average wage income of industrial workers is estimated by dividing annual earnings of factory, railroad, and mining workers by average employment. See BAE *Agricultural Outlook Charts*, 1950, p. 3. This particular group of workers has an average annual wage income of about 90 percent of the U.S. nonfarm average in most recent years. In 1945, however, the average for the group was about the same as the U.S. nonfarm average.

In 1940, in contrast, the highest average regional labor income was only $565, or about 55 percent as much as the income of employed industrial workers *after* adjustment for cost of living differences. Only in the East South Central States was there no indication of relative improvement in the ratio of farm to nonfarm labor incomes between 1940 and 1945.

10. *U.S. Census, Population, Internal Migration, 1935 to 1940, Economic Characteristics of Migrants*, Tables 10 and 3.

11. See D. Gale Johnson, "Some Effects of Region, Community Size, Color, and Occupation on Family and Individual Income." Mimeo. to be published by the National Bureau of Economic Research.

12. See Herbert E. Klarman, "A Statistical Study of Income Differences among Communities," in *Studies in Income and Wealth*, Vol. VI.

13. The test was designed as follows:

A sample of 20 percent of the counties, excluding metropolitan counties, in eight Midwestern States and of 50 percent of the counties in five Southern States were drawn. These counties were ranked by the level of (gross) farm income per farm in 1940 within each state and then grouped into quartiles or thirds. For each of these groups of counties, the average annual wages per worker in retailing and in manufacturing were calculated. Even though the average farm income in the top group of counties was three or four times the bottom group, there was no discernible difference in the level of earnings in retailing and manufacturing (See Table IV).

Gross farm income is not as satisfactory a measure of farm income as one would like. It is believed that the exclusion of the metropolitan counties will exclude most of the part time farms in each of the states. (For about half of the states the counties also were grouped by gross income per farm worker. This change in classification did not affect the results.) Differences in gross farm income could possibly reflect differences in types of farming as it affects the relation between gross and net income, but casual inspection of the counties and their area distribution do not leave this impression. In any case, the difference between the gross incomes in the highest and lowest groups in each state is so large that it certainly reflects substantial differences in net farm income—at least of the order of two to one.

14. Wage rates are not, of course, the only determinant of total job opportunities that might be brought under the control of unions. "Work standards" can be and perhaps are in some cases very important.

15. This technique may only be supplementary to the first and used by the unions as a rationing device to allocate the available number of jobs among its members. Some of my friends, who know much more about unions than I do, believe this to be the case.

16. The latter group included some industries that were partially organized during part of the period. However, the significance of unionism in this group seems to have been fairly unimportant.

17. Paul H. Douglas, *Real Wages in the United States, 1890–1926*, pp. 119 and 126. The weekly earnings were $17.83 in 1890, $21.379 in 1914 and $46.22 for the union industries (p. 118), $9.27, $12.42 and $25.47 respectively for the payroll industries (p. 124).

18. *Ibid.*, p. 177. It should be noted that the building trades, unionized after 1890, showed greater relative increases in earnings than any of the groups cited in the text. With 1890 equal to 100, 1914 weekly earnings were 146 and 1926 earnings were 331. However, 1926 was a boom year for building and this may have had some effects on the results.

19. *Ibid.*, p. 564.

20. Arthur M. Ross and William Goldner, "Forces Affecting the Interindustry Wage Structure," *Quarterly Journal of Economics*, LXIV (May, 1950), pp. 270–71.

21. The value of F is 2.14. The value of F for the 5 percent level of significance is 2.85. Of course, the test of significance used may not be a reliable one.

22. If data were available, 1929 should be substituted for 1933 so that the years compared would both be ones of essentially full employment. Likewise, a year later than 1946 should be used to avoid the effects of war-time wage policy.

23. Joseph W. Garbarino, "A Theory of Interindustry Wage Structure," *Quarterly Journal of Economics*, LXIV (May, 1950), pp. 282–305, esp. 302 and 305.

24. See Richard A. Lester, "Some Reflections on the 'Labor Monopoly' Issue," in Francis Doody, *Readings in Labor Economics*, pp. 304–06.

25. The application of the U.S. Minimum Wage law to Puerto Rico apparently had very serious employment effects and would have had even more had it not been for large scale evasion. It was found necessary to adopt special minima for the island.

4

Comparability of Labor Capacities of Farm and Nonfarm Labor

When watching the action of demand and supply with regard to a material commodity, we are constantly met by the difficulty that two things which are being sold under the same name in the same market are really not of the same quality and not of the same value to the purchasers. Or, if the things are really alike, they may be sold even in the face of the keenest competition at prices which are nominally different, because the conditions of sale are not the same. . . . But difficulties of this kind are much greater in the case of labour than of material commodities: the true price that is paid for labour often differs widely, and in ways that are not easily traced, from that which is nominally paid.[1]

In the above paragraph and his later discussion, [Alfred] Marshall indicates that wage rates or money earnings, by themselves, provide an inadequate picture of the functioning of the labor market because the laborers being compared are not of equal efficiency or the conditions of sale of labor are dissimilar. Any analysis of the functioning of the farm labor market is beset by difficulties of both kinds. This is especially true if one wishes to determine whether the labor market has tended to equalize the returns to farm and nonfarm labor. In this paper only the first of the two main problems is considered—the general question of the efficiency of the farm work force compared to the nonfarm work force.

Because the word "efficiency" has so many different connotations, the term "labor capacity" is used in its place. Our measure of the labor capacity of the farm labor force is the job distribution that a random sample of the farm labor force would have in nonfarm employment if each worker held

Reprinted by permission of the American Economic Association from the *American Economic Review* 43 (June 1953): 296–313.

the job which for him had the greatest net advantage. This measure is subject to two restrictions: it relates to a given time and it assumes that farm migrants find jobs in competition with the nonfarm labor force, as it actually exists at the time, in a given labor market. The measure of relative labor capacity that can be determined is ratio of the average money earnings reflected by the (hypothetical) nonfarm job distribution of farm labor force to the average money earnings of the nonfarm work force. The measure assumes that farm people have the same set of attitudes concerning the nonmonetary advantages and disadvantages of nonfarm jobs as do nonfarm people.[2]

This study attempts to determine the relative labor capacity of farm people by analysis of the job experience of farm persons who have migrated to nonfarm areas. The materials have come largely from the 1940 Census of Population and cover persons who migrated from farms between January 1, 1935 and the time the 1940 Census was taken.[3] The general approach has been to estimate the money earnings reflected by the farm-nonfarm migrant population, standardized to eliminate the effects of age, sex and color differences relative to the parent farm population, and to compare this estimate to the labor earnings of the nonfarm population. The conclusions reached later are based upon two significant hypotheses: (1) The migrants from farm areas are representative of the parent population in all significant characteristics, excepting age, sex and color. (2) The actual job experience of farm migrants in nonfarm areas is indicative of the labor capacity of farm migrants. This implies that recent migrants from farms are able to find the jobs for which they are best suited under given labor market conditions.

It has not been possible to find evidence that would support or reject the second hypothesis. It seems probable that the first job held by many migrants is not the one best suited to his particular labor capacities. The average time the farm migrants had been in nonfarm communities was approximately two and one-half years. Consequently, the majority of them had had an opportunity to search for different jobs, though an important fraction had not. It is quite possible therefore that the dependence of the analysis upon this hypothesis has had the effect of a slight underestimation of the labor capacity of farm people.

I. Selectivity of Rural Migration

Migration from farm to nonfarm communities is a selective process with respect to certain population characteristics. The migrants included in the present study were younger than the parent population.[4] The migrant stream included slightly fewer males[5] and significantly fewer nonwhites than the rural farm population of 1940.[6]

Selectivity of migration with respect to age, sex and color is of little significance in the present instance. The effects of these types of selectivity can be reflected in our estimates of labor capacity. Migration, however, may be selective with respect to other characteristics that are related to labor capacity, such as intelligence, manual dexterity and education. The hypothesis that the farm migrants to nonfarm areas represent a random sample of the parent farm population with respect to characteristics other than age, sex and color is not rejected by three different types of evidence that bear on the issue.

1. The median number of years of school completed by farm migrants 25 to 34 years old was almost identical to that of the farm population of the same age—about 8 years of schooling. About 11.5 per cent of the migrants had some college education compared to 8.8 of the rural farm population. Almost the same proportion graduated from college (4.3 and 3.8 per cent). It does not appear that the farm migrants were appreciably better educated than the farm nonmigrants.[7]

2. If there is a positive relation between the labor capacity of farm migrants and the level of farm income in their original community, the distribution of farm migrants by the region of origin does not indicate that the farm migrants came predominantly from either the high or low farm income regions. An analysis of the data on net off-farm migration rates by states for the decade of the 'thirties as a whole revealed a weighted correlation of −0.38 between the level of net farm income per worker and the rate of migration from farms. The correlation coefficient, which is just significant at the one per cent level, indicates that the low income states had a somewhat higher rate of net off-farm migration than the higher income states. However, it should be noted that for the particular group of migrants included in our data, the regional distribution of the origin of migrants did not indicate any relationship between the level of farm incomes and the rate of migration.

3. There have been several score of studies made for individual communities in an attempt to determine if migrants represent a selected group from the parent population that is in some way superior or inferior. These studies, made by sociologists, have used various definitions of superiority or inferiority, but most of the research workers had in mind a concept that had a significant relation to economic productivity. In a review and analysis of this literature, Dorothy S. Thomas found four conflicting hypotheses: (1) Cityward migration selected the superior elements of the parent population; (2) Cityward migration selects the inferior elements; (3) Cityward migration selects from the extremes; and (4) Cityward migration represents a random sample of the parent population. She concludes as follows: "We have, then, evidence of a sort that migration selects the better elements, the worse elements, both the better and the worse, and also that it

is unselective. Even though we may decide that the evidence cited is tenu-
ous, it is not improbable that selection does operate positively, negatively,
and randomly, at different times, depending on a variety of factors that, up
to the present, have not been adequately investigated."[8]

Sorokin and Zimmerman arrived at the following conclusions from an
analysis of the literature: "There is no valid evidence that migration to the
cities is selective in the sense that the cities attract in a much greater proportion those from the country who are better physically, vitally, mentally, mor-
ally, or socially, and leave in the country those who are poorer in all these
respects [italics omitted]. There is also no evidence that the reverse is true."[9]

Though I have not been able to provide evidence denying the possibility of selectivity in the migration process, it is significant that the hypothesis that no important degree of selectivity exists cannot be rejected on the
basis of available data. Consequently, in this paper it is assumed that there
is no selectivity.

II. Occupational Experience of Rural Farm Migrants to Nonfarm Areas

The published data from the 1940 Census of Population do not provide
estimates of the wage and salary incomes of farm migrants to nonfarm
areas. The published data indicate the employment status of the migrants
and the occupational category of the employed migrants. On the basis of
certain assumptions, which are indicated below, these data are translated
into an estimate of annual labor earnings of the migrant group.

Recently the Bureau of the Census tabulated data on the wage and salary
incomes of farm migrants to urban areas for samples of migrants from the
Corn Belt and from the Cotton Belt. The wage and salary incomes relate
to 1939; consequently, some of the migrants were still living and working in
farm areas during part or all of 1939. Thus the wage and salary data cannot
be used directly but required adjustment to reflect the effect of lower money
incomes on farms as well as for the more obvious adjustments for age. In this
section, I will present the results obtained from the published data and later
show the degree of consistency between the two sets of estimates.

A. Incidence of Unemployment

These data, summarized in Table I, indicate that rural farm migrants to
urban areas had almost identically the same employment status as the urban population. Of those classified as members of the labor force, 84.4 per
cent of the farm migrants were employed while 84.7 per cent of the labor
force members resident in urban areas were employed.[10] Some difference
does exist in the case of migrants to rural nonfarm areas. Roughly 3 per
cent fewer migrants than nonmigrants were employed.[11]

Table I. Distribution of Rural Farm Migrants and Total Population in the Labor Force in Urban and Rural Nonfarm Areas by Employment Status, 1940, for U.S. (Per cent)

	Both Sexes			Males		
	Em.[a]	S.W.[a]	E.W.[a]	Em.	S.W.	E.W.
Rural farm migrants by destination						
Urban	84.4	11.2	4.4	81.1	13.4	5.5
Rural nonfarm	79.6	12.9	7.5	77.9	13.7	8.4
Rural farm	91.4	5.3	3.3	91.7	4.9	3.4
Total population						
Urban	84.7	11.0	4.3	81.4	13.8	4.8
Rural nonfarm	82.2	10.6	7.2	81.0	11.1	7.9
Rural farm	91.7	4.4	3.9	92.1	4.0	3.9

[a]Em = Employed; S.W. = Seeking Work; and E.W. = Emergency Work.

Source: Sixteenth Census, *Population, Internal Migration, 1935 to 1940, Economic Characteristics of Migrants,* Tables 2 and 8. Data relate to individuals 14 years or older.

Corroborating data are provided by a Census sample survey of the employment experience of migrants in 1948. These data cover migrations of one-year duration or less. In this case migrants had rather different rates of unemployment than the nonmigrants. However, the male migrants from farm to nonfarm areas had roughly the same proportion of unemployment (9.2 per cent) as the migrants from nonfarm to nonfarm areas (8.1 per cent).[12] These rates of unemployment were much higher than for the total male population, which was 3.6 per cent. The higher rate of unemployment among the migrants was probably due to the relatively short time interval considered.

B. Occupational Distribution of Employed Migrants

The previous section indicated that farm migrants to nonfarm areas found employment as readily as nonmigrants, if a certain time for adjustment was allowed. The question to which we now turn is the nature of the jobs which the farm migrants obtained.

Table II compares the occupational distribution for employed male and female migrants to urban and rural nonfarm areas with the occupational distribution of the nonmigrants. In the table, the occupational groups are ranked according to the medium wage or salary income of male workers employed for 12 months in 1939.[13]

It will be noted that the farm migrants to urban areas are not as fully represented in the higher income groups as the urban nonmigrants are. In the top four groups, which contain 56.4 per cent of the nonmigrants, are found only 38.0 per cent of the farm migrants and much the same situation prevails in the rural nonfarm areas—45.0 per cent of the nonmigrants fall in the top

Table II. Percentage Distribution Among Major Occupation Groups of Employed Farm Migrants and Nonmigrants, by Sex, in Urban and Rural Nonfarm Areas in 1940

	Male		Female	
	Nonmigrants (per cent)	Farm Migrants (per cent)	Nonmigrants (per cent)	Farm Migrants (per cent)
Urban				
Proprietors, managers and officials	12.4	5.4	3.7	1.4
Professional and semiprofessional	6.5	3.9	11.2	13.7
Clerical and sales	18.4	12.7	33.5	16.8
Craftsmen and foremen	19.1	16.0	1.2	0.4
Operatives	22.8	26.4	21.2	13.0
Service workers	8.9	12.7	11.6	17.1
Laborers	10.5	16.9	1.9	1.4
Domestic service	0.5	0.8	5.4	35.5
Farmers and farm managers	0.6	3.9	0.1	0.1
Farm laborers and foremen	0.4	1.1	0.1	0.6
Total	100.1	99.8	99.9	100.0
Rural Nonfarm				
Proprietors, managers and officials	12.8	6.9	6.1	3.1
Professional and semiprofessional	4.7	4.0	13.2	18.9
Clerical and sales	9.9	5.1	21.8	9.7
Craftsmen and foremen	17.6	12.1	0.8	0.3
Operatives	25.6	22.3	20.1	13.2
Service workers	5.1	14.3	11.2	17.0
Laborers	15.3	18.1	2.9	2.2
Domestic service	0.6	0.6	22.4	32.7
Farmers and farm managers	2.4	2.1	0.3	0.1
Farm laborers and foremen	6.0	14.2	1.3	2.9
Total	100.0	99.6	100.1	100.1

Source: Sixteenth Census, *Population, Internal Migration, Economic Characteristics of Migrants,* Table 10.

four groups while only 28.1 per cent of the farm migrants do. Comparison of female farm migrants and nonmigrants shows much the same thing.

The importance of the differences in the occupational distribution of the migrants and nonmigrants depends upon the differences in the wages in the various occupational groups. If the wage differences were substantial, one would expect the occupational distribution of the farm migrants to reflect substantially lower income than the occupational distribution of nonmigrants. One way of determining the extent of the difference is to weight the occupational distribution by the median incomes actually received in each occupation. The census provides information on the money income from the receipt of wages and salaries. Any nonmoney income such

Table III. Income Reflected by Occupational Distribution of Employed Farm Migrants and Nonmigrants in Urban and Rural Nonfarm Areas, by Sex, 1939

Residence in 1940 and Sex	Farm Migrants	Nonmigrants	Per Cent Migrants of Nonmigrants
	(dollars per year)		
Urban			
Male	1075	1260	85
Female	560	700	83
Rural Nonfarm			
Male	960	1160	83
Female	590	660	89

Note: Based on data in Table II and data below. Median annual wage and salary incomes by occupational groups were as follows:

	Male	*Female*
Proprietors, managers, and officials	$2,136	$1,107
Professional and semiprofessional	1,809	1,023
Clerical and sales	1,359	883
Craftsmen and foremen	1,309	827
Operatives	1,007	582
Service, except domestic	833	491
Laborers, except farm	673	538
Domestic service	429	296
Farmers and farm managers	373	348
Farm laborers	309	76

Source: Sixteenth Census, *Population*, Vol. III, *Labor Force*, Part I, Table 2. In calculating medians, incomes of less than $100 were not included to exclude most unpaid family workers. Comparisons do not entirely eliminate effects of age.

as the perquisites to farm laborers and domestic servants is not given. The income of self-employed persons is also excluded. Likewise data are available only for the United States as a whole and not for urban and rural nonfarm areas. Of the three main limitations of the data, only the first—failure to include nonmoney wage or salary income—is important. Since relatively few urban workers fall in the self-employed category, failure to include income from that source is unimportant.[14]

The results of these calculations are presented in Table III. They show that the income reflected by the occupational distribution of the rural farm migrants, had they received the median income of each occupational group, would have been somewhat less than the income reflected by the occupational distribution of urban and rural nonfarm nonmigrants. The differences are of the magnitude of 11 to 17 per cent less.

Are these differences due solely to the different innate capacities of the groups being compared? One obvious source of difference is age. The farm migrants are younger than the nonmigrants in the areas to which the farm migrant went. Farm migrants to nonfarm areas fall into age brackets that earn less than the average of either the parent or the absorbing population.

The calculations given in Table III eliminate the effect of age upon the level of earnings within an occupational group. An adjustment of the data to include the effects of age upon the occupational distribution of a population can be made by comparing the occupational distribution of nonfarm people of the same age distribution as the farm migrants with the actual occupational distribution of nonfarm people.

Such an analysis cannot be made directly from the data at hand. Census data are available showing the occupational distribution by age only for the country as a whole and not for nonfarm groups. This distribution was adjusted by omitting farmers and farm managers and farm laborers. The Census data gave the age distribution of farm persons migrating to urban areas, but not all such migrants were in the labor force. It was assumed, since the proportion of the migrants that was in the labor force was roughly the same as for urban nonmigrants, that for any age group labor force participation was the same for rural farm migrants and urban nonmigrants.

When these adjustments were made, it indicated that the income reflected by the actual age distribution of the male farm migrants was $1,230 while the male urban nonmigrants, excluding farmers and farm laborers, had an average of $1,270.[15] This would indicate a difference of roughly $40 due to occupational differences associated with the age distribution of farm migrants. If this adjustment is applied to the data in Table III, it would mean that the male farm migrants to urban areas would earn on the average about 11 per cent less than male urban nonmigrants. This would be an income level roughly equal to the average incomes of (1) operatives and (2) craftsmen and foremen.

All of the above calculations on wage earnings relate, of course, to the given set of relative wage incomes by occupational groups. A different set of relative wages combined with either the same or different occupational distribution of migrants and nonmigrants would result in a different comparison of labor earning possibilities.

Data on wage or salary incomes by occupational groups comparable to the data used in the previous analysis are available for 1950.[16] A comparison of the changes in wage or salary incomes by occupational groups indicates that the groups in which farm migrants are concentrated to a greater degree than the urban population had the greatest relative rise in earnings. This change in relative wage rates had the effect of reducing the differential between the earnings value (before adjustment for the effect of age on occupational distribution) of male urban nonmigrant and the farm migrants occupational distribution from 15 per cent to 10 per cent. After adjustment for the effects of age, the earnings of male farm migrants were about 8 per cent less than the male urban nonmigrant.[17]

Table IV. Income Reflected by Occupational Distribution of Farm Migrants to Urban and Rural Nonfarm Areas Classified by Region of Origin and Sex, 1939

Residence in 1940	Region of Origin			
and Sex	Northeast	North Central	South	West
Urban				
Male	1125	1120	1030	1280
Female	620	590	555	650
Rural Nonfarm				
Male	960	1035	910	980
Female	575	640	545	670

Source and method of calculation: see Table III. The income reflected by the occupational distribution of nonmigrants for urban areas was $1,260 for males and $700 for females and for rural nonfarm areas was $1,160 for males and $660 for females.

The conclusion that one might draw from this analysis is that labor employed in agriculture, on the average, has a labor income capacity of roughly 90 per cent of the labor income capacity of our urban and rural nonfarm populations, for similar age and sex distributions.

This estimate is for the nation as a whole. The migrants from different regions may have fared rather differently. The next section analyzes the experience of migrants by region of origin.

III. Occupational Experience of Farm Migrants from Different Regions

The farm migrants from different regions to nonfarm areas did not achieve the same occupational distributions. Table IV shows the earning value of the occupational distributions of the migrants from the four regions to urban and rural nonfarm areas. The West farm migrants to urban areas had the highest earning value, while the South farm migrants had the lowest when the migrants went to urban areas. The North Central farm migrants did the best of migrants to rural nonfarm areas while the South farm migrants did the least well.

The relatively poor showing of the South seems to be due to the color composition of the South migrants. The published Census data did not give separate data for the occupational distribution of whites and nonwhites. However, a special tabulation of data from the 1940 Census of Population permits a comparison of the occupational distribution of male white farm migrants to urban areas for the Cotton Belt and the Corn Belt.[18] A calculation of the income value reflected by the occupational distributions was the same as for the Corn Belt.[19]

Consequently, if the occupational distributions accurately reflect the capacity of the farm migrants and thus of the farm population, it may be concluded that the labor capacity of the white farm labor force in the South is approximately the same as in the rest of the nation. This conclusion is somewhat surprising in light of the significant differences that exist between agricultural labor earnings in the South and in the rest of the nation.[20] The writer considered the possibility that the farm migrants in the South came in disproportionate numbers from the rural areas with the highest farm incomes. This might then explain the relative ease with which Southern farm people fit into the nonfarm occupation pattern.

The hypothesis does not appear to be valid, however. Though data are not available to show the origin of white farm migrants by small areas, such as counties, data on all farm migrants indicate that the rate of migration is as high, if not higher, from counties with the lowest levels of living as it is from counties with the highest levels of living.[21]

IV. Wage or Salary Income of Farm Migrants to Urban Areas

Because wage and salary data were not available for all farm to urban migrants, it was necessary to derive an estimated level of earnings from their occupational distribution. This estimate was based on the assumption that the median earnings of the migrant were the same as the median earnings for the occupational group involved. Through the cooperation of the Bureau of the Census, it has been possible to obtain the wage and salary data for 9,500 migrants from the Corn Belt, 8,900 white and 4,000 nonwhite migrants from the Cotton Belt to urban areas.[22]

The wage or salary data for the sample of migrants cannot be directly compared with the data provided in the Census for any urban group. First, the wage or salary income is for the year 1939 and at least 24 per cent of the individuals included in the sample resided on farms part or all of 1939. Thus, the income data do not indicate directly the income of the farm migrants in urban occupations. Most of the individuals who lived on farms some part of 1939 either would not have had any wage or salary income while living on farms because they were farm operators or unpaid family workers or would have worked as farm laborers at a lower level of money income than the average urban wage or salary level. Second, the age distribution of migrants is substantially different from that of urban nonmigrants. Despite these two difficulties, the availability of these data provide an opportunity to check the general validity of the assumptions involved in the earlier analysis.

The migration period covered in the Census questionnaire was from April 1, 1935 to the time the Census enumeration was actually made, which was subsequent to April 1, 1940. Thus persons who migrated within an approximate 15-month period following January 1, 1939 would have spent some time on farms during 1939. On the basis of Bureau of Agricultural Economics estimates of departures from farm to nonfarm areas, and assuming that there was no greater memory bias for migrants leaving farms from 1935 through 1938 compared to migrants after that date, 24 per cent of the migrants lived part or all of the year 1939 on farms. The wage or salary income of male farm residents in 1939 was substantially below that of urban residents. In fact, for employed male farm residents without other income, but with $1 or more of income, it was $369; for the employed male urban resident, the median was $1,188.[23] Further, 30.5 per cent of the employed male farm residents without other income had no money income and presumably were unpaid family workers, while only 5.6 per cent of the urban group fell into this category. It is reasonable to assume that the male farm migrant would not have had money wage or salary earnings during the period of residence on farms in 1939 greater than 30 per cent of the amount he did or would have earned in an urban area. Consequently, I have assumed that the median earnings of the farm to urban migrants should be increased by a minimum of 10 per cent to adjust for the factor discussed here.[24]

The adjustment for the difference in age distribution was made by estimating the level of wage or salary income employed urban residents would have had if their age distribution had been the same as that of the farm migrants. This estimate was then subtracted from the earnings level reflected by the actual age distribution of the employed urban residents. The results of these adjustments are indicated in Table V.

Table V also provides a comparison between the adjusted wage or salary medians for the employed male farm migrants and the medians for urban employed workers. The results are, I believe, quite consistent with the estimate of a 10 per cent difference derived for the United States as a whole on the basis of the occupational distributions. Some of the variation in the ratios of the incomes are undoubtedly due to sampling variation. For example, the occupational distribution of the sample of Corn Belt migrants reflected an earnings value of about 3 per cent more than for all the farm to urban migrants from the North Central States. However, the combined occupational distribution for all Cotton Belt migrants was virtually identical with that for the South as a whole.

The results are, I believe, reasonably consistent with the estimate of a 10 per cent difference in the labor capacities of the farm and nonfarm labor forces for the United States. The high relative incomes of nonwhite migrants in the Cotton Belt offset the somewhat lower relative incomes for whites in the same region. These data support one of the assumptions underlying the analysis of the occupational distributions, namely, that the farm migrants in any occupational group had the same median level of earnings as the urban resident in the same occupational group (after adjustment for age and sex). If this assumption were not valid, the relative money

Table V. Median Annual Wage or Salary Income of Employed Male Rural Farm Migrants to Urban Areas and of Employed Male Urban Residents, and Adjustments for Residence and Age, Corn Belt and Cotton Belt, 1939

Area and Color	Migrants' Median Wage or Salary Income[a]	Residence Adjust- ment[b]	Age Adjust- ment[c]	Migrants' Adjusted Iincome	Urban Median Wage or Salary Income[d]	Income Ratio: Migrants to Urban
	(dollars)				(dollars)	
Corn Belt, White	954	95	126	1,175	1,218	0.96
Cotton Belt						
South, White	690	69	122	881	1,010	0.87
North, White	708	71	134	913	1,040	0.88
South, Nonwhite	412	41	30	483	465	1.04
North, Nonwhite	404	40	50	494	465	1.06

[a]From special tabulations of Census data. Includes all employed workers on private or nonemergency government work in late March 1940, including those with no wage or salary income, but excludes any worker who had $50 or more of income from sources other than wages or salaries. The two standard deviation confidence limits for the sample medians, given in the same order as in the table are: 940–968, 670–710, 693–723, 394–430 and 386–422.

[b]An adjustment of 10 per cent applied.

[c]The age adjustment reflects the difference between the wage or salary level of the distribution of the employed migrants and of the employed urban residents evaluated at the level of earnings by age for the urban residents. For the white migrants, the age distribution and earnings levels by age were those for employed male urban residents for the United States as a whole. *Source:* Sixteenth Census, *Population, The Labor Force (Sample Statistics), Wage or Salary Income in 1939*, Table 6. For the nonwhite data from Sixteenth Census, *Population, Education, Educational Attainment by Economic Characteristics and Marital Status*, Tables 20 and 35, were used after minor adjustment.

[d]For wage or salary workers without other income on private and nonemergency government work. Median calculated for employed persons including those with no wage or salary income in 1939. For Corn Belt the income figure used is for the North Central States. (Sixteenth Census, *Wage or Salary Income in 1939*, Table 5). For the Cotton Belt, nonwhite, the South as defined by the Census is used (*ibid.*, Table I). The median wage and salary incomes for the two Cotton belt areas were not available from the Census and it was necessary to estimate the medians. The median of $1,050 for white male employees for the South as a whole was reduced, in each of the two cases, by the percentage that the median family income for white wage or salary workers' families, in the states included in the Cotton Belt area, was below the similar median for the South as a whole. (See Sixteenth Census, *Population and Housing, Families, General Characteristics*, Tables 34 and 45).

incomes derived from wage data would have been substantially below 90 per cent since the occupational distributions in our area samples were very similar to those for the regions that included the samples.

The data on money wages or salaries seem inconsistent with our earlier results at one point, namely, that the white labor forces in the Corn Belt and the Cotton Belt had the same labor capacities. This conclusion was based on the comparison of the occupational distribution, assuming that the levels of earnings by occupational groups were the same in the two areas. The data given in Table V on urban wage or salary incomes indicate that this assumption is not valid since the occupational distribution of non-farm white workers in the South is nearly the same as for white workers in the nation as a whole.[25] Does the difference in the earnings of the urban residents in the South and in the North Central States reflect differences in the labor capacities of the nonfarm labor forces of the two regions? Though I cannot justify my conclusion in the space available to me, I would answer the question in the negative. Some of the difference can be explained in terms of the distribution of workers by size of community and size of plants in which workers are employed. Much of the remainder was due to the demand and supply relations in the labor market in the South compared to the rest of the nation. There is some evidence that there was a fairly substantial reduction in the labor-earnings differential between the South and the rest of the country from the late 'thirties to the late 'forties.[26]

Even if one accepts the assumption that the difference in labor earnings of urban residents does not reflect any difference in labor capacities, there is a difference in the ratio of farm migrant and urban incomes in the Corn Belt and Cotton Belt.

It should be noted, however, that the occupational distribution of the Corn Belt migrants reflected an earnings value of about 3 per cent more than for all the farm to urban migrants from the North Central States, while the combined occupational distribution for all Cotton Belt migrants was virtually identical with that for the South as a whole. Thus if one reduced the ratio of migrant to urban incomes from 0.96 to 0.93, this might be a reasonable estimate for the North Central States as a whole. While the difference that remains of approximately five per cent in relative incomes after this adjustment is probably greater than can be explained by sampling errors, it does not seem unduly large given the other adjustments that have been made in the data.

The major surprise in the results is the nonwhite results. The consistency of results for the two samples creates doubt that the high ratio can be explained by sampling variability.

V. Qualifications and Conclusions

If the occupational experience and wage or salary income of recent migrants from farm to nonfarm areas can be used as an indication of labor capacity, and if farm migrants are representative of the parent farm population,[27] then farm people have a labor capacity approximately 90 per cent of nonfarm people of the same age and sex. Two main hypotheses are indicated in this summary statement and the conclusion obviously depends upon their acceptance.

I believe that the first of the hypotheses may be subject to some reservation. The occupational and wage experience of farm migrants in the first year or so following migration may underestimate the level of their labor capacities. Such an underestimate would occur if migrants tended to improve their occupational or income situation to a greater degree than the rate of advancement normally associated with age and experience. This proposition is one that can be put to empirical test and such a study is now under way.

It should be noted that in all analyses and calculations medians have been used instead of arithmetic means. However, we know that distributions of wage or salary data are not symmetrical, but exhibit an important degree of skewness, with the median less than the means. If the ratio of the mean to median were the same for all of the wage distributions used, it would make no difference in any of the results obtained in our analysis. However, though there is evidence that the ratio of the mean to the median is generally larger in the higher than in the lower-paying occupations, given the diversity within the broad occupational groups used by the Census, there is no systematic relationship between the mean and median as the median wage income rises.[28]

Though some doubt may be cast upon the accuracy of means calculated from open-ended distributions, comparisons were made using the calculated means as well as the medians in estimating the earnings value of occupational distributions. In every case the change in the results was very minor. For example, though the ratio of the income value of the occupational distribution of male farm migrants to urban areas to the same estimate for urban nonmigrants was slightly less when means rather than medians were used, the results when rounded to the nearest per cent were identical. In general the effect of using the estimated means rather than medians would not have changed the results given in Table III by more than one per cent.

The results obtained have several important empirical or practical applications, of which two may be noted quite briefly. One is the relevance of the results to comparisons of farm and nonfarm incomes, or more specifically, to comparisons of labor returns to farm and nonfarm workers.[29] Any comparisons of returns for large and relatively heterogeneous labor groups are always suspect, unless an effort is made to determine the relative equivalence of the labor capacities of the groups. The present results, however, should not be used directly without first standardizing the labor groups being compared for age and sex distributions. If this is done, it may well be found that the farm labor force has a larger proportion of its members in the age groups with lowest earnings. However, the effects of the unfavorable age distribution are likely to be largely offset by the relatively small proportion of females in the farm labor force compared to most other groups. Using 1940 urban wage distributions by age and sex as weights, the 1940 farm labor force had an earning capacity about 4 per cent below the nonfarm labor force. Thus, if real labor returns were as much as 14 per cent lower for the average farm worker than for the average nonfarm worker, this might well be consistent with equal real returns for comparable workers.

A second application of these results is the assurance that they provide to new employers who might wish to locate in areas of low farm incomes. The evidence indicates that such employers will find individuals having or capable of quickly achieving a wide range of skills. Only in the managerial and professional categories is it likely that new employers will have difficulty in recruiting a labor force roughly equivalent to that available almost anywhere in the United States. The evidence provided by the data for farm migrants is perfectly consistent with the favorable labor productivity experiences of many firms that have located in rural areas where farm incomes have been low.

Notes

The research on which this article is based has been done on a project financed by a grant for Agricultural Economics Research at the University of Chicago by the Rockefeller Foundation. Financial assistance was also received from the National Security Resources Board. The author, professor of economics at the University of Chicago, is indebted to Jack Ciaccio and Marjorie Penniman for assistance in making the detailed statistical tabulations, and to T. W. Schultz and O. H. Brownlee for certain critical observations.

1. Alfred Marshall, *Principles of Economics* (London, 1936) 8th ed., pp. 546–47.

2. Reference is made only to the labor capacity of farm people as evidenced by the nonfarm jobs they might hold and not also to the farm jobs nonfarm people might hold. This has been done because the net transfer of labor in the United States has been from farm to nonfarm and the relevant comparison is the one chosen.

3. A migrant was defined as a person who lived in 1935 in a county or city of 100,000 or over, different from the one in which he lived at the time the 1940 Census was taken. Thus not all persons who changed from a farm to a nonfarm residence between 1935 and 1940 were counted as migrants if the move was within a county and not to a city larger than 100,000.

4. The white farm migrants to nonfarm areas had a median age about 10 years lower than the white farm nonmigrants if only persons over 14 are included. The nonwhite median for migrants was about 4 years lower than for the nonwhite farm nonmigrants. See Sixteenth Census of the U.S., 1940, *Population, Internal Migration, 1935 to 1940, Age of Migrants*, Table 9.

5. See Sixteenth Census, *Population, Internal Migration, Social Characteristics of Migrants*, Table 2.

6. See Sixteenth Census, *Population, Internal Migration, Color and Sex of Migrants*, Table 2. The nonwhite farm population was 16.2 per cent of the total farm population in 1940; only 12.2 per cent of the farm migrants to nonfarm communities were nonwhites.

7. Sixteenth Census, *Population, Internal Migration, Social Characteristics of Migrants*, Table 7.

8. Dorothy S. Thomas, "Selective Migration," *The Milbank Memorial Fund Quart.*, 1938, XVI, 403–07.

9. Sorokin, P. and Zimmerman, C. C., *Principles of Rural-Urban Sociology* (New York, Henry Holt and Co., 1929), p. 582.

10. No difference remains when the employment data are adjusted for the differences in age distribution.

11. Farm migrants to farm areas had the same employment status as the total farm population.

12. Bureau of the Census, *Current Population Reports, Labor Force*, Series P-50, No. 10, Table 4. The two figures could have been drawn from the same populations. (See *ibid.*, p. 5.)

13. There are only minor differences between the ranks of wages and salaries for males and females. (See Table III.)

14. Out of a total of 40,450,000 male workers in 1948, something less than 4,000,000 were self-employed in pursuits other than agriculture. The earnings of the self-employed professional workers in 1947 who were about one-third as numerous as the salaried professional workers, were about 40 per cent greater than the income of salaried professionals. However, most of the self-employed were in the proprietors, managers, and officials group and the self-employed earned about a sixth less than the salaried. See Bureau of the Census, *Current Population Reports*, Series P-60, No. 5, Table 17.

15. See Sixteenth Census, *Population*, Vol. III, *Labor Force*, Part 1, Table 65, for occupational distribution by age.

16. See Bureau of the Census, *Current Population Reports, Consumer Incomes*, Series P-60, No. 9, Table 25.

17. Calculations for migrants to rural nonfarm areas indicate a difference of 10 per cent at the 1950 relative wage levels.

18. For a description of the sample, see Section IV below.

19. Though the Cotton Belt does not include all of the South and the Corn Belt does not include all of the North Central States, the income value reflected by the farm migrants in the Cotton Belt farm (both white and nonwhite) migrants differed from that of the South as a whole by less than one per cent, while the difference between the Corn Belt and the North Central States was about three per cent. This indicates that the samples were quite representative of the larger areas. These calculations and those referred to in the text assume that the rates of pay in the same occupational groups were identical in all regions. The validity and significance of this assumption is discussed in Section IV.

20. For the years 1940 and 1945, the writer has estimated that annual labor returns to farm workers were as follows (in dollars):

Region	1940	1945
United States	$385	$1,135
New England	505	1,090
Middle Atlantic	565	1,105
E. N. Central	505	1,455
W. N. Central	500	1,745
S. Atlantic	260	710
E. S. Central	260	540
W. S. Central	315	760
Mountain	505	1,560
Pacific	530	2,202

These data relate to all labor in the South. Adjustments based on the relationship between the value of products sold, traded or used on farms of white farm operators and all farm operators in the South, result in estimates of labor income for white farm labor 10 to 12 per cent higher than those given above. Since some Negro labor is used on farms operated by white operators, there is a small, but undetermined, downward bias in the estimates of white farm labor income in the South.

21. The countries in the 16 states of the South were divided into quartiles according to the rural farm level of living index developed by Hagood. The average (unweighted) migration rates for the counties in each quartile were calculated for 1930–40 decade. The quartiles had the following average rural level of living indexes: 1st, 108; 2nd, 85; 3rd, 74; and 4th, 60.6. The average migration rates were: 1st, minus 13.1; 2nd, minus 13.1; 3rd, minus 15.4; and 4th, minus 13.2. Source of data: as to levels of living: Margaret Jarman Hagood, "Rural Level of Living Indexes for Counties of the United States, 1940" (Washington, 1943); and as to migration: Eleanor H. Bernert, "County Variation in Net Migration from the Rural Farm Population, 1930–40" (Washington, 1944).

22. These data were obtained jointly by the Scripps Foundation for Research in Population Problems of Miami University, the Bureau of Agricultural Economics of the United States Department of Agriculture, and Agricultural Economics Research of the University of Chicago. Mr. and Mrs. Donald J. Bogue contributed substantial amounts of their time and energy to the study. It should be noted that the sample included only migrants that remained within the same state.

23. Sixteenth Census, *Population, The Labor Force (Sample Statistics)*, Wage or Salary Income in 1939, Table 10a.

24. The method of arriving at this estimate may be indicated briefly. As noted in the text, 24 per cent of the migrants lived on farms part or all of 1939. If the movement occurred at the same rate during the 15 months through March, 1940, a fifth of this group of migrants would have lived all of 1939 on farms, while of the remaining four-fifths their average experience would have been a half year of work on farms and a half year of work (or seeking work) in urban areas. Thus taking the urban wage as one and the farm resident's salary or wage income as three-tenths, the following weighted average for all migrants is obtained:

$$
\begin{array}{r}
0.76 \ \times 1.00 = 0.760 \\
0.048 \times 0.30 = 0.014 \\
0.192 \times 0.65 = \underline{0.124} \\
0.898
\end{array}
$$

This average indicates that the actual wage or salary income of the migrant should be increased by about 11 per cent to adjust for the inclusion of income earned while a farm resident.

25. See D. Gale Johnson, "Some Effects of Region, Community Size, Color and Occupation on Family and Individual Income," *Studies in Income and Wealth*, Vol. XV (New York, National Bureau of Economic Research, 1952), p. 64. See also pp. 57–63 for a discussion of wages and labor earnings in the South compared to other regions.

26. *Ibid.*, pp. 57–63.

27. It is argued that the farm migrants are representative of the farm population only after adjustment is made for age and sex.

28. Estimates of the mean for the seven nonfarm occupational groups for males were made from the available distributions. The ratios of the estimated mean to medians for experienced workers receiving more than $100 in 1939 were: (1) Proprietors, etc., 1.22; (2) professional and semi-professional, 1.20; (3) clerical and sales, 1.14; (4) craftsmen and foremen, 1.06; (5) operatives, 1.07; (6) service workers (excluding domestic), 1.18, and (7) laborers, 1.12. Source: Sixteenth Census, *Population*, Vol. III, *Labor Force*, Part I, Table 72.

29. See L. H. Bean, "Are Farmers Getting Too Much?" and comments by D. Gale Johnson and J. D. Black, *Rev. Econ. Stat.*, Aug., 1952, XXXIV, 248–61.

PART TWO

Research, Productivity, and Supply Response

5

The Nature of the Supply Function for Agricultural Products

It is generally believed that the total output of farm products responds little if at all to changes in the average price of farm products. For example, most farmers believe this to be true for downward movements in real prices; the willingness of farmers and their representatives to accept direct control of output and marketing clearly reflects their belief that downward movements in real prices will not substantially reduce farm output. Though this belief is apparently based on the depression experiences of 1919–22 and 1929–33, it is applied without hesitation to a period when resources are generally fully employed.[1]

The best published discussion of the responsiveness of agricultural output to price changes is by Professors Galbraith and Black.[2] Their analysis is restricted to depression conditions and lacks preciseness largely because it fails to distinguish between those conditions relevant to the decision process within firms and those relevant to the nature of the factor markets, *i.e.*, the supply conditions of factors. No systematic attempt has been made, to my knowledge, to make a similar analysis for non-depression conditions.

This paper attempts to fill these gaps by analyzing the reaction of aggregate output (1) to falling relative prices under depression conditions and (2) to changing relative prices when resources are fully employed in the economy. These two sets of conditions are chosen to simplify the discussion and because they represent the important empirical conditions.

I. Agricultural Output During a Depression

Aggregate farm output has repeatedly failed to decline during depressions. Numerous explanations have been offered for this phenomenon; such as

Reprinted by permission of the American Economic Association from the *American Economic Review* 40 (September 1950): 539–63.

high fixed costs in agriculture, the length of the production process, and the competitive structure of agriculture. Several such explanations are reviewed and analyzed in this section and an attempt is then made to outline a more general explanation.

A. *The Facts*

Much of the evidence usually presented on the behavior of agricultural output has been interpreted as indicating that agriculture's output response during major cyclical declines is peculiar to it. Table I presents data of the type usually considered rather conclusive. In 1933, agricultural output was only 6 per cent below its 1929 level, and that decline was in substantial part due to poor weather in the Great Plains area. In 1932, farm output was actually higher than in 1929. Manufacturing output as a whole declined by more than a third, while iron and steel output declined by 70 per cent and machinery, by 60 per cent.

But any theory explaining agriculture's behavior must also be consistent with another set of facts, namely that there were other important segments in the economy which produced almost as much in 1932 and 1933 as in 1929. The following tabulation shows 1933 annual output as a percentage of 1929 output:[3]

Meat packing	95
Shortening	98
Canned milk	98
Cheese	95
Butter	110
Cotton goods	87
Woolen and worsted goods	87
Knit goods	100
Shoes, leather	94
Beet sugar	151
Canned fruits and vegetables	88
Clothing, women's	91
Soap	98
Petroleum refining	86

Employment in agriculture in 1929 was 8,323,000 family workers and 2,984,000 hired workers; in 1933, 8,590,000 and 2,433,000, respectively. Total employment declined roughly two per cent, from 11,289,000 to 11,023,000. Employment in manufacturing as a whole declined by 30–35 per cent and employment in certain types of manufacturing (automobiles, for example) by more than 40 per cent and hours worked, by even more. On the other hand, employment in food manufacturing declined only 15 per cent from 1929 to 1932 and rose from 1932 to 1933.

Table I. Indexes of Production for Agriculture, Manufacturing, and Mining, 1927 to 1947
(1935–39=100)

Year	Agricultural Output	Manufacturing	Durable Manufacturing	Non-Durable Manufacturing	Iron and Steel	Machinery	Mining
1927	95	94	107	83	108	99	97
28	99	99	117	85	121	106	95
29	97	110	132	93	133	130	103
30	95	90	98	84	97	100	91
31	104	75	67	79	61	66	82
32	101	57	41	70	32	43	72
33	93	68	54	79	54	50	80
34	79	74	65	81	61	69	83
35	96	87	83	90	81	83	89
36	85	104	108	100	114	105	99
37	108	113	122	106	123	126	109
38	105	87	78	95	68	82	99
39	106	109	109	109	114	104	105
40	109	126	139	115	147	136	114
41	114	168	201	142	186	221	122
42	128	212	279	158	199	340	125
43	125	258	360	176	208	443	132
44	130	252	353	171	206	439	145
45	129	214	274	166	183	343	143
46	133	177	192	165	150	240	142

Source: Statistical Abstract, 1947, p. 816 and Agricultural Statistics, 1947, p. 533.

B. A Review of the Explanations

Roughly a half-dozen explanations of the cyclical behavior of agricultural output have been put forward. This section examines each of them; the next presents a general theory constructed to explain those characteristics of agricultural output under discussion.

1. High fixed costs. The belief that high fixed costs are responsible for the failure of farmers to reduce output during a depression has achieved more general acceptance than any other explanation.

It is generally argued that farm firms have high fixed costs because the labor supply is so closely related to the firm. Labor is viewed as a resource that is fixed to the firm because the operator and family members constitute roughly three-fourths of total farm workers. The same considerations are frequently alleged to apply to land.

This argument seems invalid. The employment of hired labor in agriculture is almost as constant as family employment.[4] Yet hired labor is certainly not considered a fixed resource by the firm. Hired workers are apparently willing to offer their services at prices which the firms believe are no higher than the value of the marginal product and so continue to be employed.

In many cases, the land operated is one of the assets of a firm in the sense that the firm owns the land. Most farm firms have the alternative of renting the land to another firm to operate, even during a depression. Only in the exceptional case would such an alternative not be available. Even if farmers who own land are unwilling to consider this alternative, the behavior of firms renting land cannot be explained by treating land as a fixed cost. There was roughly as much land rented during the 'thirties as was owned by the farm firms. Presumably operators continued to use the land because the price of land fell enough to equalize demand and supply at a level of "full" employment.

Table II is pertinent to the present argument. It indicates changes in gross income, net operator income, and production expenses from 1929 to 1932. Most of the changes in expenditures are due to changes in prices rather than quantities.[5] Net operator income represents the returns to all resources owned or controlled by the operator, including labor, land, and capital. Of the total production expenses, only taxes and farm mortgage interest would have continued to be claims on current income even if no output had been planned. These two items represented, in 1929, only 15 per cent of total production expenses and only 9 per cent of the value of gross output.

Even if one were willing to assume that labor is a fixed cost to the farm firm, nothing is gained by so doing. High fixed costs are not an adequate

Table II. Expenses of Farm Production and Farm Income
(millions of dollars)

	1929	1932	1947
Gross income	13,824	6,406	34,705
Net operator income	5,654	1,715	17,087
Total production expenses	8,170	4,691	17,618
Feed purchased	919	348	3,783
Livestock purchased	461	164	1,302
Fertilizer and lime	293	125	685
Operating motor vehicles	509	384	1,505
Hired labor	1,284	584	2,791
Miscellaneous current expenses[a]	1,146	814	1,768
Taxes	641	504	705
Farm mortgage interest	582	534	222
Rent (net)[b]	1,062	343	2,300
Depreciation[c]	1,273	890	2,579

Source: Department of Agriculture, Bureau of Agricultural Economics, Net Farm Income and Parity Summary 1919–41 (mimeo.) and Farm Income Situation, August–September, 1948.

[a]Includes such items as electricity, twine, ginning fees, dairy supplies, seeds, containers, etc.

[b]Gross rents were as follows: 1929—$1,621,474,000; 1932—$668,935,000; and 1947—$3,100,000,000. About the same amount of land was rented in 1932 as in 1929, while perhaps ten per cent less was rented in 1947.

[c]Gross investment: 1929—$1,414,000,000; 1932—$290,000,000; 1947—$3,682,000,000.

explanation of the relative stability of hired labor employment nor of the constancy of the amount of land rented. Nor can one, on this basis, explain the constancy of the output of livestock products, particularly hogs. Farm management studies indicate that 75 per cent of the average cost of producing hogs is feed cost. Hog output is maintained only because feed prices fall, and fall as much or more than hog prices during the downswing of a major depression.

The constant employment of factors apparently reflects not high fixed costs but either (a) inelastic supply curves together with highly flexible factor prices or (b) changes in the marginal opportunity costs of the factors with the business cycle. The first explanation is pertinent to physical capital assets and land; the second to labor, feed, and livestock. Since most feeds are durable, why do feed prices fall low enough during a depression to clear the market of all of the current output? Given existing cost conditions during a depression, farmers are maximizing their position by producing crops, but they have the alternatives of feeding or storing the feed crops in anticipation of higher prices later. In this way, livestock output could be contracted during depression. Constancy of output of livestock implies that the supply curve of feed grains for current use has shifted far to the right. I shall consider this point later.

2. Farmers try to offset lower prices by increased output. The explanation that agricultural output is maintained (or even increased) by farmers as a means of offsetting lower prices may have a certain validity. Over certain ranges, the supply curve of operator and unpaid family labor may be backward sloping; *i.e.*, individuals work more at a lower than a higher wage. This statement implies that as income falls, the marginal utility of income increases relative to the marginal utility of leisure. Since the employment of some inputs do decline during a depression because their prices are not as flexible as product prices, farm output is probably maintained by a small increase in the quantity of labor supplied by a given number of workers.

3. Subsistence production is important in agriculture. If production is largely for the consumption of the operator family and few of the inputs are purchased, relative prices have little effect upon the firm under any circumstances. Production decisions will be based largely on resources owned or controlled (mostly land and labor), and consumption preferences.

In 1939, about one-quarter of all farms had household use as the major source of income, and in 1944, roughly 22 per cent.[6] These farms produced less than four per cent of the total farm output in 1944.[7] Roughly three-quarters of what was produced on these self-sufficient farms was consumed in the household. In 1944, the rest of the farm operators consumed only about eight per cent of what they produced.[8] Consequently, the production of the bulk of the agricultural output is so commercialized that constancy of subsistence production cannot explain constancy of aggregate production.

4. Technological factors inhibit response to price changes. The production process in agriculture is relatively long. Consequently, a decline in prices may not be followed at once by a reduction in output. Farmers will find it advantageous to complete the production process as long as price equals or exceeds the marginal cost of completing the production process as of any moment of time. This explanation can apply only to cycles of short duration when prices fall for a year or eighteen months and then start to increase. It cannot apply to the 1929 to 1933 downswing. Here farmers did have time to change their production plans and yet failed to do so.

5. Agriculture has a more competitive structure than the rest of the economy. The belief that agricultural output is maintained during depressions because agriculture is competitive is strongly held in many quarters. The belief that output in many sectors of the nonfarm economy is highly variable during a business cycle because of enterprise monopoly is also strongly held.

The data on page [84] suggest that monopoly by no means always leads to large output variations during a business cycle. By measures ordinarily used to measure degree of concentration (percentage of output controlled by a few firms), canned milk, meat packing, soap, and beet sugar are rather non-competitive; the rest of the industries, relatively competitive. Yet all had highly stable output.

In addition, it should be noted that several competitive industries produced much smaller outputs in 1933 than in 1929. Among these were all branches of mining and lumber and most products of lumber.[9]

An enterprise monopolist would restrict output and maintain prices during a depression in two sets of circumstances. First, if the supply functions for all factors were perfectly elastic *and* if the supply functions did not shift from the peak to the trough of the business cycle. In these circumstances the monopolist would find it in its interest to maintain prices in the face of declining demand.[10] Without knowing the exact nature of the shift in demand, it cannot be said with certainty that no price change would occur but it is reasonable to assume that the change in price would be relatively small and most of the adjustment would be in output.

A competitive industry in the same circumstances would react in the same fashion. Since marginal cost did not decline, price could not fall for any period of time.

Second, an enterprise monopolist producing a durable product would be more likely to maintain prices during a depression if it believed that an extra unit sold during a depression at a lower price would otherwise be sold later at a higher price during the subsequent prosperity. A policy such as this is not without costs and it is by no means certain that if such a monopolist realized lower marginal costs during the depression that it would maintain prices at the pre-depression level. A competitive industry producing a similar product with the same demand relationship in time would not react as the monopolist does for obvious reasons.

The degree of competition in the factor markets is probably more important than the degree of competition in the product market in explaining output response during a depression. It would be difficult to distinguish between the price and output behavior during a depression of an enterprise monopolist that buys factors in a competitive market and a competitive firm. Food processors did not contract output because the supply functions for at least one important factor, farm products, shifted and the factor price fell sharply. The output behavior of the firms did not seem to have been affected by the extent of monopoly and one should not have expected that it would be.

But if the factor markets are not competitive and the supply functions for the factors do not shift during the depression, both an enterprise monopolist and a competitive firm would be unable to maintain output if demand for the product declined.

The important unanswered question about price and output reactions during the 1929–33 depression is found in the urban labor market, why the hourly earnings of production workers in manufacturing declined so little (from $0.566 in 1929 to $0.442 in 1933) despite the drastic decline in employment.[11] Unions were not then sufficiently important in manufacturing to have had much influence. Given the small decline in wages, manufacturing industries which did not use an important factor or input having a flexible price could not react in the same way as agriculture.

6. Summary of the explanation. Most of the preceding explanations of the difference between the behavior of output in agriculture and in non-agriculture must be rejected. High fixed costs, the importance of subsistence production, technological conditions are clearly invalid explanations. The differences in the competitive structure of agriculture and industry in the degree of enterprise monopoly is a superficially more plausible explanation, yet I believe it, too, is invalid. An enterprise monopoly faced with the same factor supply conditions as agriculture would, in my view, react in much the same way as a competitive firm.

The belief that farm workers may work harder during periods of low income cannot be rejected on the basis of existing data, and this hypothesis is consistent with actual behavior.

C. A Possible Theory

Any theory purporting to explain the constancy of agricultural output during a depression should explain also similar behavior in the rest of the economy; it should be consistent also with the fact that agricultural output increases when the relative price of agricultural products increases and that farmers shift from one product to another as the relative prices of different products vary.

It is my view that a theory meeting these requirements is provided by the usual economic analysis of farmers as profit-maximizing entrepreneurs and that the special characteristics of the behavior of agricultural output can be explained by the characteristics of the supply functions of factors to agricultural firms.

The supply function for agricultural products is sometimes expressed as a simple relation between the quantity of output and the price of the output. However, the use of this relation obscures the complexity of the supply process determining the supply of agricultural products. The supply of agricultural products depends on: (1) Production conditions—the technological relations between inputs and outputs; (2) Supply conditions of the factors of production; (3) Price or demand conditions for output; and (4) The behavior of firms, including the objective of the entrepreneur.

The explanations provided in this paper of the behavior of agricultural output assume that firms maximize profits and that the demand for factors of production is determined solely on this basis unless a contrary assumption is made. The assumption of profit maximization implies that output behavior will be determined by the relationship between output and factor prices. For example, a greater relative increase in product than factor prices will result in increased output, and vice versa. Further, a rise in the price of one factor relative to another will decrease the employment of the first factor relative to the second. It is not necessary for our purposes that farmers actually maximize profits, but it is important, of course, that reliable predictions can be made by using the assumption of profit maximization.

Attention must be given to the supply conditions for the factors of production. These are spelled out in some detail at the relevant points in this article, but some comment on the labor supply function is required at this point. It is assumed that there is a labor supply function including all farm labor. To do this requires a strict separation of the farm firm and the labor function of the operator and other family members. This separation is required if confusion is to be avoided.

A firm is a business unit under single control within which productive resources are combined in order to produce goods and services for sale and use as a means of achieving some objective. A firm may consist solely of entrepreneurship, a business opportunity, and liquid capital. The farm operator is both an entrepreneur and a laborer. He accepts this dual role in the belief that he can thereby achieve a larger return from his energies. Otherwise, he would forego his entrepreneurial activities and hire out as a laborer. Analytically, we can divorce the supply of labor by the operator and his family to the farm firm from the farm firm itself, *i.e.*, we need not assume that this labor is a part of the firm.

During a depression, the supply function of land for use in agriculture has a price elasticity of nearly zero for a period of five to ten years. The response in quantity supplied following a price decline is related to disinvestment in land and failure to provide for maintenance. The supply function for capital equipment—to agriculture as a whole—is very inelastic whenever the demand price is below the price of new equipment. The supply function is then related solely to the existing supply of old equipment. Since such equipment does not have alternative uses outside of agriculture, there is no reservation price above the depreciation cost. Since this cost can be postponed, it may not represent an effective lower limit. The supply function of labor shifts with the level of income and employment in the rest of the economy. The marginal opportunity cost of labor falls rapidly as unemployment increases and rises similarly as unemployment declines. As the marginal opportunity cost approaches zero, the supply curve for agricultural labor becomes very inelastic. Farm workers are willing to accept lower rates of pay rather than be unemployed.

These conditions of supply would mean that during a major prolonged decline in business activity that (1) farm prices, farm wage rates, and land

Table III. Crops Planted, Labor Employed and Power and Machinery Used on Farms and Relative Farm Prices, 1919–39

| Year | Crops Planted[a] (in millions) | Labor Employment[b] | | Power and Machinery[c] | Relative Farm Prices[d] |
| | | Total | Hired | | |
		(in millions)			
1919	363	11.1	2.78	468	112
1920	359	11.4	2.88	477	121
1921	358	11.4	2.90	505	88
1922	354	11.4	2.92	496	85
1923	353	11.4	2.89	454	101
1924	353	11.4	2.87	455	101
1925	364	11.4	2.97	458	106
1926	359	11.5	3.03	460	100
1927	358	11.3	2.95	463	99
1928	367	11.4	2.96	463	102
1929	363	11.3	2.98	466	101
1930	368	11.2	2.85	471	106
1931	372	11.2	2.69	468	73
1932	376	11.1	2.50	452	62
1933	372	11.0	2.43	416	67
1934	339	10.9	2.33	391	73
1935	360	11.1	2.43	389	86
1936	360	11.0	2.56	391	91
1937	364	10.9	2.63	403	90
1938	356	10.8	2.62	419	78
1939	344	10.7	2.60	428	78

[a]1924–1939, *Agricultural Statistics, 1940*, p. 542. 1919–23, estimate by the author.

[b]Department of Agriculture, Bureau of Agricultural Economics, *Farm Wage Rates, Farm Employment, and Related Data, 1943*, p. 155.

[c]Martin R. Cooper, Glen T. Barton, and Albert P. Bradell, *Progress of Farm Mechanization*, Dept. of Agriculture Misc. Pub. No. 630 (1947), p. 81. An index number with volume measured in terms of 1935–39 average dollars, 1870=100.

[d]Calculated from *Agricultural Statistics, 1945*, pp. 430–31. Based on ratio of prices received by farmers to wholesale prices of all commodities. Index equals 100 in 1910–14.

rents would fall in about the same proportion and (2) the employment of land, labor, and machinery would not change appreciably. Condition (2) might prevail without (1) if the resources had to be used in fixed proportions or if one of the resources had a fixed coefficient of production, conditions that seem less plausible than the conditions of supply outlined above.

Tables III and IV are not inconsistent with the above conclusions, except for the behavior of wage rates between 1919 and 1921. In part, this is explained by the fact that the peak in prices received came in May, 1920; the minimum, in June, 1921. Actually, no serious drop in farm prices came until the 1920 harvest, after wage bargains had largely been made.

Table IV. Farm Prices and Wage Rates, Farm Wages and Rents Paid, and Cash Farm Income, 1919–39

Year	Prices Received by Farmers	Wage Rates[b]	Wages Paid[c]	Gross Rent[d]	Cash Farm Income[e]
				(millions of dollars)	
1919	215	207	1,515	2,226	14,602
1920	211	242	1,780	1,645	12,608
1921	124	155	1,159	1,208	8,150
1922	132	151	1,122	1,347	8,594
1923	143	169	1,219	1,501	9,563
1924	143	173	1,224	1,651	10,221
1925	156	176	1,243	1,585	10,995
1926	146	179	1,326	1,518	10,564
1927	142	179	1,280	1,648	10,756
1928	151	179	1,268	1,640	11,072
1929	149	180	1,284	1,621	11,296
1930	128	167	1,134	1,315	9,021
1931	90	130	847	906	6,371
1932	68	96	584	669	4,743
1933	72	85	512	793	5,445
1934	90	95	601	953	6,780
1935	109	103	740	1,101	7,659
1936	114	111	880	1,187	8,654
1937	122	126	1,039	1,218	9,217
1938	97	125	1,000	1,080	8,168
1939	95	123	982	1,170	8,684

[a]*Agricultural Statistics, 1945*, p. 430, 1910–14=100.

[b]Bureau of Agricultural Economics, Department of Agriculture, *Farm Wage Rates, Farm Employment, and Related Data, 1943*, pp. 3–4, 1910–14=100.

[c]Bureau of Agricultural Economics, Department of Agriculture, *Net Farm Income and Parity Report, 1943* (1944), pp. 26 and 18.

[d]*Agricultural Statistics, 1943*, p. 412.

The 1920–22 depression did not result in any significant decrease in crop acres planted; labor employment was roughly constant, employment of power and machinery increased slightly. Absolute farm prices fell by 42 per cent between 1919 and 1921, and relative farm-nonfarm prices by 22 per cent. Because of the shortness of the time period, the 1920–22 depression cannot be considered as good a verification of our hypotheses as one might like.

The 1930–33 depression is a much better test. The period was four years; sufficiently long to permit farmers to revise their production plans completely. For employment of resources, the experience and hypotheses match very well. Acres planted may have increased slightly, total labor employment remained almost constant (decreasing only three per cent), while hired labor employment declined by about 15 per cent. Part of this decline resulted from a shift for workers related to the employer from the hired to the unpaid category.

As shown in Table IV, prices received and rent paid moved down together from 1929 to 1932 and 1933, with rent falling slightly more than prices.[12] In the absence of an important change in the marginal physical productivity of land, the employment of land would not have been expected to decline, and it did not do so.

Wage rates tended to lag behind prices and employment of hired labor might therefore have been expected to decline slightly, as indeed it did.

The net income attributable to land, capital, and labor accounts for roughly 70 per cent of gross agricultural income.[13] The remainder is attributable to products and services purchased (about 15 per cent),[14] taxes (about 4 per cent), and depreciation and maintenance (about 11 per cent). Of these, only taxes requires a net outlay regardless of the level of output or prices. Current outlay for depreciation and maintenance in agriculture can be postponed almost in its entirety for as long as four years.

Current purchases of products and services from the non-agricultural sector of the economy are not made under supply conditions comparable to the supply conditions for land, labor and capital. During depressions, prices of these products and services do not decline as rapidly as prices received by farmers. Consequently, except for products and services that are limitational in character, purchases of such items should fall considerably. And this was the case. Fertilizer prices decreased by 35 per cent.[15] Fertilizer consumption declined by more than 45 per cent.[16] Farm machinery prices declined by less than 10 per cent. Expenditures on motor vehicles and machinery, including repairs for machinery, declined by more than 70 per cent. Building material prices fell 20 per cent, and expenditures by about 80 per cent.

Other current inputs of a highly varied nature, such as the cost of operating motor vehicles, electricity, twine, ginning fees, and seeds apparently declined only moderately in price and in quantity purchased. Many of these inputs such as seed, twine, ginning fees, and containers are in the category of inputs having fixed coefficients of production. Consequently, farmers would not decrease their use of these items as long as they continued to stay at full production.

Two other categories of expenses to the individual farmer need consideration, namely, feed and livestock purchases. Though data on quantities of feed and livestock purchased are not available, the data on expenditures and prices indicate that quantities of feed and livestock purchased decreased by roughly 10 per cent. The prices of feed and livestock declined slightly less than farm prices in general, perhaps five per cent less. Because of the slightly smaller relative price declines in feed and livestock for feed-

Table V. Production of Feed Grains, Hay, and Pasture, Product Added by Livestock and Hogs

Year	Production of Feed[a] Grains, Hay, and Pasture	Product Added by Meat Animals and[a] Animal Products	Hog Production[b]
1919	106	83	14.0
1920	116	80	13.5
1921	108	84	14.1
1922	107	89	16.5
1927	107	95	16.3
1928	108	96	16.2
1929	103	97	15.6
1930	94	99	15.2
1931	103	101	16.5
1932	113	101	16.4
1933	96	103	16.6

[a]1935–39=100. Source: Glen T. Barton and Martin R. Cooper, *Farm Production in War and Peace*, USDA, BAE, F.M. 53, p. 74.

[b]In billions of pounds. Source: *Agricultural Statistics, 1940*, p. 370.

ing purposes, one would expect that interfarm sales of these items would decline some but not much.

The theory that we have outlined rests on certain presumptions about the supply functions of labor, capital, and land. This theory is not complete. Certain important aspects of output behavior, such as the constancy of livestock output, are not explained by the assumptions. The theory explains why all crop land is utilized during a depression, but it is insufficient to explain why farmers sell the output of durable products or transform the feed into livestock. Why do the farmers not store such products as wheat, corn, and oats during periods of absolute and relative price declines during a major cycle?

Tables V and VI present the data indicating the constancy of livestock output. Table V requires no comment; farmers kept on producing livestock. Table VI indicates that falling prices led to only minor increases in the stocks of corn and wheat. Most of the wheat stocks were held by the government. The increase in corn stocks should be seen in their proper perspective; from 1929 through 1932, more than 10 billion bushels of corn were produced. Of this total, hogs consumed at least 4 billion bushels.

It is necessary to specify the nature of the supply function for current use or sale of the durable farm products in order to explain this behavior on the part of the farmer. The supply function for current use or sale of durable farm products shifted to the right roughly as far as did the output curve for these products. In other words, the demand for inventories by farmers or others did not increase to result in a significant difference be-

Table VI.　Prices and Total Stocks of Corn and Wheat in U.S.

| Year | Total Stocks (millions of bushels) | | Corn Price | Wheat Price |
	Corn	Wheat	(previous marketing year) (cents per bushel)	
	(Oct. 1)	(July 1)		
1919		85		205
1920		170		216
1921		124		183
1922		96		103
1923		132		97
1927	217	109	75	122
1928	92	113	85	119
1929	148	227	84	100
1930	136	291	80	104
1931	168	313	60	67
1932	270	375	32	39
1933	386	378	32	38

Source: Agricultural Statistics, 1940, pp. 10, 23, 46 and 54.

tween the physical output of the products and the quantities offered for sale or used as inputs for further production; supply prices of factors used in producing livestock fell proportionately as much as the prices of livestock, and as one would expect, livestock output was maintained.

Two explanations seem relevant for the failure of the demand for inventories to increase during depressions. First, farmers do not believe that they can estimate anticipated prices very accurately. What I feel to be the best model available to them, namely, that next year's price will be the same as this one, is consistent with not holding stocks in any large volume.[17] Second, most farmers have never had enough capital to be able to forego the current income required by a definite storage policy or to permit them to accept the large risks that arise from storage.[18]

If private storage has not acted to stabilize somewhat the price of the durable products over the cycle through storage operations, we may find that public storage will do so. If this occurs, agricultural output will behave quite differently from the way it has in the past. The output of livestock products will decline during the depression and so will aggregate agricultural output. However, government policy may also involve subsidization of the output of livestock products. If this is done, total output would be maintained.

The theory propounded above is consistent with the behavior of the non-agricultural firms that maintained output during the depression. Most of the firms—in meat packing, butter, evaporated milk, cheese, shortening and beet sugar—had one input that was extremely important from a cost standpoint, and this input had a supply function similar to that ascribed to land or to labor in agriculture. This input was an agricultural product. Since the alternative cost of these products fell sharply, the same volume of product would be available for processing at a much lower price and at a price that would clear the market.

Since other factors used by these industries did not have the same type of supply curves for other inputs, total output in any real sense probably fell. The data seem to indicate that hours worked declined more than output. Output is usually measured for such industries as a linear function of agricultural input.

II. Agricultural Output During Periods of Full Employment

The movement of farm prices relative to nonfarm prices appears to have no important influence upon total agricultural output during periods of recession or depression. The reasons for this behavior have been outlined above. It is sometimes assumed that agricultural output would be equally unresponsive to a decline in relative farm prices under full-employment conditions. This statement, however, cannot be empirically verified by the experience of the United States. There is no period in our history for which we have reasonably accurate data—which means since 1900—when high levels of employment coincided with declining relative farm prices.[19] One such period may now be emerging—starting in 1949.

Consequently those individuals who rely solely upon empiricism and eschew the use of theoretical models can find no support for the contention that farm output is not responsive to declining relative farm prices when employment opportunities are readily available in the rest of the economy. It may well be that under dynamic conditions aggregate agricultural output would not actually decrease despite a fall in relative farm prices, but such a statement is certainly not the same as the statement that relative prices do not affect the level of output. If clarity of thought is considered important, the two statements should not be confused.

Since 1900 there have been three periods of sustained full employment. During two of these periods—1900 to 1919 and from 1940 to 1948—real farm prices were rising. The only other period of sustained full employment—1923 to 1929—was a period of relatively stable real farm prices.

A. Implications of the Theory

The theory by which we have sought to rationalize the behavior of farm output during depressions has important implications for the behavior to be expected under full employment conditions. These implications can be best outlined by considering, first, the effect of changes in relative prices under given production conditions, and second, the modifications introduced by changes in production conditions.

Under given production conditions, output of agricultural products can change only as a result of changes in the quantity factors of production employed—to speak broadly, in the quantities of land, labor, or capital employed. An increase in the real price of output will raise the marginal product of factors to farmers, and therefore lead them to demand a larger quantity at previous prices. Under given conditions of supply of factors, this will lead to an increased employment of factors—unless their supply prices are perfectly inelastic—and hence to an increase in output.

A decline in the price of real output will lower the marginal products of the factors. At previous prices of the factors, farmers will demand smaller quantities of the factors. For downward movements in factor prices, the supply function for land is almost perfectly inelastic in the short run. Land is an asset with no alternative use outside of agriculture and its quantity will be affected only as depreciation and depletion exceed maintenance expenditure. A protracted decline in prices would lead to the former exceeding the latter and thus to a decline in the quantity of land supplied.[20] Capital equipment of a durable nature also has an inelastic supply function for downward price movements in the short run. A given quantity of such equipment exists and its quantity can be reduced only by depreciation since it has no alternative use outside of agriculture. The value of the existing assets would decline and new purchases from the nonfarm sector would be reduced. The marginal product of labor would decline and the demand for labor would decrease. Given flexible wage rates in agriculture, labor employment would decrease as a result of migration though unemployment would not emerge.

In a period of three to five years of declining real output prices, the reduction in farm output will depend largely upon the reduction in farm labor employment. The supply of labor to agriculture is a function of its wage in agriculture, of the wage for comparable labor in non-agriculture, of the level of unemployment, and of the growth of the farm labor force due to the excess of additions (individuals living on farms reaching working age) over withdrawals (from death or retirement). If the elasticities of the quantity of labor supplied with respect to the farm wage rate, the nonfarm wage rate and unemployment are relatively small, the reduction in labor supplied produced by a decline in relative prices may be fairly small. These are likely to be the conditions when the price decline is assumed to be temporary. If the price decline is assumed to be permanent, the elasticities of the quantity of labor supplied with respect to the relevant variables are likely to be relatively large, and the adjustment in labor supply would occur more rapidly.

In the above paragraphs, it has been assumed that the production function remains unchanged. The production function in agriculture does change as new techniques become available. Though the availability of new techniques of production is probably unrelated to the level of farm prices, the rate of adoption of new techniques requiring significant investments might be. Many types of new techniques have not, however, required important investments by farmers adopting them in the past; for example, new seeds, new feeding methods and rations, and disease control methods. Thus it may be assumed that the production function shifts at a slow rate under any circumstances and that this rate may be increased somewhat by high real output prices.[21]

If real farm prices were constant, agricultural output would gradually increase due to the autonomous shifts in the production function. The employment of farm resources in a growing economy could increase, decrease, or remain constant depending upon the annual change in demand for farm products, the technological change in agriculture, and the technological change in the rest of the economy.[22] If the annual change in demand were equal to the technological change in agriculture, which in turn was equal to the technological advance in the rest of the economy, resource employment in agriculture would remain unchanged. However, if technological change occurred more rapidly in the rest of the economy than in agriculture, constant real prices of farm products would result in a decline in relative returns to resources in agriculture and some reduction in resource use, particularly of capital and labor. In this case, constant real prices would not represent a long-run equilibrium situation.

Given technological change, falling real farm prices need not produce a decline in aggregate output in agriculture. The autonomous shift in the production function may increase the marginal physical productivity of each of the resources sufficiently to counteract the decline in resource use. In consequence, the failure of aggregate output to decline would not be inconsistent with long-run equilibrium in the factor markets.[23]

We have so far taken no account of either uncertainty or capital rationing. A change in relative farm prices for one or two years may not affect the level of resource employment because entrepreneurs do not expect such change to be permanent. Consequently, actual plans may be made in terms of expected prices either higher or lower than the market prices. Capital employment is not determined solely by profit maximization; in addition, as has been argued elsewhere, capital use in agriculture is subject to capital rationing.[24] Consequently, new investment in agriculture is a function of the liquidity position of farmers as well as of current and expected returns. However, this factor will not reverse the general direction of movement of capital employment, though the amount of investment would obviously be affected thereby.

In testing the above propositions in any empirical situation, one specific caution must be noted. The conditions stated and conclusions following therefrom are based on the assumption that the economy has been operating at a high level of employment for some time. In other words, it is assumed that labor unemployment in the nonfarm sector of the economy has not been acting as a deterrent to the migration of labor out of agriculture. At the beginning of the period of rising real farm prices if there is much unemployment of labor in the nonfarm economy, farm labor will be earning less than comparable *employed* nonfarm labor. Consequently, if the rise in real farm prices is associated with a decline in unemployment, the supply function of agricultural labor will shift to the right; farm labor employment will decline even though the return to labor in agriculture rises relative to the return to employed labor in non-agriculture.[25]

B. Tests of the Theory

1. 1900–1920. The period 1900 to 1920 was one of almost continuous high levels of employment. According to Douglas, unemployment in manufacturing and transportation—two cyclically volatile industries—exceeded six per cent in only four years out of the 22.[26] And only two of these were consecutive years—1914 and 1915.

Real farm prices apparently rose by 25 to 30 per cent,[27] and farm output, by 25 per cent[28] or roughly 1 per cent per year.

The increased farm output was associated with increased employment of all resources except labor—labor employment was roughly the same in 1920 as in 1900.[29] The employment of other resources, however, increased sharply. Total cropland increased from 319 million acres in 1900 to 402 million in 1920—an increase of 26 per cent. However, the land added was less productive (produced less rent per acre) than existing cropland. The net increase in the production capacity of the land was probably of the order of 15 per cent. The quantity of farm power, machinery and equipment increased from an index of 295 in 1900 to 477 in 1920—an increase of 62 per cent.[30] Livestock, exclusive of horses and mules, increased by 12 per cent. Total capital inputs probably increased by 30 per cent. Current operating expenses must have increased by at least 100 per cent, perhaps considerably more.[31]

The increased employment of resources was sufficient to account for most of the increased output,[32] and the changing relative prices of output and inputs, to account for the increased employment of land, capital and current inputs.[33] In consequence, no substantial change in technology can be inferred from the increase in output.

Why did farm labor employment fail to increase? The evidence available is inconclusive in indicating what rational conduct would have been for farm workers. Farm labor income per worker rose relative to the income of employed industrial workers between 1910 and 1919. The absolute difference between annual earnings, however, increased from about $335 in 1910–14 to $570 in 1919, and to $820 in 1920. A rise in relative earnings when the absolute differences increase may nor may not indicate that real returns to farm workers increased relatively. When the absolute difference is actually larger than the farm labor return, as it was in 1910–14, possible differential changes in the cost of a fixed level of living and changes in the content of the level of living make it impossible to infer with certainty whether the real returns to farm labor kept pace with the real returns to nonfarm labor.

2. 1923–29. The period from 1923 through 1929 was one of stable relative farm prices (in the aggregate). The parity index—a measure of relative farm prices—had the following values starting with 1923—86, 86, 92, 87, 86, 90, 89.

Our theory would indicate reasonable stability in the employment of all inputs, except perhaps labor. With respect to capital and land, actual experience does not contradict the expectations. Net investment in horses and mules, machinery, motor vehicles, including tractors, trucks, and the farm share of autos, and service dwellings was zero or perhaps negative.[34] The estimate made by Cooper, Barton and Brodell of farm power and machinery also indicated no change.[35] The current inputs purchased from non-agriculture increased, in real terms, by about 22 per cent. This increase was associated largely with the shift from farm produced to mechanical power.[36]

It is not clear how labor employment should have been expected to behave. If it is assumed that the labor market was in long-run full-employment equilibrium in 1923, there should have been no appreciable change in labor employment. And this is what actually happened according to Bureau of Agricultural Economics estimates. Total farm employment declined by less than one per cent.[37]

Between 1923 and 1929, the labor income of farm workers increased somewhat relative to the income of employed industrial workers—from about 30 per cent as much in 1923 to about 35 per cent in 1929.[38] During the seven-year period, there was a net movement of 4,260,000 people off farms—an annual average of 630,000 or roughly two per cent of the population.[39] This movement, large as it was, was only sufficient to stabilize the quantity of labor supplied.

The changes in farm production and in resource use are not inconsistent with an essentially unchanged production function. Calculations similar to those in footnote 32 indicate that one-half of the approximate eight per cent increase in farm output can be explained by increased inputs. The rest of the increase could be due to weather and other natural changes, though there seems to be no evidence that weather changes were important.[40]

3. 1940–48. Following 1940, real farm prices rose rapidly—from 80 to a peak of 121 in 1946 and then fell slightly to 115 in 1948. It seems clear that the demand for all inputs would rise in these circumstances. And such increases in demand did occur. More capital was employed and net investment would have been even greater had items been available at quoted prices. Between January, 1940 and January, 1948, the quantity of power and machinery increased by about 40 per cent. Acreage of land harvested increased by about 10,000,000 or 3 per cent. Current operating expenses (except livestock and feed purchased and short-term interest) increased by 60 per cent. Farm output increased by 20 per cent.

The farm population declined from 30.3 millions to 27.8 millions by January 1, 1949, and the level of farm employment by slightly more than 4 per cent.

The change in farm employment is consistent with our theory. It is quite clear that labor in agriculture was not in a position of long-run equilibrium in 1940. The supply of labor was large in agriculture because of the heavy rate of unemployment that had prevailed previously in the rest of the economy. As unemployment declined, the supply of labor to agriculture also decreased. The differential in earnings was such as to induce movement out of agriculture at a given level of unemployment. Agriculture had a net migration during the last half of the 'thirties of 2,770,000 or 555,000 per annum,[41] despite unemployment ranging from 14 to 20 per cent.[42]

Agriculture would have lost more of its labor force had it not been for the rapid rise in returns to agricultural resources. During the early part of the period, labor income in agriculture rose more rapidly than labor income in the rest of the economy, yet the movement out of agriculture was at a fantastic rate—net civilian migration of 1,920,000 in 1942 and 1,146,000 in 1941. This was due to the large absolute difference in real earnings that still persisted. In 1941, labor returns per worker on commercial farms was $700 less than the wage income of employed workers—a difference of slightly less than 50 per cent. By 1946 the absolute difference had narrowed to $450 and the relative difference to 20 per cent,[43] and these differences remained roughly the same in 1947 and 1948.

In the three years 1946, 1947, and 1948—there was probably a rough equilibrium in the allocation of labor to commercial farms. Consequently, no significant change in farm population or labor force was to be expected during these years, except from the return of veterans. Apparently about one-half of the farm veterans returned to the farm and stayed there.

The change in the production function between 1940 and 1948 was probably not as spectacular as is frequently believed. A large fraction of the increased output—perhaps half—can be attributed to increases in resource inputs. The remainder can be attributed to changes in the production function and natural factors, such as weather.

C. Conclusions

On the whole, the simple theory that we have used gives a reasonably accurate indication of the response of resource employment and output to changes in relative farm prices during periods of high-level employment, at least as judged by the three "tests" we have been able to make.

The attempt to judge the shift in the production function in the periods surveyed is significant in understanding the probable response of farm output to falling relative prices during periods of high-level employment. If during a period of rising relative prices, a considerable fraction of the increased output reflects increased inputs, a subsequent period of falling prices may well result in an actual decline in output. As farm prices decline, net investment will fall, the level of current inputs will diminish, and labor migration will increase. The decline in output may not be large, but it does not need to be in order to re-establish equilibrium in the factor markets because of the price inelasticity of demand for agricultural output.

An important policy conclusion can be drawn from this analysis. Maintaining farm price returns at levels above market prices during periods of high-level employment will make farm output higher than it would otherwise be.[44] Such induced increases in output will increase the difficulty of maintaining farm price returns, necessitating direct controls of output if governmental expenditures are not to exceed the amounts that even a "generous" Congress is willing to appropriate. In fact, it can be argued that increasing farm price returns by governmental action during periods of full employment does a real disservice to farm people, unless agriculture is to be permanently subsidized. If the price props are withdrawn (or a depression occurs), returns to agricultural resources would be lower than otherwise because too many resources have been retained in agriculture.

The attainment of equilibrium in the factor markets in agriculture during a period of generally falling prices or after the establishment of a lower level of agricultural prices may take longer than seems desirable. If this is so, the appropriate policy action would not seem to be higher support prices *which would prevent the resource adjustments from occurring*. Rather, the slowness of the adjustment process would seem to call for direct measures to increase the outmovement of labor.[45]

Notes

The author is associate professor of economics at the University of Chicago. He expresses appreciation to Milton Friedman, F. V. Waugh, T. W. Schultz and O. H. Brownlee for suggestions and criticisms.

1. See Secretary of Agriculture Brannan's *Statement before a Subcommittee of the Committee on Agriculture and Forestry, U.S. Senate, Eighty-First Congress, First Session, on S. 1882 and S. 1971 (July 7, 11, 12, 13, 14, 15, 18, and 19, 1949)*, pp. 50–52.

2. See J. K. Galbraith and J. D. Black, "The Maintenance of Agricultural Production during Depression: The Explanations Reviewed," *Jour. Pol. Econ.*, Vol. XLVI (June, 1938), pp. 305–23.

3. *Source:* Solomon Fabricant, *The Output of Manufacturing Industries, 1899–1937* (New York, National Bureau of Economic Research, 1940). The reader may object that output behavior of the manufacturing groups included in the tabulation do not support the statement that output in some segments of the economy behaved much the same as agricultural output. Each of the industries, excepting the last, process an agricultural product and since agricultural output did not decline, it seems evident at once that the output of the specified processing groups would not decline. The output behavior of the processing groups can be explained by the same theory developed below, namely by the shifting of the supply functions of factors or by the price inelasticity of the supply functions for one or more of the factors. Inspection of the list will disclose several industries for which the cost of the agricultural raw material represents only a small fraction (less than a fifth) of the total manufacturing cost. In these instances, the supply prices of other factors declined sufficiently to permit the firms to maintain output.

4. Between 1929 and 1933, hired labor employment declined by 15 per cent, but an unknown fraction of this decline was due to the shift from the hired to the unpaid family labor category.

5. See below.

6. *United States Census of Agriculture, 1945.* Volume II, *General Report*, Chap. X, Table 4. Data based on 1945 classification by type of farm.

7. *Ibid.*, Chap. X, Table 26.

8. *Ibid.*

9. See Solomon Fabricant, *The Output of Manufacturing Industries, 1899–1937*, Appendix B and Harold Barger and Sam H. Schurr, *The Mining Industries, 1899–1939*, Appendix A. (New York, National Bureau of Economic Research, 1944). Walter F. Crowder summarized an analysis of the output behavior of 407 products from 1929 to 1933 and 1933 to 1937. He concluded as follows: "If eight or ten products which decreased more than 90 per cent are not given undue weight, the logical inference would seem to be that the changes in quantity output of the great mass of manufactured products between 1929 and 1933 were not related to the concentration ratios of the products." For the 1933–37 period, he stated, ". . . it cannot be said that manufactured products in the 'low' concentration group exhibit any outstandingly different behavior pattern from that of products in the 'high' concentration group." (T.N.E.C. Monograph No. 27, *The Structure of Industry*, pp. 350–51 and 354.)

10. This seems to be the assumption made by several writers who have dis-

cussed the relation between enterprise monopoly and cyclical price rigidity. Boulding, for example, assumed a relatively flat marginal cost curve and then argued that a decline in demand during a downturn of the business cycle would lead to price maintenance and output restriction. This explanation fails to indicate why the marginal cost curve does not shift during the cycle. If the cost curve shifted downward, regardless of how flat it is, price would be permitted to fall unless the demand curve shifted in a very peculiar manner. (See Kenneth Boulding, *Economic Analysis* [New York, Harper and Bros., 1948, 2nd ed.] p. 557.) For an analysis similar to that contained in this paper, see T. De Scitovszky, "Prices under Monopoly and Competition," *Jour. Pol. Econ.*, Vol. XLIX (Oct., 1941), pp. 663–86.

11. Farm wage rates declined by more than 50 per cent in the same period. See *Statistical Abstract, 1947*, pp. 199 and 210.

12. The following tabulation indicates the relationship more clearly than Table IV:

Year	Prices Received	Cash Farm Income	Wage Rates	Wages Paid	Gross Rent Paid
1929	100	100	100	100	100
1930	86	80	93	88	81
1931	60	56	72	66	56
1932	46	42	53	45	41
1933	48	48	47	40	49

13. D. Gale Johnson, "Allocation of Agricultural Income" [repr. as chap. 5 in this volume].

14. Includes only products and services purchased from non-farmers.

15. *Agricultural Statistics, 1945*, p. 429.

16. *Ibid.*, p. 467.

17. See D. Gale Johnson, *Forward Prices for Agriculture* (Chicago, University of Chicago Press, 1947), Chap. VI.

18. *Ibid.*, Chap. X, esp. pp. 156–61. The path of price movements during and following the 1929–33 depression indicates that farmers were relatively wise in not accumulating stocks. For example, if a farmer had stored wheat in 1930 and if the costs of storage were 10 cents a year, he would have been unable to have made a profit at any time until the present. If he had stored wheat in 1931, he would have realized a substantial gain within two years. The same circumstances were true in corn. But what basis would a farmer have had for knowing that he should have stored in 1931 but not in 1930?

19. The late 'twenties does not seem to constitute such a period. As indicated in Table III, except for the single year 1925, relative farm prices were stable from 1923 through 1929.

20. Unless the government subsidizes maintenance through payments for "soil conservation" or increases the available supply of land through irrigation and reclamation projects.

21. The evidence on the relation between output prices and technological change is admittedly conjectural. It cannot be assumed that because output per worker rises more rapidly during periods of high or rising prices than during periods of low or falling farm prices, the production function is shifting more rapidly

in the former period than in the latter. High or rising prices induce more invest-ment per worker, which is not the same thing as a technological change. It is argued below that the rate of technological change in agriculture has been much less than is generally assumed. The same may be true in the rest of the economy. Increased capital per worker over the years may be as important, if not more important, than technological change in increasing output per worker.

22. The change in the demand for farm products is assumed to be a function of the change in per capita income and of the change in population. Technological change is measured by the increase in output from a fixed quantity of resources.

23. The share of total national output produced by agriculture would decline.

24. See D. Gale Johnson, *Forward Prices for Agriculture*, Chaps. IV and V.

25. There is, of course, an increase in real farm prices that would increase the demand for farm labor by more than the supply decreased and thus result in an increase in farm employment. The argument in the text implies only that an in-crease in real farm prices and in farm labor returns (wages of hired labor and labor income of unpaid workers, including the operator) relative to the wages of em-ployed nonfarm workers need not be inconsistent with a decline in farm employ-ment in a specific situation.

26. Paul Douglas, *Real Wages in the United States* (Boston, Houghton Mifflin Company, 1930) p. 445.

27. See Bureau of Agricultural Economics, *1949 Agricultural Outlook Charts*, p. 1.

28. Martin R. Cooper, Glen T. Barton, and Albert P. Brodell, *Progress of Farm Mechanization*, Dept. of Agriculture Misc. Pub. No. 630, p. 81.

29. Barton, Cooper and Brodell estimate farm employment at 11.4 millions in 1900 (*ibid.*, p. 5). The Bureau of Agricultural Economics estimates the 1920 farm employment as 11.4 and 1910 as 12.1 (*Farm Wages, Employment and Related Data, 1943*).

30. See Barton, *et al.*, *op. cit.*, p. 7.

31. From 1910 to 1920, total current operating expenses, excluding feed and livestock purchased and short-term interest, deflated by the index of prices paid by farmers for items used in production, except feed, increased by 69 per cent. Similar data are not available for 1900 to 1909.

32. We do not have the data that would permit an accurate estimate of the production function for agriculture. However, a crude estimate of changes in the function is possible if we assume (a) that there are constant returns to scale for agriculture as a whole, and (b) that the marginal productivities of resources are equal to the average net productivities. (As defined by Joan Robinson, *Imperfect Competition* [London, Macmillan, 1934], p. 239.)

Using estimates of average productivities for 1910–14 (See Johnson, "Allocation of Agricultural Income" [chap. 1 in this volume]) and assuming two different func-tions that meet the requirement of constant returns to scale (one linear and one linear in the logs), the following functions were obtained:

(1) Log output = .44 log labor + .27 log land + .17 log capital + .12 log cur-rent expenses.

(2) Output = .44 labor + .27 land + .17 capital + .12 current expenses.

All variables were 100 in 1900, while the 1920 values were 100 for labor, 115 for land, 130 for capital and 200 for current inputs. Substituting these values in the functions gave 1920 estimates of output of 118 for the log function and 121 for the linear function. Actual output was 125. Roughly three-fourths or more of the increase seems to have been explained by increased inputs.

33. Between 1910 and 1920, the average price of current inputs increased from 100 to 188, while farm prices increased from 102 to 211. Comparable data are not available for 1900 through 1910, but during this period the wholesale price of farm products increased by 47 per cent and the wholesale price of nonagricultural products increased by 17 per cent.

34. Based on values expressed in 1910–14 dollars. Data taken largely from Bureau of Agricultural Economics, *Net Farm Income and Parity Report, 1943*.

35. *Op. cit.*, p. 81. The index was 454 in 1923 and 466 in 1929.

36. *Ibid.*, p. 90. The cost of farm produced power which is payment for inputs produced by farmers, declined by $300,000,000 measured in dollars of constant purchasing power. Costs of operating mechanical power increased by 390 million dollars.

37. See Bureau of Agricultural Economics, *Farm Wages, Farm Employment, and Related Data*, p. 155.

38. Based on estimates of labor income by writer, "Allocation of Agricultural Income" [chap. 1] and Bureau of Agricultural Economics, *1949 Agricultural Outlook Charts*, p. 7.

39. Bureau of Agricultural Economics, *Farm Population Estimates, 1910–1942*, pp. 1 and 2.

40. See United States Department of Agriculture, *Crops and Markets*, 1949 edition, p. 6 for estimates of yields per acre. Yields for field crops were identical in 1923 and 1929.

41. Bureau of Agricultural Economics, *Farm Population Estimates, 1910–1942*, p. 2.

42. Bureau of Agricultural Economics, *1949 Agricultural Outlook Charts*, p. 4.

43. See Johnson, "Allocation of Agricultural Income" [chap. 1] and Bureau of Agricultural Economics, *1949 Agricultural Outlook Charts*, p. 7.

44. The term *farm price returns* is used in this paragraph instead of *farm prices* because farm prices can be supplemented by direct subsidies to farmers.

45. Walter Wilcox has indicated a contrary view in "High Farm Income and Efficient Resource Use," *Jour. Farm Econ.*, Vol. XXXI (Aug., 1949), pp. 555–57. He argues that farm migration is increased by high farm family incomes due to increased mechanization and the aid high incomes give to migration. He used population changes for Iowa, South Carolina and Tennessee for 1930 to 1940 in support of his position. A more detailed study of farm migration for the same decade made under the direction of the writer indicates that state differences in migration rates were unrelated to the level of farm income. The two factors which were most closely associated with migration were a measure of population pressure (the excess of new entrants to the farm labor force over deaths) and changes in the level of farm income from the late 'twenties to the 'thirties. The changes in farm population in the Pacific Coast States since 1930 clearly contradict Wilcox's position, as well.

PART THREE

Agricultural Policy in High-Income Countries

6

Contribution of Price Policy to the Income and Resource Problems in Agriculture

In the many governmental programs affecting agriculture prices have played an important role. Prices have been used, however, in the dual roles of ends and means, with little or no recognition of the incongruous situation which arises because of this mixture of function. At some points, prices have been used as goals which various programs were supposed to achieve. At other points, prices have been manipulated to direct consumption and production. At times, particularly during the war, prices have been used as both means and ends at the same time. Various administrative agencies have been charged with the responsibility of attaining parity prices, of holding down prices and of using prices to direct production. It would be a surprise if all of these magic feats were attained at one and the same time.

The motivation for any economic or social policy is the desire for change from a present unsatisfactory position to a situation that constitutes an improvement, namely, from "what is" to "what ought to be." The underlying problems for which solutions are required are evident only from a comparison of the present and a more desirable situation. The major function of analysis is to present an understanding of the nature of the present situation and of the ideal which is desired. Viewed in this light, the major problems arise as discrepancies between the two circumstances. In this article we endeavor to show the most important points at which the present agricultural situation is most unsatisfactory and to indicate the contribution that price policy can make.

A preliminary but workable delineation of these problems is possible from a description of the functions of an economic system. This description does not provide information as to the actual content of the problem, but rather provides the basis of a useful analytical separation. Any eco-

Reprinted by permission from the *Journal of Farm Economics* 26 (November 1944): 631–64.

nomic system must perform at least three functions—the allocation of resources in production, the functional distribution of incomes to resource owners and the division of the supply of final products among consumers. In an enterprise economy, these functions, for the most part, are performed by the price system.

If various conditions exist, including free mobility of resources and accurate knowledge, the enterprise economy approximates the best allocation of resources within any given distribution of the ownership of factors. Each factor owner by using his resources where their marginal earnings are greatest produces a part of a total output which provides a maximum satisfaction for all concerned.[1] In the real world, however, mobility of resources and knowledge are not perfect and resources are not always, or even predominately, applied where their marginal contribution is greatest. Because of lack of knowledge firms do not reach their most favorable position and because of forces restricting mobility resources do not always find employment at the most favorable earnings. A different allocation of resources than that occurring in the real world would result in a total output that would be considered superior to the output currently being produced. *The nature and extent of the defect in the allocation of resources constitutes what we shall call the resource problem.*

The personal distribution of incomes in an enterprise economy is the result of the functional distribution of payments for the use of resources the distribution of personal capacities and the ownership of other resources among the individuals in an economy.[2] Under perfect competition the prices of the factors would equal their marginal contribution, though in actuality there may be important differences between the price and the value of the contribution. However, an income problem can exist even if all resources are employed at their best alternative if ethical considerations are given weight. Though an ethical criticism is valid and has been ably made,[3] recognition of a problem requiring general attention need not be based on such grounds. Unless society is willing to accept the type of income distribution which results from the receipt of personal incomes from returns of the resources as owned, it is necessary to distinguish or separate the effects of policies on the use of resources and on the personal distribution of incomes. It is increasingly evident that democratic societies are unwilling to accept the present unequal distribution of personal incomes. The progressive income tax, unemployment compensation, and the organization for the care of the indigent; death, inheritance and gift taxes, and a large number of other regulations and institutions are evidence that the distribution of incomes to persons in accordance with the ownership and productivity of factors does not lead to the realization of adequate democratic welfare standards. *The divergence between the actual distribution of personal incomes resulting from resource prices and the existing ownership of re-*

sources, and that fulfilling adequate democratic social welfare criteria constitutes what we shall call the income problem.

Though the allocation of resources and the personal distribution of incomes are inexorably interdependent, economic analysis and policy formulation require an analytical separation of the two.[4] Any policy actually introduced will have important effects upon both even though the major end of policy is to affect only one or the other. Yet fundamentally, as we try to show below, measures required to improve the use of resources and the personal distribution of incomes are sufficiently distinct as to require separate formulation. A partial justification for an analytical separation of the two problems is the woefully inadequate political and lay thinking evidenced in recent years. In dealing with resources and incomes, two opposite and equally unrealistic propositions have been evident. On the one hand, there has been the prevalent assumption that if full employment existed, the income problem solves itself. Growing out of this belief was the idea implicit in the AAA legislation that the agricultural income problem could be solved if, by one means or another, product and resource prices could be increased sufficiently. On the other hand, there has been the belief, particularly among the more naive socialists, that resource use is bad because incomes to persons are distributed so inequitably, believing that, if the total income pie were more equally divided, the resource problem would take care of itself!

Rational economic policy points to a course of action lying between these two extremes. The first position ignores the fact that many individuals in society simply do not have sufficient resources to provide an adequate income, even if employment can be found under the most satisfactory conditions. The second position disregards largely the relationships between income and incentives for production and the direction of investments. The intermediate area, however, provides a number of fruitful alternatives. In many cases, improving the allocation of resources will have positive effects upon income, by increasing the aggregate and improving its distribution. Similarly, improvements in the distribution of income among individuals may have beneficial effects upon resource use, if the improved income distribution results in warranted investments in the human agent thereby making it a more valuable resource and improving its mobility. Progress toward a solution of the income problem may also reduce the differential effects of capital rationing, which are particularly important and "costly" in agriculture.[5]

It is, however, an oversimplification to consider that there is a single resource or a single income problem and that any single policy measure will be sufficient to solve either or both. Both areas are highly complex and consist of important subparts which must be distinguished and recognized as important issues in themselves.

In this article an endeavor is made to discover the nature of these problems and to indicate the nature of the contribution of price policy. In addition, in a general fashion some other measures are presented which might be effective as complements to price policy in the areas of its greatest limitations.

In what follows the assumption is made that a forward price system would be an integral part of any price policy adopted.[6] This assumption is made to permit presentation of the contribution of price policy to the resource and income problem in the most favorable light. The assumption is also made that the announced forward prices are essentially accurate and that the determination of the prices is not influenced by political considerations. The analysis is limited in its application to a comparison of conditions existing in a market economy essentially free from governmental interference with an economy in which there is an active price policy and other programs designed to improve resource allocation and the distribution of incomes.

I. Resource Problem

An analysis of resource use implies a comparison of the actual use of resources with an ideal. For our purposes this ideal may be stated in terms of two simple conditions. The first condition is that the value of the marginal product of a factor is equal to the price of the factor. The second condition is that the value of the marginal product of a factor should equal the marginal opportunity cost of the factor or the factor's marginal value product in its highest alternative uses. These two conditions provide a basis for dividing the resource problem in agriculture into manageable parts. Departure from the first condition is largely within the firm, while departure from the second may exist at any point where alternative uses of a factor present themselves.[7] Though the alternative modes of use are innumerable, a few stand out as of fundamental importance. For the purposes of this analysis it is convenient to classify the resource problem in agriculture as follows:

A. Allocation of resources within agriculture
　　1. Within firms
　　2. Among firms
　　　　a. Among firms in different geographical areas
　　　　b. Among firms in the same geographical area
　　3. Allocation of resource use in time
B. Allocation of resources between agriculture and the rest of the economy
　　1. Human
　　2. Capital

A. Allocation of Resources Within Agriculture

The first step is to study the way the resources employed in agriculture at any one time are allocated. If we ignore for the present the query whether the proportion of the nation's total resources in agriculture is the proper one, we may examine several different points in agriculture where resource use could be improved. In our outline we have delineated four: (1) within firms, (2) among firms in different geographical areas, (3) among firms in the same geographical area, and (4) the allocation of resources over time. The second and third categories are separated because of differences in degree of mobility of resources, particularly of the human agent.

1. Within firms. The individual farm operator is confronted with many obstacles preventing the reaching of maximum returns from the resources he controls at any one time or might control through hire or purchase. Fundamentally, however, most of these obstacles grow out of imperfect knowledge or uncertainty.[8] The repercussions of uncertainty upon the farm are of two kinds. First, the farmer is unable to determine the kinds and amounts of outputs which would have given him maximum returns when reviewed from the vantage point of the future. The best he can do is to make his decisions on the basis of anticipations about the nature of the future that will exist when his productive efforts come to fruition. Second, the farmer is confronted with reactions of others to the particular situation in which he finds himself. These reactions are embodied in the terms and amounts of capital which he can obtain from sources external to the firm, which is the essence of external capital rationing. Closely related to this is the effect of uncertainty upon the farmer's conception of the amount of capital resources he should own or hire which in many cases leads to self-imposed capital rationing.

Inaccurate expectations are largely responsible for the types of misdirection in the use of resources which result in periodic output cycles in hogs, beef cattle, poultry, potatoes, cotton, wheat, and others, and in the failure of private individuals to adjust storage stocks in view of fluctuations in supplies and changes in business activity. Capital rationing, both external and internal, affects not only the size of the firm, but also tends to place an emphasis upon resources that are highly flexible contractually and technically. Because of past experiences, most farmers are unwilling to commit themselves to heavy and fixed contractual payments which must be met regardless of the returns of his production. Of the resources used in agriculture, the labor of the operator and his family most nearly approach this requisite degree of flexibility since this labor can be treated as a residual claimant. As a result there seems to be a distinct tendency to employ too much labor and too little nonhuman resources, resulting in a marked discrepancy in the marginal value productivity of capital and its money cost.

Price policy could result in a marked improvement in the utilization of the resources of the nation's commercial farmers. Forward prices could provide the necessary benchmarks to permit the attainment of an allocation of resources among alternative products on each farm which would closely approximate the most profitable for the farmer and the best allocation from the standpoint of society. If forward prices were announced for only one production period in advance, price policy would have some effects in reducing the importance of capital rationing. Investments in capital assets of relatively certain profits of the prospective year would undoubtedly be encouraged. Extensive recombinations of resources which would lead to any marked increase in the size of the firm and the use of a larger proportion of capital resources with a long life would occur only if there existed a forward price system covering several production periods.

For that part of the nation's farms, perhaps one-half,[9] who are only loosely related to the price system, price policy can do very little. Most of the agricultural resources controlled by these farmers ordinarily do not find their way into the market place. Consequently, relative prices are not a complete guide to action and measures to improve the functioning of the price system would not markedly affect the allocation of these resources. There is unquestionably an important quantity of underutilized and undeveloped human resources on these 3,000,000 farms. Whether these resources can be fully developed and utilized involves social programs of a broad scale, including education, health improvement, and nutritional advancement as well as, in the vast majority of cases, migration and employment at other pursuits.

2a. Among firms in different geographical areas. In a competitive equilibrium a specified amount, say a man-year, of any factor should make approximately the same marginal contribution regardless of where it is employed.[10] Not having the data which would permit the determination of the marginal productivities for the factors employed in the different agricultural areas, other measures must be used in determining whether there is a major departure from an equilibrium position in the allocation of resources. Two such measures have proven to be useful, both as separate indicators and when used jointly. One is the average value produced per farm or per worker. The other is the relative combination of resources in agriculture in the various geographical areas, with resources being divided into labor and capital, including land. Both measures are subject to certain limitations, but when interpreted with some caution the results obtained should not be misleading.[11]

It is, of course, common knowledge to any agricultural economist that

Table I. Average Value of Products Sold, Traded, or Used by Farm Households, Number of Workers per Farm and Work off Farm by Farm Operators, 1939

Region	Index of farm income[1]	Average value[2] (in dollars)	Workers per farm[3] (man years)	Per cent operators working off farms[4]	Average days off farm[3]
Pacific	380	2,647	2.02	39	171
Mountain	260	2,168	1.82	33	135
New England	240	1,793	1.85	42	179
Mid. Atlantic	250	1,727	1.80	34	158
W.N. Central	210	1,716	1.48	24	100
E.N. Central	205	1,510	1.43	29	144
W.S. Central	155	1,013	1.90	27	124
S. Atlantic	150	915	2.01	28	152
E.S. Central	100	604	1.71	27	121
United States	185	1,309	1.73	29	135

[1]H. C. Norcross, State Estimates of Expenses and Net Income from Agriculture, 1929, 1939–1942, Bureau of Agricultural Economics, U.S. Department of Agriculture, preliminary (mimeo.), May, 1944. Net farm income is net income of operators exclusive of government payments plus wages, rent and mortgage interest.

[2]Sixteenth Census, 1940, Agriculture, Farm Characteristics by Value of Products, pp. 30–35. Table 1.

[3]Derived from BAE, USDA, Farm Wage Rates, Farm Employment, and Related Data, January, 1943, pp. 157–166.

[4]Sixteenth Census, 1940, Agriculture, Vol. III, General Report, p. 342, Table 8. Average days off farm is the average for those operators reporting work off farm.

marked differences in the average output of farms in the nine geographic regions, whether measured in terms of value of products sold or net farm incomes, do exist.[12] Value of products sold per farm is lowest in the East South Central States (about $600 in 1939), while the Pacific States were highest (with four and one-half times that amount). Net farm incomes showed a slightly smaller absolute and relative dispersion, but the range was still large. The differences in productivity of the farm units cannot be explained by differences in the size of the labor force as the largest labor forces were found with the greatest and smallest outputs per farm. Likewise, the differences in productivity cannot be explained by greater off the farm employment in the regions with the lowest productivity. For the three southern regions the off farm employment of farm operators is approximately the same as for the United States.

The reasons for the differences in the regional productivities must lie elsewhere. There are three possible explanations: (1) that the particular year used, 1939, gives unreliable results because of yields and prices that were atypical. An examination of data for four other years (1929, 1940, 1941, 1942), shows no important differences in the relative variation or rank, except that as farm prices increased the North Central States tend to move ahead of the North Eastern States due to the fairly obvious characteristics of the farm business in the two areas. (2) That the variations in

Table II. Value of Product Sold, Traded or Used, Value of Land and Buildings Per Farm and per Worker by Regions, 1939
(In dollars)

Region	Value of product per farm	Value of land and buildings per farm	Value of product per worker	Value of land and buildings per worker
Pacific	2,647	11,720	1,310	5,800
Mountain	2,168	7,623	1,190	4,200
New England	1,793	5,478	960	2,950
Mid. Atlantic	1,727	5,858	960	3,300
W.N. Central	1,716	8,065	1,160	5,450
E.N. Central	1,510	7,289	1,060	5,100
W.S. Central	1,013	4,308	530	2,300
South Atlantic	915	3,099	460	1,500
E.S. Central	604	2,272	350	1,300

Source: Sixteenth Census, 1940, Agriculture, Farm Characteristics by Value of Products, pp. 32–35, and Table I above.

output arise from differences in the quality of the labor employed. Though there is no empirical way of measuring the differences in quality, it can hardly be questioned that much of the labor in the three southern regions would compare unfavorably with that of the rest of the nation. One must doubt, however, if the average worker in the Pacific States could perform work four times as valuable as that performed by workers in the East South Central States, if the two groups of workers were in identical situations. (3) That the marked differences in productivity are due to important differences in the quantities of cooperating factors combined with labor. This is a fairly remarkable circumstance since approximately 65–70 percent of the net farm income is attributable to labor if capital is valued at the going market rate of interest.[13]

The data in Table II substantiate the third explanation. The comparison of value of product with the value of land and buildings per farm indicates a very marked positive relationship between output and the value of land. The last two columns present the same data on a per worker basis, and in this comparison the relationship between output and value of land is even more direct. In almost every case, the larger the value of land per worker, the greater is the output per worker. The inclusion of cooperating factors other than land does not change the picture. Except for labor, land and the attachments to it is the most important resource used in agricultural production.[14] In addition, the quantity of other capital factors employed is closely related to the value of land. On a per worker basis, the value of machinery ranges from $90 in the East South Central States to $540 in the

Table III. Value of Product, Value of Implements and Machinery, Value of Livestock, and Value of Total Capital per Worker, 1939
(In dollars)

Region	Value of product per worker	Value of machinery per worker	Value of livestock per worker	Value of capital per worker
Pacific	1,310	420	450	6,670
Mountain	1,190	425	935	5,550
W.N. Central	1,160	540	765	6,760
E.N. Central	1,060	495	625	6,210
Mid. Atlantic	960	470	540	4,260
New England	960	295	370	3,610
W.S. Central	530	180	320	2,810
South Atlantic	460	100	190	1,830
E.S. Central	350	90	205	1,625

Source: Sixteenth Census, 1940, Agriculture, Vol. III, General Report.

Table IV. Indexes of Net Farm Income per Worker, 1939 to 1941, and Total Value of Capital per Worker, 1939, by Regions

Region	Index of net farm income per worker[1]	Index of capital per worker (per cent)
Pacific	335	410
W.N. Central	265	415
E.N. Central	255	380
Mountain	250	340
Middle Atlantic	220	260
New England	190	220
W.S. Central	145	175
South Atlantic	120	115
E.S. Central	100	100

[1]Harry G. Norcross, State Estimates of Expenses and Net Incomes, 1929, 1939–1942, BAE, USDA.

West North Central. The value of livestock varies from $190 in the South Atlantic States to $935 in the Mountain States. In general, the southern worker has less than a third as much machinery and livestock as do farm workers throughout the rest of the nation.

These three categories of capital can be brought together to give a fairly adequate measure of all the cooperating factors. The last column in Table III does this. For the nation as a whole, the regional data indicate a marked consistency in the relationship between output per worker and the total value of cooperating factors.

In Table IV an attempt has been made to obtain a net product per

Table V. Average Resource Combinations Required to Produce $1,000 Output in 1939, by Regions[1]

Region	Adjusted[2,3]		Unadjusted[3]	
	Labor (man-years)	Value of Land and buildings	Labor (man-years)	Value of land and buildings
W.N. Central	1.80	8,300	1.75	7,500
Mountain	1.85	7,200	1.85	6,200
E.N. Central	1.85	7,000	1.80	6,300
Pacific	2.00	9,000	2.00	8,800
Middle Atlantic	2.05	5,250	2.05	5,250
North East	2.10	5,000	2.10	5,000
W.S. Central	2.30	4,100	2.25	4,900
East S. Central	2.30	4,200	2.25	4,000
South Atlantic	2.55	3,700	2.55	3,700

[1]Sixteenth Census, 1940, Agriculture, Farm Characteristics by Value of Products.

[2]Adjusted data differ from unadjusted data due to attempt to remove differences in relative underestimation of products sold by regions as reported to the Census and to include estimates of costs of capital maintenance on buildings.

[3]The expenditures subtracted from value of products sold were for feed, fertilizer, petroleum products, and livestock. The estimate for livestock expenditures by regions was from BAE data, distributed among the value of products groups in proportion to feed purchases. The composite firm was found by estimating two simple linear regressions: (1) between labor and value added, and (2) between value of land and buildings and value added, for four value of products groups lying between $600 and $1,999. The data did not permit including other capital than land and buildings in the calculation.

worker which would show the relationships which would exist in a period of reasonably full employment. Value of product as a measure of productivity has tended to overemphasize productivity in the Northeastern States because of heavy purchases of partially processed raw materials, particularly feed. In addition price relationships in 1939 were probably in favor of milk and beef and against hogs, wheat, and cotton relative to the relationships which would exist if the economy were operating effectively. As a result the three years, 1939 to 1941, have been used.

This change in the measure of productivity again makes but little difference, though the results are somewhat more consistent with our implied hypothesis that output per worker increases almost proportionately with the quantity of cooperating factors. The rank of each region in net product per worker and capital per worker is the same except for a single case—the West North Central States.

The analysis thus far presented may now be summarized. The evidence indicates unmistakably that output per worker is closely and positively related to the other factors with which his efforts are combined. This positive relationship between output and quantity of cooperating factors is what one would expect from the theory of the firm unless there is perfect substitution among the factors. The substitution of labor for capital in the capital poor areas has apparently reached the point at which it takes a very large amount of labor to replace a small amount of capital. Thus addition of more labor in the capital poor areas would increase output only slightly, while large deductions from the labor force will not markedly reduce output even if capital remains constant.

The preceding analysis has been based on a comparison of output per worker in average firms of greatly different size. As a result, the analysis could be strengthened if output per worker and the relative combination of resources for firms of essentially the same size, measured in value of output, could be estimated for the different regions. The data previously presented confounds two important factors which bear upon the marginal productivity of capital and labor: (1) the combinations of factors, and (2) the scale or size of the firm. By holding the latter constant, greater insight may be obtained as to the differences in factor combinations and variations in worker productivity.

Capital costs, per dollar invested, are approximately the same throughout the agricultural economy. Any marked difference in the amount of capital employed in producing a particular output must be reflected in a lower marginal productivity for labor. The sample study of the 1940 agricultural census provides fairly adequate data for determining the combinations of resources required to produce a particular output. A third measure of productivity has to be introduced. This measure is what might be called value added by manufacturing and is found by subtracting certain expenditures from the value of products sold. For each region a composite firm producing an output of $1,000 was constructed from the data.

Because of imperfections and limitation of the data, it is impossible to conclusively verify the hypothesis that there are consistent regional differences in the relative combinations of resources required to produce the same output. In general, however, the data indicate that in the areas of lowest output per worker, there is appreciably less capital and that as the amount of capital increases the labor inputs decline. With the exception of the Pacific and West South Central States, the combination of labor and capital required to produce $1,000 output is in line with the hypothesis. A comparison between the two extremes is interesting. In the South Atlantic States 45 percent more labor and 55 percent less land is used than in the West North Central States to produce the particular output.

The evidence points to three related conclusions: (1) there is a marked difference in the productivity of the resources that go to make up farms in the various regions of the nation; (2) after consideration is given to the difference in capital costs, the major difference in productivity is between the Southern States and the rest of the nation;[15] (3) the dearth of capital and the surplus of labor in the South are all too evident.

What role could agricultural price policy play in effecting a distribution of resources which would more nearly equate the marginal productivities of resources in the different regions? The role would be minor, but still of importance for commercial agriculture. Correction of the present inequalities depends largely on the movement of the human agent. This movement is controlled by many factors and relative earnings are probably of minor influence. In terms of a better allocation of resources, price raising measures would hinder rather than help. Any measures which attempt to improve incomes by generally increasing the price per unit of output would only distort further the inter-regional differences in the marginal productivities of resources and would lead to additional capital investments in areas where capital is now most plentiful. The region with the largest firms, because their incomes increase most both absolutely and proportionately, would be in the best position to expand. In all areas the higher prices would constitute a factor inhibiting the movement of labor out of agriculture.

Price policy, through increased price certainty, could reduce capital rationing and bring in more capital resources (particularly machinery, livestock and investments *in* the land) into certain of the areas involved. This reallocation would not affect the farms that are principally self-sufficient because the increase in price certainty would affect only a small part of their total output. Additional capital inputs, however, with the immobility of human agents and inelastic demands for major products, might actually result in greater discrepancies in the regional marginal productivity of labor. For instance, large amounts of additional capital in southern agriculture might well result in greater unemployment or underemployment in agriculture. If the output of its major product, cotton, were materially increased, the long-run effect would be to lower the share of the national output received by that area.[16] Additional capital investments can be more easily integrated into agriculture in the rest of the nation because of the greater mobility of the human agent and because in the other areas incomes may be high enough that part of the increased productivity may voluntarily find its outlet in leisure, whereas in the South only greater underemployment would result.

2b. Among firms in the same region. Thus far in our comparisons of the productivity of resources we have dealt only with regional averages. The variations within regions, which are somewhat more homogeneous than the nation as a whole, are likewise of interest. It is evident from an analysis of the intra-regional data, that possibilities for the improvement of resource use is not confined to the South. Even though population mobility is a function of distance, within agriculture the migration of people to capital resources is inhibited by other forces. The existence of capital rationing

Table VI. Value Added and Labor and Land Supply for East North Central States and
South Atlantic States, 1939

Region	Value added	Labor supply (man years)	Value of land and buildings (dollars)	Value added per worker (dollars)	Value of land and buildings per worker (dollars)
East North Central	490	1.58	4,190	310	2,650
	690	1.64	4,980	420	3,050
	994	1.78	6,170	560	3,450
	1,410	1.93	8,230	730	4,250
	1,825	1.98	10,425	720	5,250
	2,570	2.17	13,950	1.180	6,450
	3,070	2.55	21,775	1,500	8,550
	6,142	2.91	30,920	2,110	10,600
	15,987	5.90	41,810	2,710	7,100
South Atlantic	565	2.14	2,370	260	910
	740	2.25	2,775	330	1,100
	1,015	2.53	3,725	400	1,250
	1,140	2.87	5,340	490	1,450
	1,860	3.20	6,450	580	1,850
	2,480	3.53	8,170	700	2,000
	3,670	4.95	12,940	740	2,450
	5,550	6.80	22,260	820	3,250
	17,235	17.30	49,200	1,000	2,850

Source: Sixteenth Census, 1940, Agriculture, Farm Characteristics by Value of Products.

and other factors (including inheritance and past income) affecting the distribution of ownership or control of factors of production do not permit
the achievement of equality of marginal productivity of comparable resources.

The differences in the combination of resources within each region have
marked effects on the productivities of the resources. This point can be
readily illustrated by drawing on material from one northern and one
southern region. After excluding the bulk of the subsistence farms (those
with value of products less than $600 in 1939), and those farms with over
$10,000 value of products sold, the following relationships seemed to exist.
(See Table VI.) Going from the smallest to the largest units, the productivity per farm increased by 13 times with less than a doubling of labor and
an eight-fold increase in the value of land and buildings in the East North
Central States. In the South Atlantic States the productivity increased
12-fold with a tripling of labor supply and a 14-fold change in the value of
land and buildings. The increase in the value of land and buildings per

worker was about four-fold in both regions. The data definitely show that as output increases on a per farm or per worker basis, the worker has considerably more co-operating factors.

Unquestionably differences in the marginal productivities of both capital and labor exist within the regions. It is not at all clear, however, that output would be increased by a reshuffling of factors so that there was approximately the same combination of factors. There is some weight provided for the argument that there is increasing returns to scale within the range of farm sizes considered. The evidence is not clear-cut, however, since as the size of farm increases the relative combinations of the factors also change in the direction of more capital per worker.[17] It may well be, however, that in terms of improving resource use that the answer lies in leaving the larger units as now constituted and increasing the size of the firm and relative amounts of capital per worker on the now smaller firms. This would involve the movement of human resources out of agriculture and capital resources in, if total output is maintained.

The contribution of price policy must be considered from two viewpoints. Considering only the commercial farmers, greater price certainty would likely improve the position of the smaller farmer relative to the larger. The smaller farm units are most seriously affected by capital rationing and the discrepancy between the marginal productivity of capital and the rate of interest is much greater than for the larger farms. Thus reducing price uncertainty would permit a larger relative expansion in the use of capital factors for the small farmer. Price policy, however, would do little to aid the non-commercial farmer for the reasons pointed out above. It may be noted that if the goal of price policy is merely that of higher prices, there would be an adjustment in the allocation of resources within an area that would be inferior to that now prevailing. The major advantages would accrue to the larger units in which the allocation is now nearest the ideal.

3. Allocation of resource use in time. In allocating resources in time, two sets of major decisions arise, one dealing with the rate at which stocks of raw materials are transformed into final goods and services and move into consumption; and the other, with investment and disinvestment of agricultural resources. A major limitation of a free market economy is what might be called its short-sightedness, an orientation toward the present and near future which results in goods flowing through the production channels too rapidly at times, regardless of the effects on producer incomes and consumer satisfaction. The unwillingness to store a larger part of the output of large crops grows out of the nature of price expectations and the financial position of most farmers. The financial position of most farmers as well as uncertainty as to future price movements makes impossible foregoing the present income or bearing the risks involved in holding. Forward prices would provide a satisfactory mechanism for a better time distribution in the flow of favorable products into processing and consumption. Forward prices would accomplish this because of the improvement in collective expectations and by providing farmers as a whole with a satisfactory alternative (storage) to immediate sale for consumption or processing.

The other important aspect is the way investment and disinvestment in agricultural resources, particularly land, is occasioned:[18] (1) by radical changes in farm prices and incomes, and (2) the nature of the farm tenure system.

When the pressure of fixed obligations upon the farm, usually in terms of fixed contractual relationships but sometimes in terms of obtaining enough production for subsistence, becomes rather sharp, there usually occurs a type of disinvestment in soil resources which is manifested by erosion and depletion. If price policy is effective in evening out the cyclical swings of prices to a considerable extent, it will be effective in reducing a part of the waste and economic hardship arising out of inflated land values and consequent contractual payments which impinge heavily upon the farm family. In addition, it is likely that a more even income flow which would result from forward prices would mean somewhat less necessity for the exhaustion of human and material resources in a vain attempt to meet the direct obligations of the family, other than indebtedness.

Many of the forces leading to soil erosion and depletion, however, are beyond what can be affected by price policy. As recognized by all students of the subject, land rental under the primitive American tenure system is conducive to socially undesirable disinvestment of the land and generally does not lead to economically profitable investments. Under most existing tenure arrangements, the tenant is not forced to pay the value of the disinvestment nor is it possible for the tenant to receive payment for any unexhausted investments. Within the framework of our tenure laws lie some of the most important divergences between private and social costs. Farm tenure reform and a soil conservation program (embodying education, community and group agreements and limited subsidies) probably are necessary to cope with the problem.

B. Allocation of Resources Between Agriculture and the Rest of the Economy.
It is difficult to find an adequate basis for comparing differences in the effectiveness with which resources are used in agriculture and the rest of the economy. A comparison of average returns for the same resource is of little value, since the problem involved is that of transfer which requires a comparison of marginal returns. If there is unemployment above the so-called minimum level in the non-agricultural sector, nothing would be gained by moving out the under-employed in agriculture and adding them

to the industrially unemployed. In a situation of unemployment, the marginal return is essentially zero for the labor transferred out of agriculture into industry, and this remains true regardless of any differences between average labor earnings in the two sectors of the economy. Actually, one has to deal with a rather hypothetical situation—namely, if resources were fully employed in the non-agricultural sector, would the economy gain from a transfer of resources between the two groups?

If the problem is posed in this way, the nature of the movement can be indicated rather simply. On the one hand, the human agent should move out of agriculture. On the other hand, many non-human resources should be moved into agriculture. The approximate extent of the movement of people out of agriculture is difficult to determine because we do not know how many people are in agriculture, in a significant sense. At least half of the nation's tracts of land that are defined as farms by the census make no significant contribution to the market supply of farm products. Some 10 percent of the operators of the 6,000,000 farms do not even classify themselves as farmers. In addition, in 1939 almost 30 percent of the nation's farmers reported work off of their farm and the writer has estimated at least half of these receive more income from wages than the value of products sold or used. Probably three-fourths of this number would receive more of their net income from wages than from agriculture. In this group is probably a fairly large number that have a satisfactory combination providing an effective use of their resources and talents.

One can only hazard a guess as to the number of people that would move out of agriculture if there were employment opportunities for them elsewhere and relative earnings were the only factor affecting mobility, but an estimate of the workers from 1.5 to 2 million farms does not seem to be too extreme. In any case, the number would be large.

The type of agricultural price policy that is instituted will probably have little to do with this movement. Though it might be possible to construct a policy that will deter the movement, it does not seem possible to do much by price policy alone to aid the movement.

The comment that non-human resources will move into agriculture as human resources move out requires some consideration since the conclusion is based on certain presumptions about agriculture. In the theoretical model of the firm and industry under competition, the demand for one factor of production (say capital) will be increased if the price of the other factor (say labor) increases if the elasticity of factor substitution is larger than the elasticity of demand.[19,20] If the elasticity of demand is less than unity as it apparently is for total agricultural output, only a moderately high elasticity of substitution would be required for the demand for capital to increase.

In addition to the influence of the elasticities of substitution and demand for the product, two other interrelated forces would be of importance in affecting the demand for capital. Capital rationing and the level of past incomes and savings are both important in affecting the demand for capital. If capital rationing exists the marginal value productivity of capital will exceed the interest rate and agricultural proprietors will have an incentive to expand the use of capital. However, the only important source of funds to expand capital use by purchase is through saving once the full ration allowances are used. Consequently, any change in the level of farm incomes per worker will, independent of other forces at work, increase the demand for capital. The out movement of labor to the extent necessary to reach the best use of labor resources would obviously raise the returns of those staying in agriculture. At the same time steps might be taken to reduce the effect of capital rationing through price policy.

An empirical analysis of certain agricultural data has been attempted to determine the marginal productivities of capital and labor in American agriculture. Census data for four agricultural regions were used; the regions were selected because the data appeared most adequate. Labor was measured by man-years; capital by value of land and buildings, and output by value added. Nine value groups were used in each region, giving 36 cases in all. A Cobb-Douglas production function was fitted without forcing the sum of the exponents to be unity. The regression equation found was $P = 1.5 \, L^{.88} C^{.74}$.[21] This equation at first sight may appear to be incongruous. It indicates that for these four regions, there are increasing returns to scale and that the total payments to the factors of production exceed the total value of the product. The latter discrepancy, however, is based upon the assumption that the factors of production are paid at rates equal to the value of their marginal productivity. There is no reason that this should be true in an uncertain world. Capital rationing by limiting the size of the firm can force a discrepancy between the rates of payments and the value of the marginal productivity without excess profits existing. Similarly the existence of capital rationing can create increasing returns to scale since it prevents firms from reaching what would be the equilibrium position in a situation where single valued anticipations existed.

Some of the implications of the production function may now be indicated. The marginal productivities, at the mean values, are 15 percent for capital and $480 a man-year. The marginal productivity of capital is about three times the interest rate, while the marginal productivity of labor is about a third higher than U.S. farm wage rates for 1939. These marginal productivities, however, are higher than what would be achieved on the average for these four regions.

In fitting the production function, each value added group was given equal weight. On this basis, the average output of $1380 was almost 70 percent higher than the weighted mean of approximately $800. At an output of $800 the marginal productivities are 12 percent for capital and $390 for labor. The latter marginal productivity is almost identical with the average farm wage rate for 1939.

Returning now to the probable direction of capital movement, it can be said that it is likely that capital will move into agriculture as labor moves out. A reduction of 20 percent in the labor supply would reduce the marginal productivity of capital from 12 percent to a little less than 10 percent.[22] The differential between the rate of interest and the marginal productivity would remain very wide and any reduction in capital rationing or increase in savings would result in the use of more capital.

It might be argued, on the basis of the above analysis, that more capital should be employed in agriculture regardless of any change in labor. If the elasticity of demand for agricultural products is unity or less, the employment of further capital will either reduce or leave unchanged total agricultural income; however, a larger share of the total and a greater aggregate amount would be paid to capital thus still further reduce returns to labor and further aggravate the disparity in the marginal value productivity of labor in agriculture compared with workers in the rest of the economy. An improvement in the use of capital would be offset by an opposite effect upon labor. However, if labor moves out of agriculture to any considerable extent, further capital investment would be desirable.

An improvement in resource allocation between agriculture and the rest of the economy can be aided by price policy if other measures are taken to improve the mobility of labor. Price policy would have two effects if greater price certainty were achieved. First, with greater price certainty the additional investment would be larger than it would be if the only force at play were the larger income resulting from less labor and a smaller output. Greater price certainty would provide the incentive for more nearly equating the marginal value of productivity to its cost. Second, the distribution of the ownership or control of the additional capital would undoubtedly be improved by a reduction of capital rationing. An important consequence of capital rationing is that its impact is variable, falling least on these individuals in the best income or capital position. Greater price certainty would give many farmers with meager resources a possible chance to acquire the additional capital.

II. The Income Problem

The income goals of American agricultural policy have been unfortunately narrow because the focus has been on commodity price parity. Interest has focused not only on the share of the national income going to agriculture, but also upon the income accruing to particular commodity producers. This narrow conception has tended to straight jacket American agriculture and has precluded adjustments in both the allocation of resources and distribution of incomes that are necessary and desirable over the long pull. To some extent attention has been given to the strictly relief or poverty aspects of the income problem, but in terms of personnel or funds involved it is clear that predominate attention has centered on commodity price parity and the proportion of the national income pie going to agriculture.

The analytical techniques for dealing with the problems growing out of the level and distribution of the national income are much more primitive than those at our disposal for the treatment of resource use. In fact, we must look for our guides in the area of social values with a certain limited use being made of tools available in economics proper. In large part one must view the income problem in terms of an enlightened, social policy and, in our case, analyze the contribution that price policy can make toward reaching the desired goals. Though much of what is said here is applicable to the whole economy, the discussion draws wholly upon materials relating to agriculture.

The major income goals for agriculture are here considered to be four in number. The first is the goal of achieving a certain minimum scale of living for all members of society on the basis of social welfare criteria which can be spelled out rather specifically in terms of health, education, nutrition, and housing. A second is the reduction of differences in income distribution in agriculture. A third, which was mentioned above, is the raising of income to people in agriculture to a par with the incomes in the rest of the society. The last is the attainment of greater stability in the aggregate level of incomes, which may be desired both for itself and as an aid in the stabilization of employment and incomes in the rest of the economy.

In the analysis of the resource problem, we did not find it necessary to make any distinction in the application of price policy which would depend upon whether or not resources were being fully employed. The purpose at that point was to analyze the contributions and limitations of price policy in attaining a full use of resources. However, in the analysis of the income problem, striking differences in the nature of the income problem and the contribution of price policy exist in a condition of optimum use of resources and in a situation where many resources are partially or fully unemployed.

A. Income Distribution in Accordance with Social Welfare Criteria
The exact extent of the poverty problem in agriculture obviously cannot be determined. Poverty is a relative concept which varies through time as the social values change with a rising average level of real income in a society. Not only is there the difficulty of determining what constitutes the nature of poverty, but the data on income distribution in agriculture are quite meager. The only reasonably satisfactory data were those obtained

during a year of relatively low incomes, 1935–36. During that year about 10 percent of farm families were receiving relief and of the non-relief families 35 percent received less than $750 income from all sources, non-agricultural as well as agricultural, while 18 percent received less than $500.[23] In 1939, almost 45 percent of all farms produced less than $750 worth of all products, which the writer has estimated would mean a net income, including house rent, of no more than $550. The two estimates, however, are not comparable since the latter does not include income from non-agricultural sources.

Granted, however, that there is an important problem of a large number of families with sub-normal incomes and that our society, through its government, were to establish certain minimum standards of livelihood, would price policy be an effective means of meeting the goal? The answer must be negative. The means for attaining some of the minimum standards, such as education, can be satisfactorily supplied only on a community, state, or national basis. Improving nutrition involves better knowledge as well as income to purchase or produce the necessary products. Further, it is unlikely that health standards would be met except by an organized program, involving free or low cost medical attention and education.

The interrelationships between achieving a better use of resources and the solution of the poverty problem can be indicated. In the first place, even if full employment is achieved there will still be large numbers in agriculture that lack sufficient resources to permit the earning of the income required to meet the minimum standards of life established in accordance with social welfare criteria. An enterprise economy, without the collective efforts of society, does not have the institutions which permit the types of investments which must be made in the human agent. A human being is born without productivity; he achieves a particular level of productivity only through investment which usually must be made by the family in which, by chance, he is born. If the productivity or income of that family is very low, the nature of the investments made may be small and value of that individual to society as valued by his ability to produce will be meager.[24] In the second place, investments in terms of improved nutrition, health, and education which may be made by society have a continuing influence upon the use of resources and the distribution of income. Mobility, which is certainly required, becomes possible only as the individual has something to sell which is more valuable when applied elsewhere.

In an economy which is operating at full gear the types of investments which society would find profitable would decline in importance unless the social welfare criteria continually expanded. Because investments in the human agent are largely made by the family, the improvement in the status of formerly poverty stricken families will generally continue into the next and succeeding generations. However, just as the educational standards have grown, it is likely that the other aspects of the good life would expand through time.

In this development the role of price policy is fundamentally that of providing the necessary climate. As we have pointed out throughout this analysis, price policy has an important contribution in achieving a better allocation of resources, but that it must be supplemented at various points both to do the job on the resource side and especially to achieve better results in income distribution.

If resources are not fully employed, the poverty issue becomes of much greater magnitude. One of the major tragedies of the fluctuations which occur in output and income is the unequal incidence. The unemployed and partially employed suffer sharp reductions, while many who are fortunate enough to remain fully employed at monetary rates of earning that are sticky receive increases in real earnings. The techniques which must be applied for tackling the poverty problem in agriculture in such circumstances must be almost wholly outside the price system.

In those instances where increased incomes are required, it is questionable if the price system is an adequate or proper means of providing the income. In many cases, the farm families involved receive such a small proportion of their total income from the market that several fold increases in prices would be required to provide the necessary income. Using the price mechanism (without price discrimination with higher prices for lower income farms) to accomplish this goal would lead to maluse of resources, overinvestment in certain segments of agriculture, and unjustifiable windfalls to the commercial farmer.

Programs which would improve the health and education of farmers and their families would make an important contribution to resource allocation, if the programs are not "financed" through the price system. A large part of the immobility of poorer farm people is due to ignorance, lack of skills, poor health, and insufficient funds to make movement possible. A program properly designed and executed could well pay large dividends to society, but the price system is not an adequate procedure for directing the endeavor. It is important to bear in mind at this point that the price system affects incomes according to the productivity and ownership of factors; and this important fact is generally lost sight of in most popular and political discussions. Poverty, ill health and poor education are negatively correlated with the productivity and quantity of resources owned. Paying more for factors of production through the price system simply fails to affect the great bulk of the families in need of assistance on the basis of social welfare criteria. To further such goals, the program must be directly related to people, not to factors of production as such.

B. Relative Incomes in Agriculture

Whether the income distribution in agriculture has an undesirably wide dispersion can be determined only on the basis of social values. In line with the trend in social values, however, we are probably on safe grounds if we accept the hypothesis that any policy which tends to increase the dispersion in a relative or absolute sense is undesirable. The impact of full use of resources would be generally considered to be desirable. If it is possible to attain a socially desirable minimum standard of living for most farm families, probably the most important single aspect of the dispersion over and above this minimum level can be viewed with much less concern if the great bulk of the families have attained that level. In addition, the techniques for changing this part of the distribution are well established. The progressive income tax provides the technique for any sort of narrowing of the personal income distribution which society may desire and it is obviously the most acceptable device.

Except as price policy is directed toward an improvement in the use of resources, the effect on income distribution is to increase the dispersion. Insofar as price policy, say through forward prices, reduces the impact of capital rationing and makes possible the recombination of resources, income differences will depend less upon the chance factors which affect the distribution of the ownership of factors. However, attempts to improve incomes in agriculture by raising prices or by any type of conditional payments, such as those made by the AAA, which are tied to non-human resources, will increase the dispersion of incomes.

C. Relative Share of National Income Going to Agriculture

One of the worse of the pseudo-statistical myths perpetrated upon the American public has been that of agricultural income parity. The belief that any period can be used to measure the fair share of the national income which should go to agriculture or any other sector of the economy is contrary to all experience and it has no support in economic analysis. Changes in degree of relative unemployment, tastes, technology and differences in mobility are all factors that influence the part of the national income that any one group receives. The fact that this particular myth has been embodied in legislation makes it even more deleterious. Even worse yet is the consideration that it has led to steps that do not in any way correct the underlying conditions which may result in agriculture being depressed.

Comparisons of income per worker in agriculture and per worker elsewhere on the basis of available information lead to highly erroneous conclusions unless extreme caution is used. Income to farm people from agriculture as defined in legislation ignores perhaps a third of the total income accruing to individuals that are called farmers in the legal (or census) sense.

Even when the income accruing to farmers from all sources is estimated,[25] problems of comparing dollar incomes would still be surrounded with a whole set of difficulties. Though the farmer normally receives only 50 percent or less of the retail value of food, farm products consumed by the household are valued at farm prices. In 1939, if home consumption were valued at retail prices, the net income to persons on farms from agriculture would have been increased by more than 20 per cent. Likewise, comparison of rental value of homes is difficult, if not impossible. The rental value of farm dwellings in 1939 was estimated at $110 a year per farm. Could comparable facilities, on the average, be obtained by the average non-farm dweller for $9 or $25 a month? No one knows, though the Bureau of Labor Statistics estimates for rents of workers from 1935–39 indicates an average cost of $20 to $35, depending upon the city.[26]

Any comparison of per capita incomes in agriculture and for the rest of the economy is grossly misleading for another reason. The fact that farm families are considerably larger than non-farm families may bear upon the desirability of income transfers to farm people for rearing and educating children, but it does not prove that the income of each individual on farms (including non-workers as well as workers) should be the same as for non-farm, nor that the income transfers should be made through the price system. There is certainly no fashion by which the price system can be used to insure that the income transfers accrue to the "right" families or are used for the desired purpose.

Basically the share of the national income going to agriculture is related to several fundamental conditions: (1) the rate of technological advance in agriculture, (2) the relatively low price and income elasticities of demand for food and fiber, (3) the high birth rate among agricultural people, and (4) the level of industrial employment. Even when the industrial economy is operating at full gear, agriculture will offer gradually diminishing employment opportunities for labor. Because of the necessity of moving people out of agriculture, and experience has shown this to be a difficult and time consuming task, agriculture may well show many of the characteristics of a "depressed" industry.

Given these underlying conditions, even with full employment, agriculture will probably never receive its "fair share" of the national income in the market place without governmental interference. The fact that workers must continually leave agriculture means that at any one time earnings are lower in agriculture than elsewhere. Only if mobility were costless, which it obviously is not, would the necessity for the lower earnings in agriculture be obviated. But price raising measures per se are useless to correct the situation over the long pull. Raising agricultural incomes in this fashion merely leads to keeping more people in agriculture under reasonably similar conditions than if prices were left alone. Some short run gains can be made, but such gains would soon be dissipated by the induced increase in farm workers.

Fundamentally there are only two types of measures which will assure agriculture that over a period of time that economic opportunities in agriculture will be essentially comparable to those in the rest of the economy. One is to make mobility as easy and costless as possible. This involves employment opportunities in industry, education and training, effective employment agencies and various other steps which increase worker productivity. The other is the transfer of at least part of the costs of supplying the nation with its population growth from farm people to society as a whole. Agricultural people have long made large investments in their children and much of the returns from these investments have accrued to society as a whole and not predominately to farm people. The nature of the investments made in children is closely related with the ease with which excess farm population can be absorbed into the rest of the economy and thus with the problem of mobility. With the declining emphasis in our mechanized economy on the need for unskilled workers, it is obviously a matter of "good business" for society to interest itself in the quality of one of agriculture's most important products—surplus population. What the exact character of the assistance to farm people should be is outside the scope of this paper, but low cost medical service, adequate educational opportunities, nutritional programs and family allowances immediately suggest themselves.

When non-agricultural resources are not fully employed, agriculture is faced with an unduly critical situation. The nature of price movements in modern economies is such that agricultural prices fall first and most, while the high rate of increase in population in agriculture forces more people to remain there. These two factors, combined also with technological advance, result in a sharp drop in total and per capita income in agriculture. It is not to be unexpected that the income position of farm people during an economic depression should cause alarm. The role of price policy for dealing with these problems is discussed in the next section.

D. Stable Agricultural Incomes

The goal of stable agricultural incomes, as enunciated by farm people, is largely a search for security. Given the primitive credit arrangements existing in agriculture, the wide fluctuations in agricultural incomes and asset values have resulted in hardship of a most acute nature. High incomes have too often found their way into inflated capital values which have later been deflated with resulting dispossession and loss. Stable agricultural incomes should also be viewed in terms of effects upon the maintenance of full employment in the rest of the economy. Because of the slowness with which expenditure patterns change and the sluggishness with which resources move, a sharp decline in incomes in agriculture can greatly intensify the severity of an economic collapse. Even though agriculture is not sufficiently important in the economy to be a factor in motivating a depression or recovery, the sharp changes in farm incomes do have important reinforcing consequences in either direction.

During periods of full employment, total agricultural income, income going to various products, and the income of individual farmers will vary. Total agricultural income is likely to vary but little under such circumstances, unless full employment is being maintained by inflationary means. The income for various products, however, will still vary as supplies change due to weather or inaccurate price expectations. The income of individual farmers will vary for the same reasons, though to a greater extent. Price policy alone cannot eliminate these income variations. It has been suggested that forward prices should fluctuate inversely with the size of the crop. There are a number of disadvantages in this procedure, including its possible application only to products that move directly into consumption without further farm processing, but the most important is that stabilizing the total value of a crop does not stabilize the income of the individual farmer. A high degree of income stability can be achieved, however, by a forward price system with storages and crop insurance applied to the individual farm unit. Income uncertainty in livestock production due to the incidence of disease would still remain, but this uncertainty is of smaller magnitude than that arising out of crop fluctuations.

The more important issue is whether the price system should be used to stabilized aggregate agricultural incomes during a complete business cycle. There are several techniques for doing this, but the discussion of their relative merits cannot be undertaken at this point. We will merely assume that the choice is between adopting the best method of using the price system to stabilize aggregate income at some predetermined level or other techniques not operating through the price system. It is also assumed that as a part of an anti-cyclical program, the government will apply measures to maintain purchasing power in the rest of the economy and that there will be no discrimination against or in favor of agriculture.

Payments to agriculture during a decline in national purchasing power can be based on two related criteria—the relief of hardship and maintenance of purchasing power. Hardship grows largely out of the incidence of the depression in forcing incomes to a level insufficient to maintain a satisfactory scale of living,[27] and the loss of assets arising out of foreclosure due to tax delinquency and failure to meet charges on indebtedness. Maintenance of purchasing power must be considered not only in terms of the total purchasing power in agriculture, but attention must be given to the manner in which the funds are expended by the recipients. In addition to these two criteria, which are special aspects of income distribution, the effect of any payments on resource use must be considered.

As a means of reducing hardships, the price system does not provide an entirely adequate procedure for distributing funds to people engaged in agriculture. Of the total payments made, less than fifteen percent of all farmers would receive sixty percent or more of the payments, while half of the farmers would receive no more than ten percent. So marked a regressiveness would be both undesirable and publicly unacceptable. Part of the impact of a depression (on agricultural people) is the loss of earning opportunities of individuals requiring outside employment to supplement their agricultural incomes. For the great bulk of these families payments through the price system would afford only partial and inadequate relief. The halting of the movement of people out of agriculture by unemployment in the industrial sector is also an important cause of hardship. There is no reason to believe that the incidence of this added burden on many farm families would be so distributed that payments through the price system would be effective in cushioning this burden. Though a large proportion of the nation's farmers sell only a small share of their total output, the loss in cash income is of major importance to them. Payments through the price system to meet the needs of these families would have to be large in the aggregate because of the small volume sold by these families.

If maintenance of purchasing power and the effects on resource use are given consideration, the price system has certain advantages. For encouraging maintenance or increase of employment in the industrial sector, the total sum of payments made to agriculture is probably of no greater importance than the way the funds are expended. A major aggravating factor in a deflation is the difficulty of moving resources from the production of one product to a rather different one. If the funds paid to agriculture are spent in about the same way as in a more prosperous period, the contribution of the payments to recovery will be greater than if the funds flow into areas that would not continue once the payments were halted. Payments through the price system will also have advantageous effects on resource use as compared to other types of payments. Producers would continue to guide resource use in line with incentives provided through the price system. Resource use would certainly be superior to that which would occur as a result of payments such as those made by the AAA. There is no justifiable reason for using payments during a depression to limit or redirect production other than would occur from shifts in relative prices.

Provisionally it may be said that payments to agriculture during a depression should be of two types. One would be a payment made directly to farmers as people as a means of directly relieving hardship. The function of these payments would be to maintain consumption expenditures and investments in the human agent (education, medical care, etc.) at a reasonable level. The other payment could be made through the price system as price supplements. Their purpose would be to encourage proper direction of the use of resources and a continuation of expenditures on durable investment goods (machinery, repairs, etc.).

Notes

The writer is indebted to T. W. Schultz for numerous criticisms and suggestions for improvement, and to J. M. Letiche for help on a couple of troublesome problems.

1. Certain discrepancies, which may be called Pigou's divergencies between private and social costs, may exist. These will arise if a firm's output results partly from resources for which no remuneration is paid for under existing property laws, if consumers receive satisfaction from scarce goods for which no remuneration is made, or if the operations of a firm result in a disservice to other individuals as producers or consumers. Cf. A. C. Pigou, The Economics of Welfare, p. 174.

2. Throughout the exposition personal capacities or labor services will be included in resources and ownership of resources will include the personal capacities of individuals as well as all other resources.

3. See F. H. Knight, Risk, Uncertainty and Profit, pp. 178–181, and Ethics of Competition, pp. 54–58.

4. Though economists have always analytically separated resource allocation (value theory) and income distribution (distribution theory), the recognition of the fundamental importance of this distinction in the analysis of policy issues has not been general. Failure to generally recognize this important distinction may have been due to the unwillingness of economists to make value judgments. Pigou (in his monumental Economics of Welfare) broke away from the tradition, and T. W. Schultz was probably the first to prove its usefulness in analyzing a specific set of public policies ["Economic Effects of Agricultural Programs," *American Economic Review*, Vol. XXX (February, 1941), pp. 127–154]. See also T. DeScitovsky, "A Note on Welfare Propositions in Economics," *Review of Economic Studies*, Vol. IX (November, 1941), pp. 77–88.

5. Capital rationing implies that the firm is unable or unwilling to borrow as much as desired at the existing rates or interest. Two types of capital rationing may be distinguished. External rationing is the limitation which individuals outside the firm place upon the funds available to the firm. Internal rationing is self imposed in the sense that the entrepreneur is unwilling to borrow more than a particular amount because of his attitude toward the risks involved in undertaking such fixed obligations. For treatments of capital rationing see T. W. Schultz, "Capital Rationing, Uncertainty, and Farm Tenancy Reform," *Journal of Political Economy*, Vol. XLVIII, pp. 309–324, (June, 1940), and A. G. Hart, "Anticipations, Uncertainty and Dynamic Planning, Studies in Business Administration," The School of Business, University of Chicago, 1940, esp. Chap. III.

6. Forward prices mean simply the announcement by a governmental agency of prices in advance of planning operations. For further analysis, see T. W. Schultz, Redirecting Farm Policy, Macmillan, New York, 1942, and the series of articles by G. S. Shepherd in this journal (Aug. and Nov., 1942, Nov. 1943, and Aug. 1944). The writer intends in the near future to publish an article on the theory of forward prices, which will deal with certain administrative problems as well as the more general theoretical problems.

7. The second condition may appear to apply directly to the operations within a firm, but it should be noted that for any given bundle of resources used by a firm the fulfillment of the first condition necessarily requires the fulfillment of the second.

8. Uncertainty is used to describe a situation in which expectations are not single valued (certain). It is thus broader than most definitions of uncertainty, but it is the writer's belief that any distinction between uncertainty and risk breaks down when applied to the individual firm. Insurance, for example, does not eliminate uncertainty; it only reduces the impact of uncertainty on the firm but can never completely eliminate it. If one insures a crop against destruction by natural causes, an element of uncertainty remains. If the crop is destroyed, the farmer may be forced to purchase feed at prices higher than he had anticipated, particularly if the crop disaster were widespread. Furthermore, insurance seldom reflects full loss because of "moral hazards"—and the cost of insurance is always greater than the actuarial cost due to administrative costs. Thus the decision to insure involves the same considerations as any other decision involving uncertainty.

9. That one-half is not an overestimate is indicated by certain data revealed by the 1940 Census. First, almost 2,000,000 farmers, or about a third of the total, gave household use as the major source of income from their farms in 1939. These farmers had a total value of product of $700,000,000 (14.5 per cent of the total for all farms) of which 60 per cent was consumed in the farm household. Second, 46.5 per cent of all farmers, employing 36 per cent of the farm labor resources and 18 per cent of the value of land and buildings, produced less than 11 per cent of the total value of farm products. Third, of this latter group of 2,700,000 farms, only 1,600,000 list farm household use as the major source of income, leaving about 400,000 farms falling in higher value brackets (over $600) which cannot be considered as fully commercial farms. Fourth, almost a sixth of the farm operators probably received more income from nonagricultural pursuits than from agricultural pursuits, though there is an important overlap with each of the above categories. These data are presented to indicate the large proportion of the agriculturally employed that are relatively detached from relative prices in agriculture.

Sources: Sixteenth Census of the United States, 1940, Analysis of Specified Farm Characteristics for Farms Classified by Total Value of Products, Cooperative Study, United States Department of Commerce and Department of Agriculture, 1943, pp. 3, 30, 58, and 102.

10. The relationship between physical quantity of resource and value of product is used explicitly in place of the usual statement made in terms of the contribution of the amount of a factor which can be purchased for a dollar. If mobility is imperfect, factor prices, particularly for labor, may differ greatly from one area to another.

11. The 1940 agricultural census, including the special sample study made by the census, provides fairly recent data for comparing regional differences in the employment of various factors and for determining a measure of relative productivity. In addition, the recent estimates of net farm income by states made by the Bureau of Agricultural Economics provides a fairly adequate measure of average farm productivity by states. Though the measure of productivity provided by the BAE study is superior to any measure that can be derived from the census, it has been found necessary to rely upon census data for some of the comparisons. In order to maintain comparability, census data have been used throughout much of the analysis. In each case when possible, the comparisons have also been made with the BAE income data, though the results are not always shown. Because the BAE data are preliminary and additional unpublished data were generously supplied by Mr. O. C. Stine, indexes are used rather than actual values.

12. With the exception of the West North Central and Mountain States, there are only minor relative discrepancies between value of products sold and net farm income. The reason for the larger discrepancies in these two groups of states is apparently due to the smaller degree of underestimation of sales in statements to the census enumerators.

13. The allocation of net farm income between capital and labor obviously rests on certain assumptions. Two approaches are possible; both, however, lead to approximately the same result. In one case, estimates of the total value of farm property may be assumed to earn a particular rate of return and the residual attributed to labor and management. In the other, labor is valued at the going rate and capital is assumed to be the residual claimant. Since it is difficult to obtain a wage factor to include the activities of the operator, the first method was used to obtain our estimate. The rate of return for capital was assumed to be 5 percent. On this basis, the residual going to labor was 62 percent for 1910–14, 65 percent for 1929, 66 percent for 1938–40, and 71 percent for 1941–42. Net farm income includes, in addition to the net income of farm operators, wages, mortgage interest, and rent to non-farm landlords. (Source of data: BAE, USDA, Net Farm Income and Income Parity Summary, 1910–41, mimeo., and Norcross, *op. cit*). These estimates are fully consistent with those made by Heady ("Changes in Income Distribution in Agriculture . . ." this journal, August, 1944, p. 440). In our case we have attributed the return to management to labor rather than as a separate category. Non-land capital is about a quarter of the total capital and if his proportion going to land is increased by that amount, the proportion left for labor is about the same as derived here. For 1910–14, for example, both methods leave 62 percent of the total for labor. No other periods are comparable.

14. In 1940, for example, the Bureau of Agricultural Economics estimates that real estate accounted for 80 percent of the total value of farm property. (BAE, USDA, Net Farm Income and Income Parity Summary, 1910–41, Mimeo, July, 1942, p. 28.)

15. For example, using the adjusted data in Table V, the New England farm uses 0.3 man-years more labor and $3,300 less capital than the North Central farm. If interest is charged at 5 percent, the difference in capital costs is $165 per year, or the equivalent of an annual wage of $550. Capital maintenance charges on buildings can be ignored because they are included in the adjusted figures.

16. If the demand for cotton is inelastic, the share of the national income going to Southern agriculture would certainly decline. Furthermore, the absolute and relative shares going to labor would fall, while the relative and perhaps absolute shares going to capital would increase. Even if the elasticity of demand is unity, labor's absolute and relative shares would both decline.

17. If increasing returns to scale did not exist, this would imply that capital use was not subject to diminishing returns throughout the relevant range of the data. This conclusion is untenable since it would mean that a withdrawal of labor from agriculture would increase total output.

18. It has been impossible to deal adequately with this topic in this paper. For good discussions of the second point see Pigou, op. cit., and A. C. Bunce, Economics of Soil Conservation, Ames, 1941.

19. See R. G. D. Allen, Mathematical Analysis for Economists, New York, 1939. Pp. 369–74.

20. The concept of elasticity of substitution may be defined as the proportional change in total physical product divided by the proportional change in the marginal physical productivity of factor A when factor B is varied and the factor A held constant. If A and B must be used in fixed proportions, the elasticity of substitution is zero since for a decrease in B, the marginal value of A is zero. If one unit of A will replace one unit of B regardless of the relative proportions of the factors, the elasticity of substitution is infinite since a reduction in B will reduce the total product without changing the marginal productivity of A. A unit elasticity of substitution occurs if the proportional change in total product is the same as the proportional change in marginal productivity of A as B varies.

21. P is the value added, L is labor in man-years and C is the value of land and buildings. The regions included were Middle Atlantic, East North Central, South Atlantic, and East South Central. This equation given is only one determined in a more extended analysis of the production function in agriculture. Using data based on state averages for the same year, the following equation was found $P = 1.4 \, L^{.88} \, C^{.77}$ which is a very close correspondence despite the entirely different nature of the data. The corrected coefficient of multiple determination for the function given in the text was 0.925 and for the other was 0.983. It is hoped that further results of this analysis will be ready in the near future.

22. This statement assumes no change in agricultural prices following the withdrawal of labor. Since physical output would fall by perhaps 10 percent, depending on the areas from which labor migrated, the marginal value productivity of capital would fall only slightly due to the price rise for agricultural products. The extent of the price rise is unimportant in the argument, however.

23. National Resources Committee, Consumer Incomes in the U.S., pp. 25, 101 and 103.

24. Marshall's treatment of this issue is particularly acute. In discussing investments made in children, Marshall points out that in the lower income ranks, investments are not made because of the "slender means and education of the parents" and that many children of the working class are poorly fed, housed and educated and "they go to the grave carrying with them undeveloped abilities and faculties." He then continued: "But the point on which we have specially to insist now is that the evil is cumulative. The worse fed are the children of our generation, the less

they will earn when they grow up and the less will be their power of providing adequately for the material wants of their children, and so on to following generations. And again the less fully their own faculties are developed, the less will they realize the importance of developing the best faculties of their children, and the less will be their power to do so. And conversely any change that awards the workers of one generation better earnings, together with better opportunities of developing their best qualities, will increase the material and moral advantages which they have the power to offer to their children." (Alfred Marshall, Principles of Economics, 8th Ed., pp. 562–563).

25. The Bureau of Agricultural Economics' estimates of income to farm people from non-farm sources indicate that such income is equal to a third or more of the income from agricultural sources. See *Farm Income Situation*, August, 1944.

26. The impression should not be left, however, that with careful and competent work that some income comparisons could not be made that have economic significance. Further studies should include analysis of income distributions for farm workers and other groups of workers on a regional or state breakdown. Such studies would provide an improved basis for analyzing resource allocation and comparative economic alternatives in agriculture and elsewhere. Such data would certainly be of great value to both economists and policy makers, and would have made our task in this paper much easier. Such studies would probably act to reduce some of the faith now held in the statistical averages on income, which would be a healthy result in itself.

27. It is assumed at present that the type of welfare measures suggested in Part 2a have not been adopted as a permanent program. If a permanent welfare program existed the emphasis on this aspect of the problem would be less.

7

Government and Agriculture:
Is Agriculture a Special Case?

A national news magazine recently headlined a story on the farm problem: "5 billions for Farmers." While the figure substantially exaggerates the net costs of the many farm programs, it does highlight the extent of the involvement of the government in farm affairs. The outflow of federal government money to farmers will probably rise rather than decline over the next few years as the so-called soil bank program, with its authorization of annual payments of $1.2 billions, comes into full scale operation.

While agriculture has received some type of special attention from government since the establishment of the federal government, programs of the present scale date from 1929 with the establishment of the Federal Farm Board and from 1933 with the emergence of a large number of New Deal agencies and programs. The land grant system of agricultural and mechanical colleges (1862), research grants (1887), extension service (1914), federal land banks (1916) and the federal intermediate credit banks (1923), all antedate the Federal Farm Board. Federal costs for these programs and other long time activities of the U.S. Department of Agriculture, such as crop and livestock estimates, inspection, disease and pest control and certain general administrative costs amounted to a little more than 6 per cent of the total costs of agricultural programs in fiscal 1955.[1] The rest of the costs were for programs started since 1929 and were largely the consequence of price support operations and the disposal of agricultural products accumulated through price support programs.

The purpose of this paper is not to appraise the present farm programs or those of the recent past. Instead, I wish to determine whether there are grounds for the special and extensive attention which the federal government is now lavishing upon agriculture. In other words, are there one or more aspects of agriculture that differentiate it from other types of eco-

Reprinted by permission from the *Journal of Law and Economics* 1 (October 1958): 122–37.

nomic activity and thus may warrant governmental intervention? With the present degree of government interference in many phases of economic activity, this is not an easy question to answer. Even if we ignore the tariff, there are many techniques now in use as a means of influencing the development of various sectors of the economy. Some are found in the federal corporate income tax structure—accelerated amortization, depletion allowances of mineral industries. Shipbuilding and ocean shipping are subsidized by various means. A limited amount of housing is subsidized. In recent years the stockpiling of certain minerals and raw materials has acted to maintain market price and encourage production.

The growth of governmental intervention in agriculture is not a unique American phenomenon. Almost all of the major industrial nations have adopted programs, especially since World War I, designed to aid agriculture in one manner or another. Perhaps the extreme case is that of the United Kingdom where in recent years the net governmental expenditures on agricultural and food subsidies have approximately equalled the net income of farm operators. For the U.S. to reach the same relative position would imply an increase in net governmental costs to about four times the level of the 1955 fiscal year.

In this essay I shall restrict my comments to the particular position of the United States and its agriculture. Seven major lines of argument have been used to justify various types of governmental aid to agriculture. The seven will be considered in turn. They are:

1. Depressions are farm-led and farm-fed.
2. Industry and labor are organized; agriculture needs governmental aid to offset their monopoly powers.
3. Almost all farms are small; there are certain activities that these farms cannot carry on efficiently.
4. Farm people are confronted with a great deal of economic instability.
5. Agriculture is and will continue to be a declining industry, as measured by the level of labor employment.
6. Farm incomes are too low.
7. A large and productive agriculture is required for military and strategic reasons.

I. Depressions Are Caused by Agriculture

A former Secretary of Agriculture, Mr. Charles Brannan, declared:

> Most depressions have been farm-led and farm-fed. Farm prices traditionally go down before, faster and farther than other prices. On the down swing of the business cycle, farm people are

the major early victims of a squeeze. As their income and, therefore, purchasing power is cut by low prices or production failure,
industrial producers find a contracting market for their production. This throws workers out of jobs. They in turn spend less for
farm products, which in turn forces down farm prices, and farm
purchasing power is further cut.

I don't mean to say that declines on farm prices are the sole
cause of depressions, but they certainly contribute greatly and
would do so more now than in the past because agriculture has
become a bigger customer of industry.[2]

The view that the rest of the nation cannot be prosperous unless agriculture is prosperous goes back to the beginnings of our nation and has been
and still is widely held. Many presidents have accepted this view, from Jef-
ferson to the present incumbent of that office. Businessmen and financial
advisers have similarly accepted it. Bernard M. Baruch said in 1921:

Agriculture is the greatest and fundamentally the most important of our American industries. The cities are but the
branches of the tree of national life, the roots of which go deeply
into the land. We all flourish or decline with the farmer.[3]

The most extreme form of the argument that national prosperity depends upon agriculture is perhaps the following:

Unless Congress recognizes the simple fact that each $1 of
gross farm income generates $7 of national income, theory and
legislation resulting from theories can easily legislate the United
States into bankruptcy and chaos. On the other hand, if Congress
will use it as a yardstick, there is no reason why the United States
should ever have a depression.[4]

The latter statement rests upon the fact that for a fairly long period of
time (from about 1929 to 1950) national income was approximately seven
times as much as gross farm income, with only moderate departures from
this ratio from year to year. Thus a statistical relationship was transformed
into a cause and effect relationship, with the cause being farm income and
the effect being national income. The exact converse could be deduced,
namely that for each $7 change in national income, gross farm income
changed by $1. It could also be pointed out that certain other sectors of
our economy have exhibited a similar close correspondence between the
income of that sector and national incomes. One such sector happens to
be the fishing industry. If the same line of reasoning could be applied it
obviously would be much easier to maintain national prosperity by subsi-

dizing the fishing industry since its gross income constitutes less than one per cent of the national income.

One could present many statistics that would, at least for some, throw doubt on the view that changes in agricultural prices and incomes are responsible for changes in national income. For example, there has been a fairly steady decline in the percentage of gross farm income of national income from about 23 per cent in 1910 to almost half that today, yet national income, in real terms, has trebled. But the argument cannot be proved or disproved by simple statistical comparisons. This is true because we live in an economy with strong interrelationships among the various parts of that economy. Thus a decline in national income will almost certainly be accompanied by a decline in farm income and in the income from the production of steel, automobiles, movies, and confetti. In some cases, the relative declines will be more than in the national income; in others, less, and in some, almost the same. But, depending upon the particular characteristics of the commodity, there will be a fairly consistent pattern of changes relative to national income. By the criterion used above in the $1 to $7 analogy, many sectors of the economy could be found to be strategic factors in causing depression or prosperity. This is why simple comparisons of statistical series will not provide us with much insight concerning the role of agriculture in national economic stability.

Let us try to trace through the effects of important changes in agricultural incomes upon the rest of the economy. These changes can occur in a relatively short period of two or three years either because of changes in production or changes in demand. Compared to most other parts of the economy, farm production in the aggregate is remarkably stable. In the last 45 years, the largest change in the volume of farm marketings and home consumption from one year to the next was 12.5 per cent; only two other changes exceeded 10 per cent. Over 90 per cent of the year-to-year changes have been less than 5 per cent, and these are small year-to-year changes compared with steel, automobiles or other durable goods. In the United States changes in farm production are too small to have any significant effect on general business activity.

Farm prices, responding mainly to changes in demand, do change a great deal from year to year. Most of the changes in gross farm income are due to changes in prices since the quantity sold changes little from year to year. It is at this point that a crucial difference between agriculture and many other parts of the economy becomes apparent in terms of effects of declines in income upon the level of incomes in the rest of the economy. When farm prices decline, as they did by about 55 per cent between 1929 and 1932, output changes little if at all. Thus the economy as a whole has as large a volume of farm products available for use after prices have dropped as before. The major change is that consumers pay less for the same amount of food and farmers receive less. The loss to farmers is approximately equal to the gain to consumers. Thus it is correct to say that no income is lost in the process. Farmers have less to spend; the rest of the population has more to spend after purchasing the same amount of food.

When the demand for housing (or any other investment good) declines, the effects are very different. Housing construction is not paid for out of current income; it is paid for out of accumulated assets and expected future savings. When housing demand falls, the income from housing construction and in industries supplying building materials falls. Contrary to the case of farm products, there is no significant offsetting gain in consumer expenditures for other products. This is because the housing is not being paid for out of current income. Thus a decline in the demand for housing can start a cumulative process which can lead to falling national income and rising unemployment.

The above argument with respect to housing should not lead to the conclusion that every decline in the amount of housing construction will lead to unemployment and declining national income. Housing construction declines may be offset by increases in construction elsewhere in the economy, as was the case in 1956 compared to 1955. The conclusion is valid only if the decline in demand for new housing is not offset by any increase in demand for any other type of investment.

It should also be noted that a decline in farm income will have disadvantageous effects upon those producers in the economy who receive much of their income from selling to farmers, such as farm machinery manufacturers and machinery dealers. To a large extent, however, these losses are offset by gains of firms that produce for or sell to the non-farm population which has more income to spend upon non-farm products.

There is little or no basis for government aids to agriculture on the ground that depressions originate in agriculture. The reasoning underlying such a position is surely faulty. However, this conclusion does not imply that there is no basis for some special measures for agriculture as a part of a general program of anti-depression measures. While major reliance should be placed upon general monetary and fiscal measures to prevent depressions and inflation, if a depression should occur and governmental expenditures are used as a means of increasing the overall level of demand, there is as much basis for channeling some of the expenditures directly to farm people as there is for paying unemployment compensation or engaging upon a public works program. But nothing said here should imply that there is any basis for a program designed to prevent any and every decline in farm income, and especially to prevent those declines that represent adjustments to changes in demand or in relative costs of producing agricultural products during periods of full employment.

II. Industry and Labor Are Organized; Agriculture Needs Government Aid To Offset Their Monopoly Powers[5]

This is a persuasive argument, at least to many people. The farmer is pictured as selling in a market dominated by the large meat packing firms, canneries, flour mills, and other firms presumed to possess monopoly powers, while buying in a market dominated by large farm machinery firms, fertilizer companies, and General Motors. The farmer has no control over the prices at which he sells, while elsewhere in the economy, producers set their prices at the level they desire. Many farmers are also convinced that labor legislation has allowed organized labor to obtain more or less any wage desired, resulting in higher costs of processing agricultural products and thus lower prices for farm products and in higher costs and prices for the products bought by farm people. The answer seems to be that since farmers cannot organize themselves to obtain similar monopoly powers, the government should provide the necessary assistance to permit the farmers to create that power. Price supports, acreage limitations and marketing quotas may be viewed as devices to give farmers a degree of monopoly power; so can marketing agreements in many fluid milk markets which are supervised and enforced by the federal government.

One counter argument to this view is that the appropriate way to fight monopoly in one part of the economy is not to create another monopoly, but to destroy the monopoly which exists. To use an analogy, nothing is gained by creating an additional monopoly since the farmers may find that, though they may be getting a larger piece of the total national income pie, the pie has so shrunk in size that the actual piece of pie is no larger than before. This analogy, like most analogies, has its defects. The larger share of the smaller pie may, in fact, be greater than the smaller piece of the larger pie. The analogy also implies that it is possible to break up monopolies elsewhere in the economy; farmers may well remain skeptics on this point.

There are two other lines of argument that are more appropriate, though I am not sure that they are any more convincing. The first is that the degree of monopoly which exists in the rest of the economy is greatly exaggerated. The second is that there is no evidence that farmers can gain economic benefits over a period of time from the measures adopted to offset the presumed monopoly powers elsewhere in the economy.

1. First, the power of labor unions to raise wages cannot be demonstrated by a comparison of wages in the unionized and the non-union sectors, except perhaps for unions that involve no more than 4 million workers. Second, since the relative wages of union workers have not increased relative to non-union, the second major goal of unions—the maintenance

of full employment—is clearly inconsistent with labor receiving a larger share of the total national income than would be true in the absence of the unions. Third, most of the markets in which farmers sell their products, with the exception of fluid milk markets and perhaps tobacco, are quite competitive with a large number of buyers, including buyers representing markets from all over the world. The agricultural processing industries have not been singularly profitable in the past decade. Over the past several decades the so-called Big Four of the meat packers have suffered a substantial reduction in their share of total livestock slaughter and two of these firms have paid no dividends on their common stock for several years.

2. The second point, namely that farmers have not been able to gain from efforts to give them monopoly power, is the most weighty one. The nature of agricultural production is such that efforts to create a cartel under guidance and subsidy from Washington will almost certainly lead to disappointing results. There is no way to restrict entry into agriculture. Thus if a program were to result in higher returns for agriculture through price supports and acreage restrictions, potential producers will attempt to enter the field. One way that this can be done is to buy land which has attached to it the right to produce the particular commodity. After a fairly short time land prices will be bid up and new producers will find that it is no more profitable to produce this particular crop than a number of others. This is not a hypothetical case, but is essentially what has happened in tobacco, where acreage controls, marketing quotas, and price supports have been maintained for two decades.

It should be remembered by those who want greater emphasis upon price supports and production limitations that these measures were used during the recent years of declining farm income. While these measures may have stemmed the fall of farm incomes somewhat, they certainly have not provided the panacea that was hoped for.

III. Farms Are Small; Government Must Undertake Activities That Small Firms Cannot Maintain

American farms are small, have been small in the past, and if present trends continue will remain small in the future. Small is, of course, a relative term. Large scale farms have been defined as farms with over $25,000 in sales and in 1950 there were 103,000 such farms; these constituted less than 3 per cent of all commercial farms.[6] The average total payment for hired labor for the year 1949 was about $9,000 for these 103,000 farms.[7] In April of 1950 less than 16,000 farms had 10 or more hired workers; the majority of the commercial farms had no hired workers.[8]

The small size of farms justifies governmental support of research on farm production techniques, information activities about marketing conditions, and the maintenance of an extension service to bring available knowledge to the attention of farmers and help them to utilize that knowledge effectively. Measured by willingness to adopt new methods of production, American agriculture has been extremely progressive. Its record compares favorably with that of manufacturing. Much of the credit for the rapid advances achieved must be given to the experiment stations and agricultural colleges supported by the state and federal governments to undertake research and train individuals in the sciences especially important to agriculture. The extension service has also played a significant role in making the information available to farmers. The individual farm is too small to organize research on most problems that are important to it. The research has to be done by an agency larger than the farm, and the particular innovations adopted in the United States have paid very large returns for the relatively small expenditures involved. These expenditures have meant more and cheaper farm products; as consumers, we have all gained.

IV. Farm People Are Confronted with Much Economic Instability

The income of an individual farm family varies a great deal from year to year. This is due to two things. First, farm prices fluctuate over a much wider range than do wage rates or the prices of most other products, except some mineral raw materials like copper. Second, the output of a given year is subject to many influences beyond the control of the farmer—the weather, diseases, insects and pests. Sometimes the gods of chance smile on him and a large production coincides with favorable prices; other times the opposite occurs. The higher degree of instability makes planning of farm operations and family consumption difficult.

I am not certain how much farm people dislike the risk which confronts them. We do know that federal crop insurance, which is designed to remove part of the risks from natural hazards affecting yields, has not been widely accepted by farmers even though the premiums have never covered the actuarial risks involved. On the other hand, many farmers do insure against certain specific risks, such as hail damage.

Available evidence supports the view that when the price risks are reduced by a government price support program farmers do produce more than they would for the same average price in a free market. This stems from the fact that with the reduction of price risks greater quantities of some production items, such as fertilizer, are used and investments are made in machinery that would not be made if the chances of loss were greater. Several years ago I outlined a method of reducing and transferring the price risks of farming, calling the method forward prices.[9] The basic idea was that the government would estimate the price of product prior to the time the most important production decisions were made and then to guarantee to the farmers a large fraction of that estimated price when the product was actually sold several months hence. There are economic difficulties with this scheme, especially arising out of errors in the estimates of future prices. Nonetheless, I felt then and I feel now that such a scheme could result in more efficient agricultural production.

One basic objection to the proposal is political; the pressure to estimate prices at levels that are too high would be ever present. Whether any methods can be devised for protecting the administrators of such a program from these pressures is uncertain. If prices are set too high, the gains from reducing the price risks would soon be offset by inducing too much production of agricultural products, which is the situation to which price supports have now brought us.

The instability of farm incomes provides grounds for some kind of governmental intervention in agricultural markets. But the gains to the economy are not so large that they cannot soon be dissipated by misguided efforts to "try to help the farmer" through overdoing a good thing and setting prices at levels above those that would equate supply and demand. I should point out that I do not see any substantial political support for the views expressed in the previous paragraphs.

V. Agriculture Is a Declining Industry

Since about 1915 farm employment has been declining more or less continuously, and while employment has declined by more than 40 per cent, output has increased by about 80 per cent. Thus decline in employment has not meant stagnation, but it has meant that the migration from farms to cities has had to continue at a rapid pace throughout the past 40 years. Since 1910 the net migration from farm to non-farm areas has been about double the present farm population of 20,000,000. The high rate of migration has been required more by the relatively high fertility of the farm population than by the decline in the number of profitable job opportunities in agriculture. But the two factors taken together have meant, and will continue to mean, that if farm earnings are to come into balance with non-farm earnings for people of comparable skills and capacities net migration rates of perhaps 5 per cent per annum will be required for some time to come.

An industry in which employment is continuously declining is always confronted with difficult adjustments. Many people must change to very different types of work. If a declining industry is located in a diversified urban area the adjustment problem is simplified by not hiring new workers and thus employment decreases with retirements and quits. But agriculture, organized as it is around family farms, has a more or less built-in training system that trains almost every male youth in the farm families as a farmer. Thus the rather impersonal process of not hiring new workers does not function, and farm youth must make a positive decision not to be a farmer. Furthermore, when a farm person decides to change occupations, he usually must change location and move away from familiar surroundings which may increase his reluctance to move to more remunerative employment.

The necessity for a continual movement of workers trained for farming to other occupations means that the returns to workers in agriculture must remain somewhat below the level in occupations requiring comparable workers. Because of the inherent difficulties of measuring and comparing farm incomes with non-farm incomes, we do not know how large a differential is required to induce a million people a year to move away from farms.

I would argue that the government might well play a valid role in minimizing the difference in earnings required to achieve a given rate of migration. Perhaps the most appropriate role that the government could perform would be that of supplying adequate and reliable information. Under present circumstances, information about job opportunities in the non-farm economy is extremely difficult for a farm youth to obtain; the most common source is from relatives or friends who have moved to a certain area. The state employment services, as now organized, are essentially designed to meet the needs of urban people. Undoubtedly, their efforts could be expanded to meet more fully the needs of farm people. However, this would require close liaison between services of several states since much interstate movement is involved. Whether additional information alone would be adequate to reduce the difference between farm and non-farm earnings to an acceptable level, is uncertain. Finding jobs before a person leaves the home community; loans or grants to finance the movement for families might also have merit.

My colleague, Professor T. W. Schultz, has dramatized the need for some measures to speed up the flow out of farming by suggesting that the government establish homesteads in reverse by making a substantial payment to a farm family now actively farming which moves to a non-farm area and takes a non-farm job. The homestead idea was used to settle a large part of the United States. Perhaps it might also be used to "unsettle" agriculture. In any case, it seems clear that it would be much more economical to approach the farm problem by this route than by the present expensive, but almost wholly ineffectual, methods.

VI. Farm Incomes Are Too Low

Probably the most frequent reason given for special governmental programs for agriculture, especially the price support and allied measures, is the low income level of farm population compared to non-farm population. Since World War II, farm per capita income has ranged from 45 to 63 per cent of nonfarm per capita income. The lower figure was for 1956 and the higher for 1948. In 1957 the corresponding figure was 49 per cent. What does such a comparison mean? It means very little as an indication of whether farm incomes are high or low. An obvious, but incorrect, implication of such a comparison is that the per capita incomes should be the same if farm people were to be as well off as nonfarm people. Such a statement has about as much meaning as the statement that families with a head aged 65 or more ought to have the same income as families with a head aged 40 to 50 years, or that manual laborers in manufacturing should have the same incomes as craftsmen and foremen in manufacturing.

Any comparisons of the incomes of two or more groups always implies certain assumptions about the degree of comparability of the groups and the income that they receive. Certain questions must be answered before any legitimate comparisons can be made.

1. Will each of the dollars of income that are being compared actually purchase the same amount of goods and services? In other words, does the income which is earned by the farm population have more purchasing power per dollar than does a dollar earned in a city? Farm income, as presently calculated, includes an estimate of the value of home-produced and consumed food. This food is priced at the price which the farmer would receive if he sold it. At the present time, the farmer receives less than 40 per cent as much for his products as the city consumer pays at retail. Thus if the farmer purchased these same foods at retail, he would have to pay approximately two and one half times as much for them as the value placed on them in estimating his income. A careful study made by Mr. Nathan Koffsky, based on data collected in 1941, indicated that a dollar of income of the farm population had a purchasing power of about 25 per cent more than a dollar earned in an urban setting.[10] While this result is probably not exactly applicable to 1957, it would seem to be a reasonable approximation.

2. Are the groups comparable with respect to age and sex and composition, the percentage of the population actually working, and education? We know that persons in the age group of 35 to 55 earn more than do persons older or younger. Likewise, on the average, men earn substantially more than women, and earnings increase with the level of education. Thus, if the two groups being compared are substantially different in these respects, the differences in incomes may be due to the differences in these and related characteristics. The farm population has a larger fraction of its working population in the age groups under 20 and over 65 than does the non-farm population. It also has a substantially smaller percentage of its total population in the labor force, 33.5 per cent compared to 41 per cent. The farm group has, for each age group, about two or three years less

schooling. Considering only persons in the labor force, differences in age, sex, and education would indicate that the average member of the non-farm labor force would earn about 15 per cent more than the average member of the farm labor force.

3. Are the payments of direct taxes, principally income taxes, approximately equal for individuals of the same income level in the two groups? The federal income tax, which is the most important direct tax paid, taxes only money income. Thus the nonmoney income of the farm population in 1955 equalled 18 per cent of the total net income of the farm population and 27 per cent of the net income of the farm population received from agriculture. Other groups in the population, principally home owners, also receive nonmoney income that is not subject to tax, but such income is much more important to the farm population. For 1953 personal income taxes were 11.3 per cent of the non-farm personal income and 6.9 per cent of the net income of farm families.[11]

4. Is the relative importance of labor income in total income approximately the same? Contrary to general opinion, more capital is used per worker in agriculture than in the rest of the economy. In agriculture about 40 per cent of total income is return on capital, while in the economy as a whole, after payment of corporate income taxes about 20 per cent of total income is return on capital. Thus a larger share of the income of the farm population is required as compensation for capital used than is true of the rest of the economy.

Taking all of these factors into account, if the per capita income of the farm population was roughly 65 per cent of that of the non-farm population, the returns to workers of similar age, education, and sex would be approximately comparable. While 65 is substantially above the 48 per cent that prevailed in 1957, the discrepancy is substantially less than when one naively assumes that the two incomes must be equal.

Even if one believes that average returns to farm people are now lower than for comparable resources in the rest of the economy, as I do, this belief cannot be used to support just any measure designed to increase farm income. There is a great deal of diversity in farm income from area to area in the United States, a much greater degree of diversity than exists in urban incomes. In 1949 there were perhaps a million farm families with able bodied male heads with net money incomes of less than $1,500. Most of these families are concentrated in the South and live on small farm units, deriving much of their income from cotton, tobacco or subsistence farming. Their market sales are small and changes in the level of prices have rather small absolute effects upon their incomes. There is no question that the low incomes of these million families represents an important economic and social problem. But there is also no question that their low incomes should not be included in an average which is then used as evidence for the need of a program whose benefits go primarily to farmers who have substantially higher incomes.

In 1949 the 40 per cent of the farms making 88 per cent of the total farm sales, and which would receive at least that large a share of the net gains from a price support program, had average incomes somewhat higher than the average incomes of non-farm families. Since that time the average income of these families has increased by about 10 per cent, while the incomes of non-farm families have risen by 30 per cent.[12] But if one takes into consideration the differences in the purchasing power of income, the income spread is now not very large.

While the concentration of low incomes in certain rural areas may well justify governmental programs to meet the problems of the farm people in those areas, the existence of these low incomes is not an adequate basis for the present expensive farm programs which are not designed to solve their problems. The incomes of the farm families that receive most of the income gains from price supports and the soil bank are not so low that they can be said to constitute an important social or economic problem.

VII. A Large and Productive Agriculture Is Required for Military and Strategic Reasons

There can be little question that many nations in the world have subsidized their agriculture as a means of insuring a supply of food in case war disrupts the normal channels of trade. The very solicitous treatment that the English farmer has received from his government since World War II is probably in large part based on the precarious position in which Britain might find itself in another war with respect to its food supply if it still relied as heavily upon food imports as during the first third of this century. Almost all of Europe, including the Soviet Union, has related its actions in agriculture to its actual or presumed military needs.

The position of the United States is such that there seems little or no grounds for special agricultural measures to increase total agricultural production for military or strategic reasons. With or without a system of price supports and related aids to agriculture, our farm output is going to continue to expand. The use of more fertilizer, advances in animal nutrition, supplemental irrigation in the relatively non-humid areas, improvements in plant varieties, more effective methods of controlling plant and animal diseases and other changes not now known will contribute to an expanding output. As we have discovered in the past, a decline in the farm labor force is consistent with increasing output as farmers substitute power and machinery for labor.

The United States is a major exporter of a number of food and fiber products—wheat, rice, cotton, tobacco, and a wide variety of other products. We are a major importer of sugar and a considerable number of subtropical and tropical products such as coffee, bananas, and spices. We are also a large importer of wool and jute. The variety and overall excellence of our diet, especially the heavy emphasis upon livestock products, means that the population could be relatively well fed even if aggregate agricultural output fell sharply as a result of war. Synthetic fibers provide reasonably satisfactory substitutes for almost any natural fiber now in use.

In adopting measures designed to increase agricultural output for military or strategic reasons, we must always recognize production somewhere else is reduced. The farm worker who is induced to continue producing cotton could and probably otherwise would contribute to the production of airplanes, guided missiles, or related items. Thus before concluding that more farm output is a good thing, we must consider the cost in terms of the alternative products that are lost in order to get more farm products. When this comparison is made, my conclusion is that there is no justification for producing more agricultural products than would be provided in the ordinary course of events through equating the demand of consumers and the amount farmers would be willing to supply.

There may be some basis for maintaining stockpiles of certain agricultural products as a safety measure. But it seems unlikely that the size of these stockpiles would be as large as the quantities of certain farm products now held by the government. Government owned or controlled stocks equal one and one-half years of domestic food and seed use of wheat, a third of a year's corn production, and about 7 months domestic mill use of cotton. And it should also be remembered that when we add to our stockpiles of these products, we are sacrificing the output of other products that may have equal or greater value for military and strategic reasons.

It is worth noting that our present and past policies which have led to the accumulation of large stocks of many agricultural products and a continuing excess of production relative to market demand at the supported prices is a negative strategic factor at the present time and is likely to continue to be so. These programs prevent normal economic and trade relations with many nations, particularly the underdeveloped nations that we would like to attract to our political position in the present contest between communism and the free world. This is true for two reasons. First, our price supports and accumulated stocks have led us to engage in large scale export dumping of wheat and cotton and the use of foreign aid funds to dispose of significant quantities of these and other agricultural products. These actions have limited the access of many nations to markets that might otherwise have been theirs. Examples are Canada for wheat, Egypt, Mexico and Brazil for cotton, Burma and Thailand for rice, and Greece and Italy for dried fruits.

Second, these policies have meant that we cannot, for economic or political reasons, provide a market for the agricultural products of many underdeveloped nations. It is quite possible that even if there were no price supports on rice, for example, the U.S. would still be a net exporter of rice. But there is no question that without the price support our production of rice would be less and the world market price somewhat higher.

The Soviet Union's economic and political position has been strengthened as a consequence of our domestic agricultural policies. This is probably a higher price than we want to pay for the continuance of these programs.

Notes

1. F. D. Stocker, Governmental Costs in Agriculture, The Concept and Its Measurement, U.S. Dept. of Agric., A.R.S. 43–28, p. 25 (May, 1956). The total cost for all agricultural programs was about $2.5 billions.

2. See Agricultural Adjustment Act of 1949, Hearings, Subcommittee of the Senate Committee on Agriculture and Forestry on Sen. 1882 and Sen. 1971, 81st Cong. 1st Sess. 31–2 (1949).

3. Some Aspects of the Farmers' Problems, 138 Atlantic Monthly, 112, (July, 1921).

4. Testimony by Carl H. Wilken, 1949 Extension of the Reciprocal Trade Agreements Act, Hearings, House Committee on Ways and Means on H. R. 1211, 81st Cong. 1st Sess. 377 (1949).

5. Since this section was written, Prof. Kenneth Boulding has prepared an illuminating article that arrives at essentially the same conclusions as expressed here. See Does Absence of Monopoly Power in Agriculture Influence the Stability and Level of Farm Income, in Joint Economic Committee, U.S. Cong., Policy for Commercial Agriculture, pp. 42–50 (Nov. 22, 1957).

6. Table 19, U.S. Census of Agriculture, 2 General Report 116 (1950).

7. Table 1, ibid., at 1124.

8. Ibid.

9. D. Gale Johnson, Forward Prices for Agriculture (1947).

10. Nathan Koffsky, Farm and Urban Purchasing Power, in Studies in Income and Wealth (vol. 11) National Bureau of Economic Research 151, 156–72 (1949).

11. Non-farm estimated by author; farm from F. D. Stocker, The Impact of Federal Income Taxes on Farm People, U.S. Dept. of Agric., A.R.S. 43–11, p. 13 (July, 1955).

12. See Johnson, Comparisons of Farm and Non-farm Incomes, Farm Policy Forum, pp. 2–7 (Spring, 1956); Koffsky and Graue, The Current Income Position of Commercial Farms, in Joint Economic Committee, U.S. Cong., Policy for Commercial Agriculture, p. 87 (Nov. 22, 1957).

8

Government and Agricultural Adjustment

When the new round of GATT negotiations start next month, negotiations on tariff and nontariff barriers on farm products will have a central and critical role. That role may be so critical that if little or no progress is made toward liberalizing trade in farm products, the negotiations will fail. Indeed the tide of isolationism and protectionism may well become so strong that recent trends toward increasing interferences with trade and other international economic relations, evident in the United States for the past five years, will win the day. It cannot be stressed too much that the strongest political support for trade liberalization in the United States comes from certain farm commodity interests. If these interests are thwarted once again, as they feel they were in the two previous GATT negotiations, they also are likely to turn inward.

On many of these issues Canada and the United States find themselves in a similar situation. Both are industrial nations with less than 5 percent of their gross domestic product originating on farms. But because of the organization of their agricultures, based on highly effective divisions of labor and functions, served by efficient input industries and effective marketing services, the percentage of gross domestic product produced on farms greatly underestimates the trade significance of the production of farm commodities in each of the economies. Further, because of enormous amounts of land available, at least on a per capita basis, and the efficiency of their farms and the industries that are associated with farm commodities, in each of the two high income industrial countries farm products have a distinct comparative advantage. Thus both countries stand to lose if there is trade liberalization for nonfarm products and services and none

Reprinted by permission from the *American Journal of Agricultural Economics* 55, no. 5 (December 1973): 860–67.

for farm products since they would not have the opportunity to exploit the major comparative advantage that they possess in agriculture.

So far as the GATT negotiations are concerned, the major international markets for the farm products of North America are in Western Europe and Japan. Unless Western Europe and Japan significantly modify their tariff and non-tariff barriers affecting farm products, many of the possible gains from agriculture's comparative advantage in North America will not be realized.

It is not my purpose to repeat the numerous objections that North Americans have to the trade and farm policies of the Common Market and Japan. Nor do I intend to emphasize the sophistry that North Americans have used in previous trade negotiations, when we have asked for reductions in trade barriers affecting our export products but have not offered significant modifications in our own protectionist measures such as high dairy supports, Section 22, Crowsnest freight rates, high sugar prices, subsidized grain storage, or export subsidies.

My purpose is to examine the fundamental issue that makes it difficult for governments and those organizations that represent farm interests to confront rationally the implications of free or nearly free trade in farm products. I believe that resistance to free trade in farm products is based primarily upon fear—an undocumented fear that free trade in agricultural products would require agricultural adjustment of enormous magnitude and severity.

This fear grows out of two considerations or, to put it differently, there are two ways of interpreting developments in agriculture since the end of World War II. First, in an industrial economy with increasing real per capita incomes agriculture has to (and does) adjust to changing circumstances at what appears to be a frenzied pace, as evidenced by the large annual decline in the farm labor force and in the number of farms. It is argued that if the adjustments required by free trade were added to the already existing requirements for change, they would impose burdens greater than the social and political fabric could withstand.

Second, and quite ironically, the presumed failure of existing farm programs to reduce significantly the income differentials between farm and nonfarm families leads many policy makers in Japan and Western Europe to the conclusion that the lower farm prices that would result from free trade would create great economic distress among farmers.[1] There appears to be an acceptance of the conclusion that farmers in Western Europe and Japan are so inefficient that free trade would result in the virtual disappearance of agriculture and the forcing of millions of farm people off the land and into cities that are already severely overcrowded. Thus, in effect, the failures of current programs and policies are presumed to provide evidence that a very different approach should not be tried, or for that matter, even seriously studied.

As I shall develop more fully later, the agricultural policies of the industrial nations—with their emphasis upon price supports, trade barriers, subsidies, and supply management—have not attacked the problems that are important to the economic welfare of farm people. The primary determinants of the welfare of farm people are the level of national income per capita, the rate of economic growth, the way that growth is managed, and the extent to which farm people are integrated into the economy and society.

Governments have done little to assist farm people in adjusting to changing conditions. Almost all that governments in the industrial countries have done to improve the circumstances of farm people has been outside the framework of national agricultural policy. Thus it is not surprising that national agricultural policies appear to have failed to achieve their objectives. In general, the means used have not been appropriate to the objective, such as increasing the relative level of farm to nonfarm family incomes or reducing income inequality in agriculture.

Ignorance and Agricultural Policy

How real is the basis for the fear that adjustment problems resulting from free trade would impose unacceptably high costs upon farm people? Unfortunately, we have to say that our evidence on this point is rather limited. This limitation is perhaps not too surprising since governments have made almost no effort to determine what effects the existing farm programs have had upon such important variables as net farm income, the return to farm labor, the return to and value of land, the level of farm employment, and the output of farm products, separately and in total. It hardly seems too much to ask governments that spend billions of dollars of either taxpayer or consumer funds to support farm incomes to spend a few millions to determine these effects. Not only have individual governments failed to undertake such studies, but repeated efforts to induce international organizations such as OECD or FAO to undertake them have failed.

What is perhaps even more surprising is that rather little is known about the level and distribution of income among farm families in several major industrial nations. The European Community has recently estimated that about a fourth of the labor force living on farms has nonfarm employment. It is, however, a very rough estimate since for the two countries with approximately 70 percent of the total farm employment in the Community of Six—France and Italy—virtually no data are available. The European Community has made no real effort to determine the economic position of farm families. The reason, so I was told by an important Community official, is that the topic is politically sensitive and for the Community to study independently the incomes of farm people would be considered interference in the affairs of the national governments. Yet the Community proclaims that it must continue the Common Agricultural Policy in order to have satisfactory farm incomes while admitting that it does not know the average level or the distribution of farm family incomes.

In all frankness we must also say that while we have quite good data on the income of farm families, from both farm and nonfarm sources, for Canada and the United States, we are not able to indicate with acceptable accuracy the major income effects of the farm programs of the two countries. While I believe that most agricultural economists now agree that the major income benefits of current farm programs go to the owners of farm land, we have not yet convinced policy makers of this nor have we been successful in showing how small a fraction of the benefits, if any, go to human resources.

The Elements of Agricultural Adjustment

The conclusion that agriculture, under the best of circumstances, must undergo rapid and continuous adjustment in a growing economy is well established. Why this is true is well known: an income elasticity of demand for output that is less than unity and declining as real per capita incomes increase, the quite direct applicability of knowledge to the farm production process, and the substitutability of inputs, including purchased inputs, for land. These relationships, when taken together, result in a potential growth of supply of farm output that is greater than the potential growth in demand for farm output.

A consequence of the potentially more rapid growth of supply than of demand is that if returns to resources are to remain unchanged, certain resources must be withdrawn from agriculture. Since only a small part of the land is likely to be withdrawn for nonfarm uses, most of the resource withdrawal must be labor. It is conceivable, of course, that part of the reduction in resources could occur through reduced employment of purchased inputs, but it appears that this has not been the case in any industrial country for which data are available. In fact, the opposite has occurred and purchased inputs have increased in importance, both in absolute amounts and as a percentage of total farm inputs.

This simplified exposition of the causes or sources of agricultural adjustment ignores a major determinant of the rate of adjustment required and, I believe, generally achieved. The excluded element was hidden in the phrase, "to keep the returns to resources unchanged."

What this phrase excludes is what my colleague, T. W. Schultz, discussed a year ago at these meetings—the increasing economic value of human time [12]. If the value of human time did not increase, there would be little or no need for agricultural adjustment since the growth in real per capita income and the increasing value of human time are intimately and causally related.

Agricultural adjustment is not a response to a constant real return to labor, but rather is a response to a continuously moving target—the rising value of human time. If the value of human time did not increase in agriculture but did elsewhere, two results would be evident: (1) the annual rate of reduction of the farm labor force in industrial economies would be at about half of its recent rate of 4 to 6 percent, and (2) the disparity in returns to farm and nonfarm labor would grow by 2 to 3 percent annually. Much of the cost of agricultural adjustment is the price we pay for keeping returns to comparable farm and nonfarm labor in some rough balance, after account is taken of the pecuniary and nonpecuniary costs of job mobility and/or migration.

The increase in the absolute and relative importance of purchased inputs referred to earlier is a response to the rising real wages (value) of farm labor as well as the rising real marginal product (rent) of land. Thus imbedded in the adjustment process are adjustments that serve to reduce the equilibrium employment level of farm labor since the purchased inputs are highly substitutable for labor.

The fact that agriculture adjusts to the increasing value of human time means that the conventional methods of farm programs—higher support prices, output restraints, acreage limitations, or input subsidies—will have little or no long run effect upon the returns to labor engaged in agriculture. A product price increase or a given reduction in cropland has a once and for all effect upon the demand for labor and thus upon the return for labor. Even if the elasticity of demand for farm output is infinite, as it is made so by an effective price support that is maintained at a constant level, the shift in demand for labor is a once and for all change. At some time, in fact very soon, it will be necessary for labor to transfer out of agriculture, and the effect of the increased output price (or reduction in land input) on the return to farm labor will have been reduced to zero, though the cost of the price support or land limitation will continue indefinitely into the future.

I have developed this point more fully elsewhere [6, Ch. 9], but at this time it may be appropriate to refer to three studies that are directly relevant. One is by Micha Gisser [2] in which he shows that annual rightward shifts in the demand for farm output of 1.8 to 2.2 percent, with a shift in the production function of 1 percent and a capital formation rate of 2 to 3 percent, would fall substantially short of providing for a 1.6 percent annual increase in the return to farm labor. The needed leftward shift in the supply of farm labor ranged from 0.7 to 5.4 percent annually and this is on the assumption that there was no natural growth in the farm labor force. If the long run elasticity of supply of labor to agriculture is approximately 3 as estimated by Gisser, there is no conceivable shift in the demand for farm output that would eliminate the necessity for some outmigration if the returns to farm labor are to increase at a rate comparable to the increase for nonfarm labor.

An entirely different model developed by Lee Martin [7], also for the United States, can be used to estimate the effects of price changes and reductions in cultivated area on the required reduction in the farm labor force to achieve an annual increase in the gross product per farm worker of 4.4 percent in actual prices. The latter percentage growth rate was for undeflated per capita disposable income between 1946–50 to 1966–70. If farm output prices increased at 2 percent per annum, the required reduction in the farm labor force would be 0.6 percent per annum compared to a reduction of 2.1 percent if product prices were unchanged and 3.7 percent if output prices declined 2 percent per annum.

Martin summarizes the effects of continuous annual reductions in cultivated cropland—up to 4 percent annually—on the return to farm labor and the need for reducing the farm labor force as follows: "Worthy of note in this simulation is that marginal product of labor can grow at the desired rate (0.044) only if the supply adjustment is made through the labor market. If acreage reductions are relied upon to keep supply in balance with demand, the annual increase in the marginal product of labor will not be large enough to keep labor earnings in agriculture from falling farther behind labor earnings in the nonfarm economy."

Wilhelm Henrichsmeyer has developed a model for Germany for 1950 to 1965 with projections to 1980 that shows the relationships between farm labor migration, farm prices, technical change, and farm wage rates [3]. His model indicates that for 1965–80 the farm labor force must decline by 4.5 percent annually to achieve the same rate of increase of wages in agriculture as in the rest of the economy, namely 4.9 percent annually.

One can use his model to estimate the difficulties that would be involved if migration were zero and farm wage rates were to increase at 4.9 percent annually. Assuming no control on output, real farm prices would have to increase by about 2 percent annually. But farm output would be growing by approximately 2 percent more than the growth in demand. If the market for farm output were to clear, farm output growth would have to be held to something less than 1 percent rather than more than 3 percent and real farm product prices increased by about 4 percent annually. To hold farm output growth to less than 1 percent annually would require continuous annual reductions in the amounts of land *and* capital employed in agriculture of about 5 percent.

It is sometimes argued that if the increase in factor productivity were somehow kept small, it would be unnecessary to have such a large outflow of labor from agriculture.[2] Under some circumstances there may be some validity to this view. But the validity hardly extends to the conditions that prevail in the United States and Canada because of the dependence of our two countries upon export markets for such a substantial fraction of total demand for key farm commodities. The price of slowing down productivity growth would be rising costs of production of export commodities and a gradual loss of export markets. Thus agriculture in our countries is faced with declining employment whichever way the fates turn. For some other countries that can insulate their agricultures from world markets, it might be possible for a time to slow down growth in productivity and reduce the required outflow of labor. But here the costs would be high as well. One of the major sources of productivity growth as conventionally measured is the increase in human capital per worker. Do nations wish to limit this growth and thus penalize the farm population for now and the next generation? The other cost is rising food prices and there are surely limits to how far that can go or would be permitted to go.

Labor Force Adjustment

The critical element in agriculture adjustment is the responsiveness of the farm labor force to economic opportunities. Although there is a good deal of confusion on this point, there is a growing body of evidence that shows clearly that the labor market has changed radically over the past decade or so. And the change has significantly increased the elasticity of supply of labor to agriculture.

Much of the earlier work on the farm labor market, including my own, was strongly influenced by the assumption that the choice between farm and nonfarm employment involved migration—a change in physical location—and that mobility—a change in job—requires migration. As late as 1950 in the United States only 16 percent of the labor force living on farms had nonfarm jobs: there had been little increase from a decade earlier when 10 percent of the jobs held by farm people were nonfarm. But by 1960 a third of all who were living on farms and employed had nonfarm jobs and by 1971 the percentage had increased to 45. Another, and I think rather unexpected, development was that by 1971, 38 percent of the labor force employed in agriculture had nonfarm residence [13]. Thus the isolation of the farm population from nonfarm economic and employment activities that was implicit in much of our earlier work is no longer nearly so valid an assumption.

Obviously something similar has occurred in Canada. In 1971 it is estimated that 49 percent of the total net income of farm operator families came from off-farm sources [10]. This compares with 53 percent in the

United States [14]. In the Prairie Provinces 35 percent of the income of farm operator families came from off-farm sources.

Because of the land reform imposed by American occupation authorities, but now so popular and entrenched, there has been little increase in the size of Japanese farms over the past two decades. The farm consolidation that has occurred in North America and to a lesser but still considerable extent in Western Europe has not been possible in Japan. Yet the labor force adjustment seems to be progressing quite smoothly through off-farm employment of farm family members. Out of 16 million farm and nonfarm workers living on farms in 1970, over 8 million had nonfarm jobs (either part-time or full-time) [9]. Off-farm income accounted for over 63 percent of farm family income in 1970; this was a significant increase from the 52 percent from this source in 1965 [8].

The radical changes that have occurred in the conditions of labor supply to agriculture in North America, Japan, and in much, though not all, of Western Europe have not been the result of conscious governmental policy—at least not of policy directed to that end. The change has been the consequence of many aspects of economic growth—lower cost transportation, improved communication, dispersal of employment opportunities, changes in the regional distribution of farm populations, and the near absence of serious unemployment for the past quarter century. Improvements in the education available to farm youth have also been important, though in all too many countries rural educational opportunities still lag behind urban opportunities.

These changes, however, have largely negated the long run effects of the usual farm policy measures upon the return to farm labor. As the elasticity of supply of farm labor has increased, product price increases or input subsidies have two primary effects: increasing output and increasing the return to farm land. The level of farm employment can be influenced by such measures but not to a sufficient degree to have any significant impact upon the return to farm labor.

Another change in economic parameters has occurred that reduces the effectiveness of supply management in increasing the returns to farm labor or other farm resources. As international trade in farm products has increased in importance in North America, the price elasticity of demand for farm products has increased significantly. It is not only the growing importance of exports and imports that has had this effect, but the increase in incomes in the other industrial countries has meant that our livestock sector, even though trade is relatively small, is functioning within the framework of an international market. Thus efforts to manage output, even if successful, will at best have temporary effects on the returns to all resources engaged in agriculture.

Free Trade and Agricultural Adjustment

FAO has recently published projections of the population economically active in agriculture [11]. Admittedly the projections are based largely on recent trends in the farm labor force, population growth rates, and anticipated rates of economic growth, but the results are instructive. Based on what amounted to the assumption that farm policies existing in 1970 would continue, it is projected that the economically active farm population in the EEC (the Six) would decline from 10.5 million in 1970 to 5.3 million in 1985 or by nearly 50 percent. This compares with an actual decline of the farm labor force from 21.5 million in 1950 to the 10.5 million in 1970. For North America the decline in the labor force is estimated to be rather less, approximately a third. The decline in Japan is projected from 10.8 million in 1970 to 5.4 million in 1985, also a decline of half.

Would free trade—or a significant trend toward trade liberalization— significantly increase the required rate of reduction of the farm labor force in the agricultural regions that would import substantially more farm products? In the absence of quite detailed studies of the effect of free trade upon the demand for farm labor, it is not possible to give a definitive answer to this question. Yet based on some quite preliminary explorations, it appears that the fear that free trade in farm products would require an important further reduction in the farm labor force is unfounded.[3]

There is little evidence that the particular forms of protectionism that have been adopted by the industrial countries have been designed to minimize the primary adjustment problem of agriculture, namely the need to reduce the farm labor force. If policies were so designed, protection would have favored products that were labor intensive and would have neglected, at least relatively, products that were labor extensive. With some notable exceptions, to be noted later, the farm policies of the industrial countries have favored farm products that are labor extensive and land intensive. This is shown most clearly by the Common Agricultural Policy, but it is also evident in North America. The major protection of agriculture in the industrial countries is for crops, especially the grains. Livestock products, except dairy in North America and Australia, have generally received much less effective protection than crop products.

When one looks at the composition of farm output in the enlarged EEC, the emphasis upon the protection of grain seems remarkable indeed. As of 1970, according to some calculations based upon the Flanigan report [1], the gross value of grain was $8.5 billion. The gross value of the main livestock products was almost $39 billion. Both calculations were made in terms of prices received by farmers in the EEC as of 1969–71. If the value of grain and oilmeals fed (imported and domestically produced) is deducted the remaining value of livestock output is about $30 billion or almost four times the output of grain.

If it is assumed that labor used per dollar of output is the same for grains and livestock products (net of concentrated feed costs), and the production changes projected for 1980 in the Flanigan report between a continuation of current policies and free trade for the enlarged EEC are accepted, then the demand for labor would be somewhat larger with free trade than with a continuation of present policies. This theory assumes that labor use per unit of physical output would not be affected by changes in product prices and thus somewhat overestimates what labor use would be under free trade. But it also assumes that labor requirements per dollar of output is the same for grains and livestock products, and it appears that this assumption significantly underestimates the effects of the changes in composition of output on the demand for labor that would result from free trade.

There can be no doubt that Japanese agriculture would face difficult adjustment problems as a result of free trade. But even in the case of Japan, the added adjustment problems due to free trade would be marginal to the adjustment problems that must be faced in any case, even the adjustment problems that will be required if present farm and trade policies are continued. It will not be possible, even with very high prices, to maintain farms of the present size as viable economic units—units that would provide approximately the same return per hour of human effort as will be provided by nonfarm employment 10 to 15 years from now. It is my opinion that by 1980 the major farm policy problem as seen in Japan will be how much it will cost to produce the desired level of farm output. The issue will not be whether farm people have a satisfactory income level. The conflict between continuation of present policies and free trade will not be the result of concern over the incomes of those who remain in agriculture but rather the concern as to the ability of Japan under free trade to produce enough of its own food consumption to provide adequate national security.

According to the recent projections of the Japanese Ministry of Agriculture, farm employment is expected to decline by almost 50 percent by 1982—a decline of about half in 12 years rather than the 15 years indicated in the FAO projections. But even the decline projected by the Japanese Ministry is likely to be an underestimate if real wages generally increase by 6 percent annually and real farm product prices remain constant. Higher real output prices would be required, I believe, if Japan were to maintain approximately its current level of import-dependency in food production.

My speculative comments concerning the adjustment problems in Western Europe and Japan due to free trade have put major emphasis upon the employment of labor. I should not leave the impression that all other adjustments would be of a similar marginal nature. Because most, if not all, of the net income benefits from protection have been absorbed through higher returns to land, free trade would have a major impact upon land prices. Even with a gradual transition to free trade there would probably be declines in the absolute prices of farm land, especially highly productive land now devoted to grain production. Even though there has never been, to my knowledge, an adequate reason given for governments seeking higher farm land prices, a significant decline in land prices might well have effects that many would consider to be undesirable. However, the decline in land values would in all cases be small compared to the savings that consumers and taxpayers would realize as a result of free trade. Consequently all existing land owners could be fully compensated for losses in land values, leaving a substantial net gain for consumers and taxpayers.

The Important and the Unimportant

As measured by effects upon the welfare of farm people, the agricultural policies of the industrial countries have generally concerned themselves with matters that have been relatively unimportant in determining the economic welfare of farm people. These policies have emphasized price supports, trade barriers, export subsidies, and supply management. These measures have only a marginal impact on the level of living enjoyed by farm people in industrial countries.

The primary determinants of how well off farm people are have been the level of national income per capita, the rate of national economic growth, and the way that growth has been managed. Of nearly equal importance is the extent to which farm people are integrated into the economy and society in terms of access to educational opportunity and the ease, speed, and cost of communication and transportation. The high levels of income of farm families in North America are not the result of the governmental agricultural policy measures, but are due to high levels of productivity in the economies generally.

Governments have generally done little, as a matter of conscious policy, to aid agriculture in the difficult adjustments that it must make. In fact, much of governmental policy has been motivated by a futile desire to prevent adjustment.

Almost all of what governments in the industrial countries have done to improve the welfare of farm people has been outside the framework of national agricultural policy. I refer to the support of education, the road and highway system, the encouragement of the extension of railroads, the support of agricultural research, the postal service, the provision of adult education, and the management of the economy to achieve economic growth with minimum periods of less than full employment. Had some of the resources devoted to price supports or supply management over the past three decades been positively directed toward measures to increase human capital possessed by farm people, both those who have remained in agriculture and those who found it necessary to leave, the average level of farm income would now be measurably higher than it is.

As economists we should not be guilty of the same error as has been so evident in farm policy. Free trade is not a panacea for farm income problems. The economic analysis underlying this paper says that the effect of free trade on the real income levels of farm people over the long run would be so small that it probably could not be measured.[4] The real income per farm family would be higher, but only because of the small increase in real per capita income that would result from more efficient use of all resources in an economy.

As I see it, the major advantage of free trade as a national policy is that it would clear the way for inducing policy makers and the rest of us to consider the important issues that affect the welfare of farm people.[5] And the avoidance of the enormous income transfers inherent in the existing policies would make available the fiscal resources required to permit farm people to share fully in the growth of the economy of which they are a part.

Notes

The research on which this paper is based has been supported by the National Science Foundation (GS-3071) and the Rockefeller Foundation (RF 63096). Neither is responsible for the contents of the paper. I want to express my appreciation for their helpful comments to Joseph Barse of the Economic Research Service, U.S. Department of Agriculture, and to Harry G. Johnson and Theodore W. Schultz of the University of Chicago.

1. The "presumed failure" of farm programs to reduce income differentials should not be interpreted to mean that there are substantial differences in returns to comparable resources engaged in agriculture and elsewhere in the economy. Most of the income comparisons made by governments are wholly inadequate as a basis for assessment of the effectiveness of resource allocation [6, Ch. 10].

2. Change in factor productivity is here used in a somewhat loose sense, as the change in the ratio of output to total inputs as the latter are conventionally measured. A large fraction of the increase in output per unit of input is due to inaccurate measurement of inputs—primarily the failure to reflect changes in quality of the inputs over time. The measurement of inputs fails to reflect changes in the quality of human resources, and similar measurement errors occur for complex inputs such as tractors, combines, and transportation equipment. But for present purposes it does not matter much whether the shift in the demand curve for labor is due to an unrecognized improvement in tractors or, equivalently, a discrepancy between the measured and the true price of tractors.

3. I have made rough estimates of the possible effects of free trade upon the demand for land and labor resources in North America [5] and somewhat more

detailed estimates for the United States [4]. The results indicate that even though a labor intensive product is highly protected in North America—dairy—the effect of free trade would be to increase the demand for both labor and land.

4. Land rent will change as a result of free trade, increasing in some countries and declining in others. It is not clear that rent would increase in all the countries that would export substantially more as a result of free trade since such a large fraction of the income benefits of existing elements of protection have resulted in increased rents. It cannot be ruled out that in some countries the decline in rent would be large enough to have some measurable effect upon farm family incomes, but I think that this would be the exception rather than the rule.

5. Free trade does not rule out many programs and policies designed to improve the welfare of farm people. As I have argued at length elsewhere [6, pp. 205–225], there are many measures that could benefit farm people and yet be consistent with the basic principles of free trade. These measures can be, on balance, "production and consumption neutral" and thus have little or no effect upon the value and volume of international trade. Obviously many measures will have some small effect upon production; this would be true of, say, improving education available to farm people, but some of the effects would be to increase production (greater capacity of the farm labor force) while others would be to reduce production (increased labor mobility). With some imagination, the worst effects of price instability can be minimized.

References

[1] Council on International Economic Policy, *Agricultural Trade and the Proposed Round of Multilateral Negotiations*, U.S. House Committee on Agriculture and Forestry, 93d Congress, 1st Session, Washington, D.C., U.S. Government Printing Office, 1973.

[2] Gisser, Micha, "Needed Adjustments in the Supply of Farm Labor," *J. Farm Econ.* 49:806–815, Nov. 1967.

[3] Henrichsmeyer, Wilhelm, "Economic Growth and Agriculture: A Two-Sector Analysis," *The German Economic Review* 10:310–326, 1972.

[4] Johnson, D. Gale, *Farm Commodity Programs: An Opportunity for Change*, Washington, American Enterprise Institute, 1973.

[5] ———, "The Impact of Freer Trade on North American Agriculture," *Am. J. Agr. Econ.* 55:294–300, May 1973.

[6] ———, *World Agriculture in Disarray*, New York, St. Martin's Press, 1973.

[7] Martin, Lee R., "Some Market Effects of Agricultural Development on Functional Income Distribution in Developed Countries," Dept. of Agricultural and Applied Economics, University of Minnesota Staff Paper P72–9, March 1972.

[8] Ministry of Agriculture and Forestry, Japan, *Abstract of Statistics on Agriculture, Forestry and Fisheries, Japan, 1971*, Tokyo, Association of Agricultural-Forestry Statistics, 1972.

[9] ———, Japan, *Norinsko Tokei Hyo (Statistical Yearbook of the Ministry of Agriculture and Forestry), 1970–71*, Tokyo, 1971.

[10] Porteous, W. L., "Canadian Farm Income-Levels—Outlook," Canadian Agricultural Outlook Conference, Proceedings '72, 109–121, Ottawa, 1972.

[11] Schulte, W., L. Maiken, and A. Bruni, "Projections of World Agricultural Population," *Mon. Bul. of Agr. Econ. and Stat.* 21:1–10, Jan. 1972.

[12] Schultz, Theodore W., "The Increasing Economic Value of Human Time," *Am. J. Agr. Econ.* 54:843–850, Dec. 1972.

[13] U.S. Bureau of Census and U.S. Department of Agriculture, *Current Population Reports, Farm Population*, Series Census-ERS, P-27, various issues.

[14] U.S. Department of Agriculture, Economic Research Service, *Farm Income Situation*, F15–220, Washington, July 1972.

PART FOUR

*Agricultural Policy in China
and the Soviet Union*

9

Eye-Witness Appraisal
of Soviet Farming, 1955

The trip that I made as a member of the American Agricultural Delegation to the U.S.S.R. during the summer of 1955 was a conducted tour. The particular farms that I visited cannot in any sense be considered a random sample of Russian farms. Only one of the more than 30 farms visited by our group was selected by us. The rest were selected by the Soviet authorities. Most of the farms that we visited were substantially better than the average of the U.S.S.R. However, some farms were near the average for the area in which they were located and we did travel two or three thousand miles by car in farm areas. Thus one has some possibility of putting what he saw in some kind of perspective, but one must also say that he cannot be certain as to the accuracy of that perspective.

I do not intend to give what might be called a systematic appraisal of the problems and prospects of Soviet agriculture. I shall try to give you a number of impressions that I have as a result of my five weeks' trip. These impressions represent nothing more than an honest effort to report some of the things that I witnessed or heard.

I

Many westerners have interpreted the attention that has been given to agriculture in the speeches and decrees since late 1953 and the announcement of the reasons for Malenkov's resignation as evidence that the present Russian food situation is extremely critical. If these interpretations are carried so far as to imply that the Russian people are on a near starvation diet, what I saw would not support so strong a conclusion. The people I saw on the streets, in the factories and on the farms did not appear to be underfed. Other evidence is that there seemed to be plenty of bread available in the

Reprinted by permission from the *Journal of Farm Economics* 38 (May 1956): 287–95.

173

stores. Since bread and potatoes plus other grain products supply about 75 per cent of all calories consumed by the Russians, the general availability of bread means that people are probably not going hungry. The bread is relatively cheap and any employed person can afford to buy any amount of bread that he desires, particularly if he is willing to eat rye bread. Most other food items, such as fresh milk, meat, and fruits and vegetables were not available in the state food stores I visited. However, these items were available in the free or collective farm markets at prices substantially above the state store prices. However, the margins between state store prices and free market prices, which ranged about 25 to 150 per cent above state store prices, hardly indicate a near starvation situation.

However, there does seem to be considerable concern that the over-all food situation has not improved since the beginning of the present decade. Apparently the 1950 grain crop has not been equalled, though the 1955 crop may be as large. In both 1953 and 1954 there were serious crop failures in major grain regions, the 1953 failure coming in the east and the 1954 in the European part of Russia. The amount of grain that has been fed to livestock in recent years has been very small and unless there are significant increases in grain and feed output in the near future, there is little or no prospect for the improvement of the Russian diet. When these and other facts are confronted and when it is remembered that there are about 3 million new mouths each year, it can be seen that the present rather tight food situation might be transformed into a crisis in 5 or 10 years.

II

Grain and crop production as I saw it in the summer of 1955 was better than I had expected it to be. By this I mean that the cultivated crops had been well tended and the fields were free of weeds; the small grain fields were generally free of weeds and the crops had apparently been sown on time on reasonably well prepared seed beds, and the harvesting was being done on time and there seemed to be little loss of grain from the combines. I know relatively little about cotton production, but the irrigated cotton fields that I saw compared favorably with the few similar fields that I have seen in the U.S. On one trip of about 100 miles from Odessa I did see mile after mile of small grain fields that were infested with weeds and which as a consequence would yield quite poorly. But this was the exception; perhaps in other years when growing conditions were less favorable before and after seeding this condition is more general.

As you undoubtedly know, Russia is now to become a corn country, rivaling even the United States in total corn acreage. We saw corn everywhere, along the Volga River where rainfall does not exceed 12 inches, near Moscow with its short growing season and in Siberia in dry land farming country. The corn had been planted so that it could be cross culti-vated; since we saw few planters that could plant it in this way, it was evi-dent that much of the corn had been planted by hand. The Russians are great believers in cultivation, much of the corn being cultivated a total of six times. It must have come as a shock to the Russian delegation in the U.S. to learn that our farmers were now giving up the check-row planting of corn and that they cultivated their corn as seldom as possible and hardly ever more than three times.

So far as I could tell the Russians do not have hybrid corn in any sig-nificant quantity. We visited upwards of a dozen places where hybrid corn would have been produced if such production existed—agricultural insti-tutes and colleges, experiment stations, seed selection stations and state seed grain farms—yet we did not see a single field where hybrid corn was being produced. Only at one of these places was there even a claim that they were producing hybrid corn. This was on a state grain farm that had 3,000 acres of corn for sale as seed to collective and state farms and it was said that on 100 acres hybrid seed was being produced. I didn't see it and I suspect that what was done was the crossing of two open-pollinated varie-ties. At many of the farms we visited we asked what kind of corn seed had been used and the answer was always some open-pollinated variety, frequently a variety that had come from the United States.

The blame for the lack of hybrid corn can probably be given to Lysenko. An official of the Ministry of Agriculture, who was not a devotee of Ly-senko, told me that work on the development of the inbreds required for hybrid seed was well underway in the late twenties. But Lysenko ordered the work stopped on the grounds that according to his theories this was not the correct way to conduct breeding work.

However, I do not believe that Russia will be able to obtain from hybrid corn the miracles that Krushchev has attributed to hybrid corn in the United States. Apparently, as a result of long training, Krushchev and his statisticians have grossly overestimated the gains we have made from hy-brid corn. This was done by using the drought years of the thirties as the prehybrid corn base period and 1953, the year with the highest yield on record, as the current base, in estimating the effect of hybrids on yields.

But the problem is more complex than that. In areas in the United States that are most comparable to the U.S.S.R. we have obtained very modest increases in corn yields. This has been true in North and South Dakota and, to a somewhat lesser degree, in Nebraska.[1] Russia has only about 10,000,000 acres of crop land in the Kuban that compares with any part of the Corn Belt and this to northeastern Nebraska. In areas of limited rainfall and relatively short growing seasons, hybrid corn has not worked miracles in the United States and there is not much reason to believe that Russian experience will be very different.

I do feel that crop production suffers from the failure of the Russians to specialize—to concentrate on specific crops in the areas that are best adapted to them. If more corn is desired, the corn should be concentrated in the Kuban and in the Ukraine with the best growing conditions. Instead corn is being planted everywhere and the corn expansion is not significantly greater in the areas where corn has the greatest comparative advantage. There is a general tendency to make each area as self-sufficient as possible. In part this may be due to the lack of adequate transportation, since few farm products (other than grain) move long distances from their point of origin.

The mechanization of agriculture still has considerable distance to go. Most of the field operations for the small grains seem to be rather fully mechanized. However, machinery must be used quite intensively. For example, each combine apparently has to harvest about 750 acres each year. In order to complete the harvest in minimum time, the grain is not cleaned adequately at the combine and a great deal of labor is used at the grain cleaning floors to get the grain in shape for delivery and storage. The grain cleaning is mechanized in only certain limited respects and Krushchev has stated that "considerably more labor is consumed on postharvest processing of the grain and the harvesting of straw and chaff than on all preceding operations in raising grain crops."[2] From what I saw, I see no reason to doubt this statement.

III

Although I was favorably impressed with the grain output per acre, this was not the case with livestock production. This does not mean that I saw no herds of relatively high producing milk cows or no hogs of good quality. I did. Obviously we visited a number of model farms, but with one exception, the milk yields on these farms were about the average for major dairy states such as New York and California. Most of the livestock that I saw was poor in quality and many of the cattle seemed to be underfed. It is not hard to believe that average milk yields on collective farms are about 2,300 to 2,500 pounds per year, and we in fact visited farms where the reported yields were substantially below this.

About sixty per cent of the milk cows in the Soviet Union are now owned by the individual peasants who are limited, in most areas, to one cow per family. Thus milk production occurs at the extremes of a distribution of production units in terms of scale. On the one hand, 60 per cent of the cows are in production units of minimum size, namely one cow. The rest are in units ranging upwards from perhaps 40 or 50 to several hundred cows. Thus I suspect that the Russians have the worst of all possible worlds with respect to scale. The disadvantages of the extremely small unit are well known, but some comment on the larger units may be mentioned. In several farms that we visited, special summer quarters were used for milk cows. These summer quarters were apparently required to get the cattle out on pastures and near enough to the sources of green feed. Although the capital investment in summer quarters was not as high as for year-round building of the traditional type (but as great as the milking parlor and open shed buildings now coming into use in the United States), the investment required was quite substantial. In some cases the distances from the villages were so great that living quarters for the milk maids had been built. The necessity for the summer quarters is more a function of the size of the farm as a whole, than of the dairy herd.

IV

All of the members of our delegation were surprised at the large amount of labor that was being used in conjunction with every farm operation. While I had some rough idea of the total number of farm workers in the U.S.S.R. I still found it somewhat difficult to imagine what all of these workers do. This is especially true since the tractor power available is about the same as we had in the thirties and there is a greater number of combines.

On the farms that we visited in the Ukraine there was about one farm worker for 8 or 9 acres and in the Kuban one worker for 9 or 10 acres of crop land. Only in the dry-land areas of western Siberia and northern Kazakhstan, where there is relatively little livestock, did the ratio of labor to land reach approximately the average of the U.S. In the U.S. we have about 50 acres of cropland per worker; in the Russian areas referred to there were about 40 acres per worker. But in a comparable area in the United States, such as North Dakota, there are about 225 acres of harvested land per worker.

On the whole, I have the impression that in Russia the output per worker in agriculture compared to output per worker in industry is much lower than in the U.S.[3] This crude statement assumes some concept of physical labor input and comparable outputs, weighted by approximately U.S. prices or at least by non-Russian prices. But if we accept the relative price structures of each country and calculate the labor inputs and outputs in dollars for the U.S. economy and in rubles for the Russian economy, the differences might not be very great. This is true because the relative prices of food are so much higher in the U.S.S.R.

The wages paid to workers on state farms seemed to average about 550 rubles per month, while the wages paid to factory workers was about 750 rubles. This is a very small difference, compared to differences that exist in the United States, Canada or western Europe. Part of this difference is eliminated by the greater importance of private plots for the farm workers and the absence of costs of getting to work. The employees of machine-tractor stations have higher earnings than state-farm employees and may have real earnings that approach or exceed those of factory workers.

While I have the impression that in the areas we visited the peasants who belong to a collective have incomes equal to factory workers, this is a difficult statement to prove. My impressions may be biased by the fact that all collectives visited were at least average or better in terms of productivity and income. But even on the poorest of the farms that we visited, which were probably about average for the U.S.S.R. as measured by milk yields per cow and cash income per hectare, the payments from the collective farms were equal to about a third of the average factory wages, if grain is priced at 2 rubles a kilogram. For the farm income to equal the factory income it would be necessary that the private plot income be about 6,000 to 8,000 rubles per year, since there is more than one worker per household. If the family had both a cow and a sow and raised a calf and four pigs each year, the income from the plot would be about 8,000 rubles at free market prices.

But perhaps the most convincing evidence of the near equality of farm and nonfarm incomes is that the farm population and work force has increased over the past two years, halting a decline that has lasted for more than two decades. We were told that the number of able bodied workers in agriculture increased by 1,500,000 in 1954. On the farms we visited we were told of rather significant increases in the number of families and workers. Of course, the increase has not been uniform over the U.S.S.R. In the European part, the rate of increase has been higher in the south than in the north, while the increase in the east has undoubtedly been much greater than in the west. While some of the movement of workers to the new lands area may have had an element of compulsion in it, I doubt that this was an important factor in the Ukraine and Kuban and Volga areas.

If the labor force on farms continues to increase in response to the more favorable incomes and the increased labor needs resulting from the corn and new land programs, the conflict between farm and industrial production will be of real significance. It will be interesting to watch how long the favorable impact of recent policy changes on farm incomes will be allowed to continue. One way out would be to increase nonfarm wages or to increase delivery requirements, either at obligatory prices or at the purchase prices, so that more farm products would be available at state-store prices. I am not predicting that either of these changes will occur, but neither will I be much surprised if they do sometime within the next couple of years.

V

A few comments on the level of living in the U.S.S.R. may be in order. I find it difficult to make meaningful comparisons because my own experience has been with high-income countries. I am also greatly puzzled by a special phenomenon for which I can find no explanation. When you move from one farm area with relatively high incomes to another area with low incomes in the United States, it is easy to observe the difference in terms of quality of housing, the clothing, the age and type of cars, the number of television sets, automatic washing machines and so on. But as I went from one farm to another in Russia I was quite unable to detect the large differences in apparent income in terms of the quality of housing, clothing, furnishings in the houses, or the general appearance of the people or the village. The only apparent difference seemed to be in the number and quality of farm buildings and in the size and quality of the palace of culture. The reason for these last differences is fairly obvious—a certain share of the money earnings of the farms must be invested.

Why didn't differences in payments from the collective farms of the order of four or five to one reveal themselves in a way that could be detected readily by an outsider? A part of the reason is probably that a large share of the food intake comes from the private plots on most farms. This would be true of most foods with the possible exception of the grains, which are received by distribution from the collective farms. Part of the reason may also be that the kinds of consumer goods that would be obvious to the visitor are simply not available in quantity in the U.S.S.R. Here I refer to such things as washing machines, gas or electric stoves. But clothing, which is very high priced, does seem to be available to a degree that a farmer with cash and some time for shopping should be able to acquire it.

I am also puzzled that the difference in income did not show up in the size and quality of houses. In fact, some of the poorest housing that I saw was on a very high-income and model farm near Moscow.

The only exception to the above statements was in the cotton producing areas of central Asia. But comparisons are difficult because of the different modes of living of these essentially Asiatic people and the more westernized Russians. There were more cars on the farms in this area—on one farm 10 per cent of the families owned cars while the general rule is none or one to three cars in villages of several hundred families. One also saw sewing machines in the homes of some. But since such homes, by tradition, have little in the way of furniture other comparisons cannot be made.

On the whole, I felt that farm housing compared favorably with urban housing. Except for the small proportion of urban people who live in apartment buildings constructed since the end of World War II, urban housing can only be described as miserable. Many urban families, perhaps most, still depend on community wells or hydrants and outside toilet facilities. In addition, they live in more crowded circumstances, both with respect to the amount of space inside the house and outside. Many urban families apparently also share kitchens, though I have no direct evidence of this. The relative openness of the rural housing must be attractive to many urban families, though the paucity of community facilities and shopping and the ever-present mud during the spring and fall and the isolation caused by snow and cold in the winter are undoubtedly serious drawbacks.

But to return to the general question of comparative levels of living in the U.S.S.R. and the U.S., in a rough sense I believe that aggregate output would have to increase at least four fold before the per capita content of the U.S.S.R. level of living could duplicate that of the U.S. This is based primarily upon observation of the content of consumption, upon labor productivity in agriculture, wage rates and some rough price comparisons. I doubt if the purchasing power of the ruble for consumption purposes much exceeds 6 cents. The average monthly wage of factory workers is about 750 rubles or perhaps $45. Even if the purchasing power is as high as 10 cents, which I find it hard to believe, the earnings would be $75 or about a fifth of the U.S. average. If one adds a quarter for social services that are not available in the U.S., such as medical services and nurseries, the earnings are still about one fourth of the U.S. level. And it should be noted that the Russian factory worker is probably higher up in the urban income scale than the American factory worker.

VI

I would like to comment briefly on what I saw with respect to the two great programs for expanding agricultural output—the corn program, and the virgin and idle lands program. I must admit that I was surprised at the number of acres of corn that I saw on the trip. I did not believe, and I made the mistake of saying it, that the Soviet Union would actually increase the corn acreage very significantly. In contrast to the new lands program, the corn program appeared to be a program based solely upon exhortation. This has been fairly common in previous agricultural programs, and the output has generally about equalled the input. The state was going to put relatively little into the corn program; machinery wasn't available for many aspects of corn planting, cultivation and harvesting; and the collective farms were to receive little or no aid in constructing the necessary silos. But the corn was there, apparently 40,000,000 acres or about four times as much as last year. And there were many new silos on almost every farm that we visited.

If a nation is interested in maximizing the output of feed per unit of land, the corn plant has much to commend. In temperate climates, if the entire corn plant is used, corn will probably outyield any other plant and will yield about as much starch (more protein) than the root crops. This seems to be true in cold climates, dry climates, or in relatively hot climates. But the difficulty with corn is that it requires much more labor and machinery than the small grains. As a result, we do not grow corn unless it will yield almost twice as much grain as the competing small grains.

I saw corn being produced near Moscow that would not produce ears this year; I saw corn in western Siberia that would yield no more than 3 to 5 bushels to the acre if harvested as grain. Corn acreage is being expanded everywhere. There is apparently no attempt to expand it relatively more in the areas with the highest relative yields.

Since the corn is to be harvested as silage, with the ears separated from the stocks so that hog feed will be produced, output per acre will be higher than it would otherwise have been. But the return to labor for the additional effort will be very low indeed, especially in the drier areas. Furthermore, recent press reports have indicated that the harvest is proceeding very poorly in many areas because the harvest labor requirements have exceeded the available labor input.

An additional facet of the corn program may be mentioned. Milk yields were substantially higher this year than last year. Part of this is due to the fact that last year was a year of drought in much of western Russia. But part of the increase can be attributed to the availability of corn as a green feed in July and August. Large areas of corn were cut for cattle feed during this period. I suspect that in most years cattle have not had enough feed after the spring growth of grass was eaten and that milk yields were seriously cut and never recovered during the same lactation period. Now the corn meets this deficiency. However, if the Russians allow too large an increase in livestock numbers, they may find themselves very short of feed at a different period of the year, say in the late winter months.

I did not see much that would serve as a basis for appraising the new lands program and its prospects. The area that I visited was part of a well established farming area. The newly plowed lands were similar to lands that had been farmed for several decades, but had not been farmed because of a lack of manpower and machinery. Little moisture had been received this year, though the crops were undoubtedly aided by the carryover of moisture from the previous year. This year's yields will be less than half of last year's. In northern Kazakhstan where the other part of our party visited, crops were much more affected by the drought which seems to have blanketed most of the new lands area this year. Last year the yields were about 15 bushels per acre and this year from 2 to 6 bushels per acre.

I am somewhat puzzled by what our group was told in Kazakhstan and what has since been mentioned a number of times in the Russian press, namely that the yields on new lands were much higher, say 6 bushels per acre, than on lands that had been plowed before, even twenty years before. What this may mean is that these lands soon lose their ability to retain moisture once the original cover is disturbed. If true, the yield prospects on much of this land may prove to be quite disappointing to the Russians.

VII

Russian agriculture presents one with paradoxes on all sides. You have bigness carried to the extreme in many things—farm sizes, machinery, sows, boars, barns and fields—but alongside this they have minute peasant farming on the small plots. Mechanization has been emphasized in field operations, but other aspects of grain production have been mechanized little if at all. There is centralization of marketing and production decisions, but free markets continue to have a very important role in feeding the city population and in providing incentives for rural people. Labor specialization is carried to an extreme on many farms, but there is little tendency to have specialization in crop production.

I must confess that after my brief acquaintance with Soviet agriculture that I am hard pushed to explain why it works as well as it does, rather than being puzzled by why it has performed rather poorly.

Notes

1. The average corn yields for 1920–29 and 1940–49 for the three states were as follows: North Dakota, 21.0 and 22.4; South Dakota, 24.7 and 25.8; and Nebraska, 25.7 and 28.1. In Iowa the yield increased from 40.2 to 51.2. Some of the increase in corn yield in Nebraska is probably explained by the different area distribution of corn, since the corn acreage declined in the drier parts of the state between the two periods.

2. See *The Current Digest of the Soviet Press*, May 5, 1954, p. 8 for translation from *Pravda*, March 21, 1954.

3. By this statement I mean the following:

$$\frac{\text{Agricultural output per worker—U.S.S.R.}}{\text{Industrial output per worker—U.S.S.R.}} < \frac{\text{Agricultural output per worker—U.S.}}{\text{Industrial output per worker—U.S.}}$$

10

Agriculture in the Centrally Planned Economies

From a number of viewpoints the agricultures of most centrally planned economies performed well from about 1950 to the mid-1970s. Agricultural production grew more rapidly in Eastern Europe and the Soviet Union from 1950 until the early 1970s than in Western Europe or North America (USDA 1981). Per capita production of meat and other livestock products increased, in some cases at an unparalleled pace, with most of the East European countries increasing annual per capita meat consumption by approximately 20 kilograms in the 1970s (USDA 1982a).

But since the mid-1970s the agriculture and food economies of Eastern Europe and the USSR have faltered and, in some cases, stagnated or even declined in terms of production (USDA 1981). Any pretense of self-sufficiency in food and agriculture has had to be abandoned. In one country strikes, riots, and martial law were responses, at least in part, to an unsatisfactory food situation. In another country bread rationing has had to be introduced—an almost unbelievable response in a world in which wheat is so abundant and cheap. In still another, meat has essentially disappeared from retail stores, with its distribution now largely controlled by employers and trade unions. And in almost all countries the financial costs of food price subsidies have been or are now a major budgetary element and source of actual or potential disequilibrium in the markets for food.

The sharp increase in grain imports by the centrally planned economies (CPEs) during the 1970s was associated with the rapid increase in meat production achieved in Eastern Europe, though the even more striking growth of grain imports by the USSR and China must be explained on other and more complicated grounds. For some of the CPEs the increased grain imports were made possible, if not actually paid for, by credits ex-

Reprinted by permission from the *American Journal of Agricultural Economics* 64, no. 4 (December 1982): 845–53.

tended by Western banks and governments. Now that the credit worthiness of several of the East European economies has been questioned, the growth of grain imports has slowed down and has been reversed in some countries. The enormous overhang of debt, some of which has figuratively been eaten, poses serious problems for several economies for the decade ahead.

I have for some time argued that the primary source of the difficulties of socialist agriculture is that such agricultures are found in socialist economies. I mean this—it is not intended as a facetious remark. It now seems quite evident that any form of land tenure can be made efficient.[1] And socialized agriculture, in its most general terms, is a land tenure system. Elsewhere I have discussed what would be required for a socialized agriculture to be an efficient agriculture (Johnson 1980). A set of well-defined property rights is required; in terms of efficiency, it makes little difference whether the land is owned by the state (all of the people), by the members of collective farms, or by private persons, either as landlords or operators. What matters is that each party to the tenure relationship has well-defined rights to the fruits of one's efforts or one's contribution to the output. There is no economic reason why the state cannot be a reasonable landlord who promotes efficient use of resources. Unfortunately, most policy makers in the centrally planned economies have not learned the lesson of how to be good landlords, though China may now be giving some evidence to the contrary, as does Hungary.

But it is no longer enough to have well-defined property relationships to have an efficient agriculture. Modern agriculture is very dependent upon its economic relations with the rest of the economy, through its use of nonfarm purchased goods and services and its need for efficient, reliable, and low-cost marketing services.[2] When input markets and marketing services are poorly organized, it makes little difference how agriculture is organized, be it as private, collective, or state farms.

The output performance of Polish agriculture, which is three-fourths private, has not been any better than the other Eastern European agricultures that are fully socialized. But there is no reason to expect better performance from a private agriculture unless the policy setting within which the agriculture operates is supportive and nonthreatening. But in Poland the Communist Party has been unwilling to rescind its long-run objective of a socialized agriculture or to adopt an even-handed treatment of private and socialized farms in the allocation of machinery, fertilizer, or other farm inputs. I must hasten to add that, in most of the CPEs, the policy setting of socialized agriculture can hardly be described as supportive of a low-cost and efficient agriculture.

I use the term "policy setting" in a very broad sense to include the significant relationships between farms and the state—output and input prices (the terms of trade), procurement requirements, the planning process, and other aspects of decision making. I also use it to include an intangible element—the extent of mutual trust and confidence that exists between the state, as represented by the numerous agencies that deal with farms, and the farms. Mutual trust goes beyond matters of honesty and includes the attitudes of governmental officials about the competence of farm people to carry out their functions efficiently and responsibly. Where these elements of mutual trust are circumscribed, other and more objective aspects of the policy setting may have less than expected effects upon outcomes.

Different Socialist Economies, Different Agricultures

It is not possible to progress very far in a discussion of socialist agricultures without recognizing the differences, as well as the similarities, that can be observed among different countries or groups of countries. I shall discuss three countries or groups: USSR, Eastern Europe, and China. Within Eastern Europe there have been and are significant differences in organization, policies, and performances. My comments can only highlight a few things about each.

USSR

I start with the agriculture of the USSR because it was the first of the socialized agricultures and there is now more than a half-century of experience with what can and cannot be achieved under socialism. The output record for 1950 to 1970 was a very good one compared to Western Europe or North America. By 1950 Soviet agriculture had recovered from World War II except for the impact upon the farm labor force which at that time consisted very largely of women and older men. Given the composition of the labor force, the performance of Soviet agriculture during the 1950s was quite remarkable.

With the death of Stalin, the rapacious exploitation of rural people by their government was largely brought to an end. Prices paid to farms in 1958 compared to 1952 were increased severalfold—the grains by six times; livestock by eleven times; sunflowers by eight times; and sugar beets doubled. Milk prices in 1958 were four times the 1952 prices (Strauss, p. 201).

Khrushchev undertook several bold and risky agricultural measures— the New Lands program which brought 36 million hectares of marginal land under cultivation, the corn program which increased the planted area of corn from 4 million hectares to 37 million hectares in 1962 though the maximum area harvested for grain was just 7 million hectares, the abolition of the Machine Tractor Stations (MTS), and the introduction of a single procurement price for each product. While Soviet agriculture responded positively to some of the measures, it may have been success that undid Khrushchev. In part because of measures adopted and in part due to favorable growing conditions, 1958 was a bumper crop year. It was then that Khrushchev abolished the MTS in the expectation, widely shared outside the USSR, that this step would significantly improve productivity by making the collective farms more responsible for the use of their resources and by providing greater incentives. Like all too many agricultural reforms, this one was poorly planned and executed. Repair services were not adequately provided for, and the machinery was transferred to the farms under unfavorable terms for the farms. A new burden replaced an old one, and farm incomes declined after what could have been a constructive move.

Agriculture performed far below the belicose Khrushchev's expectations—to catch up and overtake the United States in meat and milk production by 1965. Several of the goals for 1965, announced in 1958, still have not been met, including the critically important one for meat. While farm output grew by 43% between 1952 and 1958, for the next six years the output increased by just 17% (USDA 1981). At least in part due to the poor performance of agriculture and the need to import 10 million tons of grain in 1963/64, Khrushchev was replaced by Kosygin and Brezhnev in 1964. Some of the personal and politically liberalizing measures introduced by Khrushchev may have assured his physical, if not his political, survival.

The new administration carried out a major reform of agricultural policies—farm prices were increased, an enormous fertilizer production program was inaugurated, investment in agriculture was increased sharply, wages were introduced for farm workers, and a pension system for members of collective farms was introduced. These were clearly sensible measures and were expected to have resulted in a revitalization of agriculture.

But hardly any other aspect of the agricultural policy inherited from Khrushchev was changed. Moscow still maintained tight control over the minutest details of farm operations—crop areas, plowing dates, seeding dates and rates of seeding, harvesting, delivery quotas, and the annual and five-year plans for each farm. Nothing was done to achieve trust or respect of the planning or other governmental officials by farm people and thus, of course, confidence in those officials among farm people remained at a minimal level.

Given the material resources devoted to agriculture, it would have been reasonable to expect rapid and continuing output growth. For a time, from 1964 to 1970, it appeared that the program was being successful since output grew at an annual rate of 3.9%. But the 1970s saw a much slower growth rate of agricultural output—at an annual rate of 1.2%, with an even lower growth rate after the mid-1970s; 1980 output was the same as in 1973 and 1976 (USDA 1981).

The shift in resources to agriculture can only be described as enormous. During 1961–65 the percentage of national investment allocated to agriculture was 19%; during 1976–80 this percentage had increased to 27%. If the investment in agriculturally related industries is included, the percentage increases to 33 for 1976–80. Annual rates of investment increased from 9 billion rubles in 1961–65 to 34 billion rubles in 1976–80. Fertilizer deliveries to agriculture, in terms of nutrient content, increased from 6.5 million tons in 1965 to 18 million tons for 1976–80. There were significant increases in the delivery of farm machines; but, due to high scrappage rates, inventories increased slowly during the 1970s.

Some of the recent output performance of Soviet agriculture can be attributed to poor weather—grain production from 1979–81 may have been 13% less than it would have been with normal or average weather— actually 179 million tons instead of 205 million tons. But the effect of the low production of grain and other feed supplies was partially, if not wholly, offset by grain imports averaging 36 million tons for the three years. The level of grain imports was greater than the shortfall in grain production from trend levels for these years. Thus, the fact that per capita meat output in 1981 was the same as in 1975 should not be attributed wholly or even primarily to poor climatic conditions. Milk production in 1981 was below the absolute level in 1974. Milk production per cow in 1981 was 2,040 kilograms, compared to 2,260 kilograms in 1977 and 2,110 in 1970 (USDA 1982). This decline in milk output per cow is a remarkable development and can hardly be accounted for on rational grounds. Current milk output per cow is among the lowest in Europe.

The very high rate of investment in agriculture and the increase in the purchased inputs over the past two decades has not resulted in a marked decrease in the use of labor. Between 1965 and 1980 employment in agriculture declined by just 15% to an annual level of about 27 million. Quite remarkably, even this enormous number of farm workers—Soviet agricultural output is some 20% less than U.S.—has not been sufficient. In 1979 some 15.6 million nonagricultural workers were sent from the city to the countryside to help with various farm operations, primarily harvesting. This is approximately double the number of such nonfarm workers sent to the farms in 1960 and some 40% more than in 1970 (*Current Digest*, no. 8, 1982).

Nor does the immediate future look much brighter, either in terms of increased production or greater efficiency and lower production costs. By Soviet estimates the percentage increases in costs of production (excluding land costs and interest on capital) was 38% for grain, 58% for meat, and 50% for milk between 1969–71 and 1978–80 on collective farms.[3]

The cost increases have been responded to by higher procurement prices, including significant price increases in 1979 and 1982 and the extension of the 50% bonus for deliveries in excess of actual deliveries during 1976–80 to all farm products. The average procurement prices for all farm products were increased to include the bonus payments paid in earlier years.

The increase in procurement prices has occurred in the setting of a policy of holding fixed the retail prices of major food products, especially meat, milk, and potatoes. The policy of subsidizing consumer prices was introduced as an emergency measure in 1965, based on the expectation that the new agricultural measures would result in such a sharp increase in production and in significantly lower costs that the subsidies could be eliminated. But some seventeen years later, the annual cost of the food price subsidies is at least 35 billion rubles ($45 billion at the official exchange rate) and will increase in the years ahead.

The policy of fixed prices for food in the face of slowly growing supplies, relatively high income elasticities of demand for meat and milk, and significant growth rates in money wages has had results that are having seriously adverse effects upon the functioning of the Soviet economy. Meat has almost disappeared from state stores; an alternative distribution system centering on places of employment has emerged not only for meat but for many other products in short supply, including such necessities as pickles, catsup, and plum jam (*Current Digest*, no. 2, 1982). The prices of food in the collective farm market in 1980 were almost double the official prices, up from an excess of just 50% in 1970.

In October 1981 Brezhnev called attention to the central role of food in the Soviet Union: "The food problem is, economically and politically, the central problem of the whole five-year-plan." He went on to note the necessity to improve the management system (*Current Digest*, no. 46, 1981): "The collective farms and state farms themselves should have the final say in deciding what should be sown on each hectare and when one job or another should begin." But in May 1982, in announcing the much heralded new food program, there was almost complete silence about the transfer of such authority to the farms.

The food program, as announced, was shockingly unimaginative. It must have been the case that general agreement on what was required to improve agriculture could not be reached. Instead, amid a mass of detail and numerous resolutions, what emerged was the creation of two new bureaucratic levels and a further sharp increase in prices paid to farms and, given the policy of fixed retail prices, an increase in food price subsidies of almost 50%. The increase in farm prices, including some special price increases for output produced under relatively poor conditions and unprofitable and low profitability farms, will cost 16 billion rubles a year starting in 1983. Thus, the food subsidy bill will reach at least 51 billion rubles annually.

Instead of giving the collective and state farms "the final say in deciding what should be sown on each hectare and when one job or another should begin," agriculture was subjected to two new bureaucracies. One is the agro-industrial association, organized at the district or province level, which is to be a single agency to manage all enterprises and organizations of agriculture, including the processing industry and input services. A higher level organization, called agro-industrial commissions, is to be created at the republic and union levels.

But hardly a word was said about the initiative or independence of the collective and state farms. The farms now have an additional master, who may or may not replace the others to whom each farm has had to be responsible. If one remembers that in all of the USSR there are but 46,000 collective and state farms with 24 million members and employees, it would be reasonable to assume that farms of such average scale could rather well manage their own affairs if given the opportunity to do so.

While Brezhnev apparently had to bow to the bureaucracies in his effort to give greater authority to the farms, he scolded the bureaucrats for their excessive meddling: "It's necessary to get rid of administrative fiat and petty tutelage with respect to collective farms and state farms, which can rightfully be called the foundation of all agricultural production. No one should be permitted to demand that farms fulfill any assignments not envisaged by the state plan or to ask them for any information except as established by state reporting requirements" (*Current Digest*, no. 21, 1982, p. 7). At least, Brezhnev must be given credit for trying.

China

At the present time it is very difficult to describe the socialist agriculture of China. Revolutionary changes are underway, and it is not obvious that any one in China knows when and how the changes will end. It is not at all clear, at least to me, how farming will be organized and administered in 1985.

As recently as 1977 on many, if not most, communes the members were directed in their daily activities by some official. Often times people were assigned tasks that were of little productive value, perhaps so no one would be idle. There was little or no relationship between work or effort and reward. Work points were allocated on the basis of time input and not on the amount of work performed. Learning from Dazhai, a production brigade with only 83 households, was a national campaign. It is perhaps typical of the cynical attitude toward farm people prevailing at the time that, while Dazhai was claimed to have made a productive garden spot out of a hilly mountainside solely on the basis of their own resources and achieved a very high level of output, it has now been revealed that the brigade received substantial assistance from the state and that for a number of years the output data of the Dazhai brigade and the county in which it was located were falsified (Zhou).

Under the commune system and related policy positions, including a severe agricultural price scissors (Huang), grain production per capita in 1977 was the same as in 1957 and, shocking to some, as in 1936 (Johnson 1982). Contrary to generally rosy views concerning the success in eliminating poverty, it has since been revealed that 100 million Chinese suffer from malnutrition and almost every year one or more parts of China suffer from famine or food stringency (Johnson 1981). And recent analyses have revealed that the distribution of income in China is no more equal than in other developing countries. Much of the enormous inequality arises for two reasons: First, urban incomes are at least double those of farm people, and, second, there is very great income inequality within rural China— not just regionally but from one production team to another in the same commune. Mobility between farm and city, from commune to commune, or from one production team to another within the same commune is extremely limited. Mobility from farm to city is restricted as a matter of policy; mobility within agriculture is limited because there is no incentive for the better-off communes or production teams to accept new members.

What are the revolutionary changes? The most important, by far, has been the introduction of the work responsibility system. The work responsibility system consists of a variety of arrangements designed to create a relationship between one's contribution to output—marginal productivity—and one's reward. Some of the more significant work responsibility arrangements are (*a*) payment for a specific task, such as transplanting a given area with rice or harvesting a crop on the basis of amount harvested, each to be done according to an agreed schedule; (*b*) allocation of land, seed, fertilizer, machinery, draft animals, and tasks to a group, with work points allocated on the basis of the amount and quality of work, with the value of the work points dependent upon output achieved; and (*c*) contracting with a household, or occasionally an individual, for the delivery of specific amounts of output to the state and supply of services by the production team which must be paid for with the contracting party retaining all output above that delivered to state or paid to the team.[4] The last—the household responsibility system—appears to be dominating all the others. In 1981 it was said that 20% of all farm households were on this system; in June 1982 it was said that 45% of farm households were participants. Originally, the household responsibility system was to be restricted to poor and mountainous areas, but this restraint no longer seems in effect.

The household responsibility system is a tenant or rental system, with the rent calculated in terms of product rather than money. For the output delivered to the state, the household receives the basic procurement prices and, possibly, higher prices for output in excess of the required sales or delivery quotas.

The responsibility systems, which seem likely to spread to almost all of agriculture, has greatly reduced the role of the commune and the production brigade as well. However, the production team (approximately 30 households) still has substantial influence because it continues to own and control the means of production other than labor, though there is evidence that some production teams have actually sold machinery, such as the small walking tractors, to their members. The ownership of the land remains with the production team, and the team retains the authority to reallocate land among households. Land is generally allocated according to the size of family.[5] If this criterion is maintained, there must be authority to reallocate land as total population and/or the sizes of individual families change.

The role of the commune may be changed under the new constitution by removing its political functions. Until now the commune has been both a governmental and economic institution.

Other changes have occurred since the fall of the Gang of Four in 1977. Prices of most farm products were increased by 20%–25% and a bonus of 30% to 50% was to be paid for deliveries in excess of the required quotas. Of crucial significance to the quality of life in rural and urban areas was the permission to reopen rural fairs and urban free markets at which farm people could sell farm products and handicrafts; these markets had almost all been closed down during the insanity of the Cultural Revolution as representing the tails of the capitalist dog. Production brigades and teams were permitted and even encouraged to create nonfarm activities as a means of providing employment and income; such nonfarm sideline activities had been severely restricted in the late 1950s and early 1960s and almost abolished during the Cultural Revolution (Johnson 1982).

Most of the recent changes are likely to result in increasing the already high degree of inequality in rural incomes. High income production brigades and teams have the capital to invest in nonagricultural sideline activities, which are far more profitable than farming. The suburban agriculture areas have been high income in the past; the reopening of urban markets has added to their income earning potential. The more energetic, intelligent, and educated can now come closer to receiving the value of the marginal product for their labor. The income inequality consequences are recognized in China; a headline in the *China Daily* on 11 June 1982 was: "Policy of Getting Rich with Honesty Will Not Alter."

Eastern Europe
The agricultural policy and institutional setting in Eastern Europe varies so much from country to country that it is impossible to deal with each of

the countries in a brief review. It is only possible to highlight some of the important differences as well as some common elements.

Agricultural output growth for the region was 22% for the 1970s. This is a quite respectable performance. However, in Poland output at the end of the decade was just 3% greater than at the beginning compared to a 33% increase in Hungary and 23%–24% in Bulgaria, Czechoslovakia, and East Germany. The reported increase of 59% in farm output for Romania seems difficult to reconcile with the current food problems confronting the country (USDA 1981). Poland's very low gross output growth occurred in spite of the large increase in grain imports during the decade.

All of the Eastern countries, except Yugoslavia, have had a major policy in common, namely the subsidization of food prices. I have dealt with this subject at some length elsewhere, so only brief note is made here (Johnson 1981). But commitments to hold the prices of some or all food products constant in the face of rising costs of production, rising money wages, and increases in the general price level have had two significant disequilibrating effects. One has been to encourage increases in the quantity of food demanded as the real price of food in the state stores has fallen. This policy encouraged rapid growth of desired consumption of livestock products and was responsible, at least in part, for the large increase in grain imports during the 1970s. Such imports were required if the imbalance between the amount demanded and supplied at artificially low retail prices was to be kept at politically acceptable levels.

The other effect was the pressure to increase agricultural output to keep up with the growth in the amount demanded. This necessitated keeping investment at a high percentage of national investment, though at much lower levels than in the USSR. But more important, the increases in prices paid to farms had to be paid for in whole or in part from the public treasury, not by consumers. The burden of the food price subsidies increased to such levels by the end of the 1970s that most of the countries had to take drastic steps and significantly increase food prices. The outcome of the Polish price increases announced in the summer of 1980 and generally delayed until early 1982 are well known. Price increases announced in 1970 and 1976 were almost all rescinded in the face of active and occasionally violent opposition. By 1980 the price subsidies for meat exceeded the amount paid by consumers, and agricultural subsidies equaled at least 20% of the wage fund (Johnson 1981).

Hungary increased food prices by a third in 1976 and by a fifth in 1979, though even these increases did not result in any decline in the absolute level of food price subsidies. Bulgaria increased food prices by about a third in 1979; Romania announced some price increases to take effect at the beginning of 1981, but public opposition resulted in postponing the increases; finally, food price increases averaging 35% were put in effect in

February 1982. Now Romania has bread rationing, though it is possible its purpose is not so much to limit human consumption as animal consumption encouraged by a highly subsidized price. Other foods are also being rationed in Romania.

The burden of the food price and agricultural subsidies in Poland became greater than the system could support. The viability of the economy depended upon increasing retail prices and nearly eliminating the price subsidies.[6] The price increases required were enormous, with retail food prices in early 1982 being double to quadruple the average prices for 1980. But even at these higher prices nonprice rationing prevailed, either officially or unofficially. In mid-1982 meat is officially rationed, with a ration for an ordinary citizen set at 2.5 kilograms per month. A year earlier the ration was 3.7 kilograms per month and in 1980 per capita meat consumption was about 6 kilograms per month (USDA 1982b). The existence of unofficial rationing is indicated by the sharp increase in prices in the free markets. Wheat prices in early 1982 were four times the average 1980 price, with similar increases for other grains.

Price increases of the magnitude indicated were possible only after martial law was declared. While there were a number of explanations for the deterioration in the relations between the Polish government and the citizens of that country in 1980 and 1981, one important cause was the pressure that the government faced in greatly reducing or eliminating the food price subsidies. The earlier mismanagement of food price increases in 1970 and 1976 and the general lack of trust between the Polish government and many of its citizens exacerbated an extremely difficult and complex situation that it would have been hard to resolve, at best.

Lazarcik has divided the Eastern Europe agricultures into two categories—centralized and predominantly decentralized agricultures. I agree with the distinction, which emphasizes control rather than ownership of land. The centralized agricultures are Bulgaria, Czechoslovakia, and the German Democratic Republic (GDR). A possible surprise in the predominantly decentralized category is Romania, since for the rest of the economy it is the most centralized of these countries. Since either 1965 or 1970 output growth has been significantly greater in the countries with decentralized agriculture. Net product has also grown more rapidly—20% greater in 1979 than in 1965 or 1970 in the decentralized agricultures and from no change to a 5% increase for the rest.

Not too much should be made of the two-way classification. The agricultures differ in many other ways. For example, the decentralized agricultural economies had a rather modest decline in farm employment between 1965 and 1979 of 22.5%, while the other countries had a decline of 32%. The centralized agricultures were much more capital intensive, having more tractors, for example, both per hectare of land and per worker.

But recent events in Poland will mean that the agricultural performance of the decentralized economies will lag behind that of the centralized ones. Agricultural production in Poland in 1980 was almost a fifth below 1978 and declined further in 1981. What has occurred in Poland illustrates the role of the general policy setting and its importance compared to whether agriculture is predominantly private or socialized.

Concluding Comments

It is possible that in the not-too-distant future the performance of agricultures in several of the centrally planned economies will be of such a nature as to call for significant policy changes. As noted, China is currently undergoing what can be described only as revolutionary changes in policy and organization. The concern over the ability of the Eastern European economies to service their external hard currency debt will put substantial pressure upon several of these countries to undertake significant agricultural policy changes, since output is unlikely to be increased during the 1980s through expanded imports of grain. Certainly it will be very difficult to borrow to purchase grain; the heavy burden of servicing even the current debt could put pressure upon the ability of some of the countries to maintain their current levels of grain imports.

The agricultural difficulties confronting the Soviet Union and Poland are probably the most critical and difficult of solution. The recent extensive review of agricultural matters in the Soviet Union was unable to come up with any significant steps that would prevent further deterioration in the performance of Soviet agriculture. Poland is confronted with a significant reduction in per capita food consumption, something few nations at her per capita income level have even been subjected to.

But very striking about the past and current difficulties of the agriculture and food policies of most of the centrally planned economies has been the role of one policy objective—namely, that of stable nominal prices for major food items, such as bread, meat, potatoes, and milk. This policy objective has had a major role in performance of the food and agricultural sectors during the past decade. It has generated rapid growth in desired consumption, a growth in demand that could not be met through expanded production. The policy objective ruled out the use of price, at least consumer prices, as the tool for equating demand and supply in the market. Experience indicates that it is exceedingly difficult to eliminate the food price subsidies. But until the subsidies are eliminated or reduced radically, agricultural policy problems will remain high on the agendas of economic and political groups in the centrally planned economies.

The real cost of the policy of stable nominal food prices is not solely the budgetary costs. Of equal or greater importance is that increases in prices paid to farmers require budgetary expenditures, often of enormous size, as witness the increases in the USSR to become effective in 1983. This means that farm price adjustments lag behind changes in costs. In general it appears that price adjustments cover cost increases that have already occurred and take little or no account of cost increases that will occur prior to the next price adjustment. Thus, for some or all farm products, farms are caught in a price scissors most of the time.

As we know, agricultural policies in every country are subject to controversy, and in many countries the budgetary and resource costs are very large indeed. But there are important differences in the outcomes in the socialized agricultures and those of the industrial countries. In the industrial countries output is rather greater than desired and resources required to produce the actual output in terms of labor and investment are a minor fraction of the economy's total. In the socialist economies, output is less than desired and the resources used to produce that output represent a substantial drain upon the economy.

But some changes have been occurring in socialist agricultures. This is particularly true in Hungary, Romania, and China. Only time will tell if similar changes will occur in other economies. What does seem certain is that unless major changes do occur, agriculture will continue to be a major drain on economic growth and a source of dependence on others.

Notes

The preparation of this paper was assisted by grants to The University of Chicago by the Rockefeller Foundation and the Prince Charitable Trusts, as well as by a contract funded by the National Council for Soviet and East European Research. The views expressed are the author's own and are not to be attributed to any other person or organization.

This paper, as well as everything else the author has written on Soviet agriculture, would not have been possible without the generous help of his late colleague, Arcadius Kahan.

1. It may be questioned whether community or tribal property systems, such as communal grazing in Africa, can be modified to achieve an efficient use of resources without such major modifications as to leave the systems unrecognizable. With this possible exception, I believe that all tenure systems can be made efficient if that objective is given high priority.

2. A nearly identical view is expressed by Zheng, a Chinese economist (p. 117):

"Modern agriculture's level of forward and backward linkages is very high. Its level of production efficiency is also conditioned by economic activities outside of agriculture. For example, the quality, price and supply availability of production materials, the ability of credit structures to provide managers with money to buy production materials, etc., all can influence the production efficiency of agriculture.

". . . raising agriculture's production efficiency is closely related to each agriculture's organizational abilities and quality of management, but the support for agriculture from the country and the national economy's other sectors cannot be ignored. Indeed, one can say that the latter is the more important aspect."

3. On state farms the increases were 37%, 55%, and 53%, respectively. The percentage increase in costs of meat production is a rough average of the increases for beef and pork. For collective farms the increase in costs of producing beef was 66% and for pork 51%; on state farms the increases were 65% and 43%. The estimates are made by Soviet official agencies.

4. Land is allocated to the family for its use; currently the land is apparently allocated for a single year. The production team provides certain services, such as plowing the fields, and makes available or sells current production inputs to the family.

5. The rapid expansion of the work responsibility system, especially the allocation of land to families, weakened the campaign to enforce the goal of but one child per family. When income was directly allocated by the production team, payments could be made to a family that agreed to have but one child and penalties could be imposed on families that had more than one child. With the widespread adoption of the family responsibility system, the birth rate increased in rural areas. It was noted in China in 1982 that at least some production teams were supporting the one-child-per-family campaign by refusing to allocate any additional land for any child in excess of one and by imposing a significant fine for the second or any additional child.

6. Price subsidies remain for milk and milk products (USDA 1982a).

References

Huang, Da. "Some Problems Concerning Pricing." *Social Sciences in China*, vol. 2, no. 1 (1981), pp. 136–56.

Johnson, D. Gale. *Agricultural Organization and Management in the Soviet Union: Change and Constancy.* Office of Agr. Econ. Res. Pap. No. 80:26, University of Chicago, 8 Aug. 1980.

———. "Food and Agriculture of the Centrally Planned Economies: Implications for the World Food System." *Essays in Contemporary Economic Problems: Demand, Productivity, and Population.* Washington DC: American Enterprise for Public Policy Research, 1981.

———. *Prospects for Economic Reform in the People's Republic of China.* Office of Agr. Econ. Res. Pap. No. 82:7, University of Chicago, rev. May 1982.

Lazarcik, Gregor. "Comparative Growth, Structure, and Levels of Agricultural Output, Inputs, and Productivity in Eastern Europe, 1965–79." U.S. Congress, *East European Economic Assessment*, part 2, pp. 587–634. Washington DC: Joint Economic Committee, U.S. Congress, 1981.

Strauss, Erich. *Soviet Agriculture in Perspective.* London: George Allen & Unwin, 1969.

The Current Digest of the Soviet Press. Columbus OH: American Association for the Advancement of Slavic Studies, various issues.

U.S. Department of Agriculture, Economic Research Service. *Eastern Europe: Re-*

view of Agriculture in 1981 and Outlook for 1982, Suppl. 3 to WAS-27. Washington DC, June 1982a.

———. *USSR: Review of Agriculture in 1981 and Outlook for 1982.* Suppl. 1 to WAS-27. Washington DC, May 1982b.

———. *World Indices of Agricultural and Food Production.* Statist. Bull. No. 669. Washington DC, July 1981.

Wadekin, Karl-Eugen. *Agrarian Policies in Communist Europe: A Critical Introduction.* Totowa NJ: Allanheld, Osumun & Co., Publishers, 1982.

Zheng, Linghuang, "Agricultural Modernization and Agricultural Production Efficiency." *Social Sciences in China*, no. 3 (1981), pp. 104–20.

Zhou, Jinhua. "Appraising the Dazhai Brigade." *Beijing Review*, no. 16, (20 Apr. 1981), pp. 24–28.

11

Economic Reforms in the People's Republic of China

Introduction

The emphasis in this paper is primarily on rural economic reforms. Industrial reforms are treated quite briefly. The primary reason for this allocation is that the agricultural reforms have been much more sweeping than the industrial reforms and have had more striking consequences. Another reason, of course, is that China remains a primarily rural country, with 80% of the population living in rural areas. A final reason is that there has been much trial and error in urban and industrial reform efforts but nothing comparable to the rural reforms has yet emerged. This is not to say nothing has changed in the urban area; there have been changes that are important to the ordinary citizen. But these changes have represented desirable adjustments within the framework of the existing system, such as increasing the hours that stores are open or extending to the city some of the benefits of the rural reforms, such as permitting the opening up of markets where farmers can sell their products directly to urban residents.

It is an infrequent occurrence in human history when there are radical changes in political and economic institutions. When such changes do occur, they are generally the result of armed conflict and revolution, such as the 1917 revolution in Russia, the 1949 revolution in China, or the eighteenth century French Revolution. While the American Revolution had major political consequences, it had very little effect on our economic institutions. But there has occurred a peaceful revolution in rural China at least as radical in its effects as the 1952 land reform in China, the establishment of the communes in 1957, or Stalin's brutal farm collectivization program of 1929.

Reprinted by permission from *Economic Development and Cultural Change* 30, no. 3, supplement (April 1988): S225–45. © 1988 by The University of Chicago. All rights reserved.

The status quo is very resistant to change; one needs only to look at some of our own policies or institutions to confirm the statement. Let us remember how long we stayed with a misguided energy policy. Thus the peaceful transformation of economic and political institutions in rural China is a most unusual achievement with few, if any, parallels in modern history. So very much has changed in a very brief period of time, and there is no evidence that the reform process has been completed.

Since the end of the Communist Revolution in 1949, there have been few years that could be called normal. From 1949–52 was a recovery period, and substantial progress toward a return to normalcy was achieved. The high inflation rate of the late war period, which was partially responsible for the downfall of the nationalists, was successfully tamed. The land reform in the countryside was completed. About a third of the rural population was in liberated areas (liberated from the Japanese and/or the Nationalists) prior to 1949. The remaining two-thirds of the rural population underwent land reform soon after 1949.[1] Consequently, by the early 1950s most Chinese farm families had rights to their own land, though these rights were rather quickly violated by subsequent agricultural reforms.

The first 5-year plan (1952–57) is generally considered to have been an economic success with rapid growth of national income of about 9% annually. This was a period of change in the organization of agriculture. At the conclusion of the land reform, there were nearly 100 million individual farms that were owner operated. But, as has been true in most communist economies, farm families were soon pressured or forced into socialized farming units. There was first the creation of mutual aid teams with the sharing of draft power and machinery. Then cooperative farms were created, in which land was merged and farmed together. By 1957 most farms had joined cooperatives. Then suddenly all of the cooperatives were transformed into communes.

The Great Leap Period—1958–60—must rank as one of the most disastrous in the history of the world in terms of its effects on a people in time of peace. The individual farms and mutual aid teams were forced into the communes in 1958. This was the year of the emphasis on the backyard furnaces for producing iron; household utensils were melted down to produce useless materials. This period has been described by a Chinese economist, Xu Dixin, as follows:

> Beginning in 1958, however, "Left" errors, characterized by one-sided emphasis on high speed, spread unchecked throughout the country. Zhou Enlai's rational proposals were negated. The drive for a "Great Leap Forward" was launched, calling for a doubling of output within one year. Ideas like the calls for the large-scale production of iron and steel and for a per hectare yield of over 75

tons, the belief that "the more man dares, the more the land will produce," and the practice of free supply of food all gained widespread credence.

In industry, the central authorities gave arbitrary directions and set excessively high targets for production and construction, which were further raised at local levels. This led to serious imbalances in the economy.

In agriculture, the advanced agricultural producer cooperatives were hastily amalgamated to form the rural people's communes. These rapidly increased in size so that in some cases they even embraced the population of an entire county. The people's communes established three levels of ownership and management: the commune, the production brigade, and the production team. But in the initial period, distribution and accounting were carried out at one of the upper two levels, remote from the production team which actually did the work. A "communist" wind was stirred up, whereby egalitarianism prevailed and human and material resources were transferred at will without regard to the actual collectives to which they belonged. All this dampened the enthusiasm of the peasants and cadres at the grassroots.

These erroneous policies brought grave consequences. Production not only did *not* double, but the rate of growth actually declined. From the end of 1958 to the early stage of the Lushan Meeting of the Political Bureau of the Communist Party's Central Committee in July 1959, efforts were made to rectify the "Left" errors that had already been recognized. In the later stage of the meeting, however, a struggle was launched to oppose "Right opportunism," and this was quite the opposite of what was needed. Plan targets were raised even higher. The countryside again suffered under the policy of egalitarian distribution. Both agricultural and industrial production further deteriorated.[2]

A young man I met in 1983 discussed the early months after the formation of the communes. There was a brief period in which all members of a production team ate together, with the food being provided from a communal kitchen without charge—the free supply of food referred to in the quotation. For a few months, as a young boy, he found the system a good one—apparently he enjoyed eating with his friends, and the food was both good and adequate in quantity. He then stopped in his story and when he continued, he blurted out: "And then we starved." He was referring to the beginning of the great famine that affected large areas of rural China, with an estimated loss of life of 30 million.[3] There were many reasons for the famine, though bad weather was of only limited importance. An important factor was that many Chinese officials believed the outrageous claim that 1958 grain production was double that of 1957. Literally, too much was eaten in the months following the 1958 harvest, and the year's food supply was nearly exhausted in the first few months of the crop year. The system of communal feeding simply did not have the appropriate market signals to indicate that supplies were being used too rapidly until it was too late. As implied by the young man, there was more than adequate food for several months and then there was virtually no food available and millions starved in 1959 even though the 1958 grain crop was a good one, somewhat larger than the record 1957 crop.

In 1961 a start toward economic sanity was made and the outrageous production goals promulgated during the Great Leap were lowered, construction projects were curtailed, and the most repressive of measures affecting rural people were modified. The production team, rather than the commune or the brigade, was made the basic unit for production and distribution and egalitarian distribution was opposed. The family sideline occupations of commune members were restored—for centuries most rural Chinese had engaged in many household activities that earned money or permitted trading for desired objects—and the rural markets were permitted to reopen.

The effect of these and other measures was rapid revitalization of the economy—from 1963–65 the average annual increase in national income was 14.5%; agricultural output increased by 11%.

But 1966 saw the beginning of the Cultural Revolution. Xu writes: "Just as the economy was beginning to develop, yet another large-scale disruption took place in the form of the 'Cultural Revolution.' Seizing some of the leading positions in the Party and the state, Lin Biao and the 'Gang of Four' . . . wrought terrifying havoc on the national economy through their so-called 'cultural revolution group.' This disruption lasted throughout the Third Five-Year Plan (1966–1970) and the Fourth Five-Year Plan 1971–1975)."[4] After noting the sharp decline in industrial production in 1965 (13.8%) and in 1968 (5%) Xu adds: "Although the production figures for 1969 and 1970 showed some increase, they were inflated."[5] Numerous Chinese economists note that economic data for the period of the Cultural Revolution need to be interpreted with caution.

When did the Cultural Revolution end? One date was established by the arrest of the Gang of Four in 1976. But normalcy did not return until 1978. In the time between the arrest of the Gang of Four and the meeting of the Central Committee of the Communist party in December 1978, many of the mistakes of earlier periods were repeated. Xu describes policy during the period: "Excessively high targets were set and the scale of capital construction, which was already at a level beyond available resources, was further extended. All this greatly intensified the imbalances in the economy and added to the financial difficulties."[6]

Many policy mistakes were made in 1977 and 1978. In the rush to undo past errors of overcentralization, many moves were made to decentralize decisions in the urban sector. The decentralization was not well thought through and many important features of the previous period were left unchanged, such as product prices, wage rates, arbitrary labor assignment to enterprises, and the "iron rice bowl." One outcome was a further increase in the share of the national income devoted to investment—it rose to a level approaching that of the Great Leap Period. In 1978, investment was 36.5% of national income, an allocation of resources that is now recognized as being inappropriately high.

China's Agriculture in 1978

In 1978, the year that the rural reforms were under serious consideration, there were 294 million people employed in agriculture. These people were organized into 52,780 communes; each commune was organized into an average of 13 production brigades, which in turn were made up of seven to ten production teams. The total population of the communes was a little more than 800 million or an average of about 15 thousand per commune. Each production team averaged about 60 workers and 35 households. A commune covered the approximate territory of a township, approximately the same size as a township in the American Midwest.

The commune was both a political and economic organization. It held most of the local governmental functions, including police, justice, welfare, administration of family planning programs, and operation of schools and hospitals. The commune also had a monopoly of economic functions—assignment of production plans to brigades and teams, allocation of procurement quotas, and control over the leadership of brigades and production teams. The commune, including all of its units, determined who got how much pay in kind and in money and who got assigned to specific jobs, even on a daily basis. Given the numerous restrictions that had grown up during the Cultural Revolution on all forms of private activities, such as use of private plots, handicraft production, and other private sideline activities plus numerous restrictions on rural fairs and markets, the collective organizations had enormous power over the lives of rural people.

What were the economic results of the commune system, dating from 1958 over 2 decades? Grain accounts for more than 80% of all calorie consumption in China. As put by a famous Chinese economist, Xue Muqiao, "During the ten years of the Cultural Revolution, both industry and agriculture crawled at a slow pace. The per capita grain output in 1977 was roughly the same as in 1957 and total cotton output remained at the 1965 level."[7] Soybean production was less than in the mid-1950s by approximately 25% in 1978. While the absolute level of cotton output was the same in the late 1970s as in 1965, per capita production declined by a quarter.

One striking claim made by Chinese officials, and echoed by most Western visitors to China from 1960 to 1977, was that while per capita income growth was modest, the inequality of income distribution was greatly reduced. A related claim was that famine had been conquered and no one ever went hungry. On both of these points it was common to compare India with China in a very unfavorable light. After all, anyone who had been in India saw poor people, including the homeless, everywhere. But you didn't see beggars or obviously hungry or very poor people in China's cities. Thus the surface indications seemed consistent with the claims.

There was substantial reduction in income inequality in urban areas and within rural production teams. But it is now clear that the large differences between urban and rural incomes that prevailed before the revolution were not reduced nor was significant progress made in reducing income inequality among villages or from one region to another. Again quoting Xue: "For over twenty years, the ratio between the living standards of the workers and the peasants has basically stood at about 2 to 1. It has dropped a little where agriculture has developed faster and has risen where agriculture has made little progress. It exceeds 2 to 1 in most areas and is as high as 3 or 4 to 1 in some areas."[8]

With respect to changes in income differences within the rural areas, Xue has written the following: "The differences in living standards among peasants are even more pronounced than those among the workers or between workers and peasants. In the more than two decades since the completion of the movement to set up agricultural producers' co-operatives, differences between communes, brigades and teams have not narrowed but have continued to widen."[9]

As for the elimination of famine and hunger, the evidence is clear—it didn't happen. As noted earlier, China suffered an enormous famine from 1959–61, when as many as 30 million people died not so much as a result of bad weather as of inept policies and Mao's giving greater weight to national self-reliance than to avoidance of human suffering. Maintaining independence from the international community may or may not be a meritorious policy, but in no case can the deaths of millions be justified by emphasis on such a political ideal.

A few other comments about hunger and poverty are appropriate. A Chinese official reported in 1979 that 10% of the Chinese population did not have enough to eat; this would have been approximately 100 million people. It was reported that in Anhui Province, "There are many people in the villages who have not enough to eat or enough clothing to keep them warm."[10] The *People's Daily* stated that "in some parts of the country containing in all 100 million people, there has never been a good life since the collectivization and production has not picked up since the three years of economic collapse. In these regions the population has increased but the amount of grain has not increased. . . . The peasants have lost their faith in collectivization."[11]

But the picture of life in the countryside as of the late 1970s should not be painted in too bleak terms. While it is true that the real income available for expenditure by the peasants increased by only a few percent between 1957 and 1977, there is evidence of substantial improvement in the living conditions of rural people. The evidence is found in the sharp increase in life expectancy, including a large decline in infant mortality, increased availability of schooling, and increased literacy. Life expectancy at birth was 34 years during 1950–54, increasing to 64 years in 1975–80.[12] Infant mortality declined from 236 per thousand to 65 per thousand for the same periods, respectively. The number of children in elementary school in 1950 was about 27%, increasing to 67% in 1960 and to 90% in 1980. Secondary school enrollment was about 20% of the relevant age group in 1960, increasing to 40% in 1980.[13]

Agricultural and Rural Reforms

Briefly, the rural reforms started with decisions taken at the December 1978 meeting of the Chinese Communist party Central Committee. The reforms introduced in 1979 can be described as both conservative and appropriate. They were to function within the existing structure and were designed to make that structure work better. There was no hint to indicate that the commune was to disappear; it was to be reformed. The three years—1979–81—were to be devoted "to readjusting, restructuring, consolidating and improving the national economy." Agriculture was to be given priority, and in 1979 substantial increases in farm prices were made; taxes imposed on agriculture were reduced; agriculture investment was planned to increase from 10.7% of the total in 1979 to 14% in the near future and later to 18%, and increased emphasis was to be given to agricultural mechanization and input supply to agriculture. Prices of most farm products were increased by 25% and premia of 30%–50% were paid for deliveries in excess of procurement quotas. Each commune, brigade, and team had procurement quotas that they were required to meet.

Mao's policy of regional self-sufficiency in grain production, which had disastrous effects on the environment as well as on agricultural productivity, was to be abandoned and production was to be specialized according to local conditions. Cotton and soybean production, which had been neglected under the grain first policy, were to be encouraged. Village fairs were given the green light and private plots were once again to be private and other forms of private production were encouraged. The amount of grain, over and above local consumption needs, that had to be sold to the state was reduced from 90% to 70%.

The most radical of the reforms were the introduction of various forms of the responsibility system and the subsequent abolition of the communes. There is considerable evidence that some of the most striking rural reforms were unplanned and perhaps unanticipated. The responsibility systems were almost certainly created from the "bottom up" rather than from the "top down" as so generally occurs in centrally planned economies. In numerous cases, policy has simply sanctioned locally initiated experiments that proved to have positive effects on production and income.

After some local experiments, apparently without central government sanction, proved that various forms of the responsibility system greatly increased production and incomes in poor areas, the adoption of such systems were approved for application, apparently only to the poorest fifth of the agricultural areas. But once such approval for limited application was given, responsibility systems spread to virtually all of agriculture by 1982.

At the start there were many different responsibility systems. As noted earlier, in the communes there was little relationship between how hard and well one worked and one's income. The responsibility systems attempted to create quite direct relationships between productivity and reward. This was done through a variety of means—paying on a piece rate basis for particular farm operations, assigning land and production responsibility to a small group and giving the group the value of all output in excess of a specified amount, and assigning land to a household (or an individual) that was responsible for fulfilling the procurement quotas, paying the land taxes and making modest payments to the production team. There were many other forms of responsibility systems, but today most of the land in China is farmed under the household responsibility system.

What has emerged in China is a reasonable facsimile of the family farm. The families do not own their land, but they have been given a long-term right of use, generally for 15 years following a 3-year period. The commune has been abolished, and its political functions have been taken over by township governments.[14] I know of no evidence that it was anticipated in 1980 that the adoption of the household responsibility systems would result in the abolition of the communes.

One consequence of the dissolution of the communes and the implementation of the various responsibility systems has been that it has been discovered that there are far more workers in the villages than are required to do the farm work under current conditions. In villages visited in 1985, it was said that at least one-third of the farm workers were no longer required for farm activities. Thus it has been both necessary and possible to create nonfarm employment sources in the rural areas. Some of the enterprises are run by villages; some of them are run by individual households or a small group of individuals. And there has been a marked response. The output growth of sideline activities has surpassed the rapid rate of growth of farm output. Between 1978 and 1984, the annual growth rate of farm output was 7.2%. Individual or household sidelines increased at an annual rate of 11.8% while village-run industrial enterprises' gross output grew at an annual rate of about 30%. I shall later return to the problems that must be solved in terms of increasing nonfarm employment in rural areas if the current policy of virtually prohibiting migration from the countryside to the city is maintained.

There were many other and important features of rural reforms.[15] These include strong approval of local fairs and markets; approval of free markets in the cities; permitting farm households to become specialized in some particular activity rather than requiring peasants to engage in grain and crop production; encouragement of a wide variety of nonfarm pursuits such as operating eating places, repair services, and manufacturing plants; and providing transportation services (one of the most profitable enterprises in the countryside).

As noted earlier, in 1979 farm prices were increased for the obligatory procurement quotas and substantial premia (up to 50%) were paid for deliveries in excess of the quota. The quotas were at reasonable levels and a large fraction of procurements of grain were at the above quota price. But success creates problems. Farm production apparently increased at a much greater rate than anticipated. One result was that the government was procuring more grain and cotton than it knew what to do with. But Chinese officials have very good imaginations—along with the ability to rationalize even quite abrupt changes in policies. Under the procurement system with its required deliveries there was a commitment to purchase whatever amount was offered to the procurement agency. The government found itself purchasing more than it wanted at ever increasing average cost.

The official answer was that the government desired to increase the role of the market in the farm sector, and it was going to do so by abolishing procurement quotas and introducing a contract system. Obligatory cotton deliveries were halted in 1984, and in 1985 a contract system was introduced in the procurement of grain.[16] The multiple price system was replaced by a single price for grain at the average procurement price of the prior year. There was no longer an obligation for a farm household to deliver any grain or most other farm products to the government, and very important, neither was there any obligation on the part of the government

Table 1. Per Capita Incomes and Expenditures of Peasant and Worker Households, People's Republic of China, 1957 and 1978–86

	1957	1978	1979	1980	1981	1982	1983	1984	1985	1986
Income:										
Worker	235	316	–	–	500	495	527	608	752	890
Peasant	73	134	160	191	223	270	310	355	398	425
Expenditure:										
Worker	222	311	–	–	457	471	506	559	690	828
Peasant	71	116	134	162	191	220	248	274	317	368

Sources: State Statistical Bureau, People's Republic of China, *Statistical Yearbook of China,* English ed. (Hong Kong: Economics Information and Agency) 1984 and 1986; U.S. Department of Agriculture, Economic Research Service, *China: Outlook and Situation Report,* RS-84-8, June 1984.

to purchase a specified amount, either in total or from individual units. In 1984 the government purchased in excess of 100 million tons of grain; in 1985 it signed contracts for somewhat less than 80 million tons. In 1984 cotton production was 6.1 million tons, up from 2.2 million tons in 1979. For 1985 the government stated in advance that it would sign contracts for just 4.25 million tons of cotton or only 70% of the 1984 production. Cotton production in 1985 declined to 4.15 million tons.

The single price and contract system has been introduced for oilseeds. The prices of some of the less important crops have been entirely freed, and some livestock prices are now also free. Thus the free market may involve a larger fraction of farm output in China than it does in the United States, Canada, or the European community.

Effects of Reforms on Farm People

For many millions of farm families a very important aspect of the reforms has been the relaxation, if not the elimination, of restraints on how they use their resources. Farm families are now free to engage in a wide range of nonfarm activities; many farm families (more than 25 million) have become specialized farm households and are now no longer required to produce according to state plan or to deliver products to the state. Farmers are now permitted to own tractors, trucks, and farm machinery; true, most of the tractors are used primarily for transportation and not for farm work but this says more about the failure of the system to produce enough trucks than about the allocation of resources available to farm people. Farm people can sell products directly to consumers in urban as well as in rural areas; the products can be handicrafts, furniture, cloth, or garments as well as farm products.

The incomes of farm people certainly have increased since 1978, though by how much is uncertain. In current yuan the per capita income of peasants was 134 in 1978 and 425 in 1986, as shown in table 1. Prices of purchased items increased during this period, but the greatest uncertainty concerns the changes in evaluation of income in kind. A reasonable guess is that real incomes of farm people doubled between 1978 and 1986.

Has the large gap between urban and rural incomes been closed? The per capita income of workers' families increased from 316 to 608 yuan between 1978 and 1986. These raw figures imply that the income gap narrowed substantially. In 1978 worker income was 2.35 times the peasant income; in 1984, the ratio was 1.7 times, implying a substantial narrowing. But by 1986 the worker-peasant income ratio had increased to 2.1. However, none of the differentials adequately measure the differences between urban and rural incomes. Urban workers receive a wide range of subsidies that are not available to farm people. Urban housing, transportation, medical care, pensions, vacations, the cost of visiting certain family members, energy, and food are heavily subsidized. Most of these subsidies are not available to peasants. Ma Hong, a very influential economist in China, reported that for employees in state enterprises such subsidies equaled 82% of the wage received in 1978.[17] Lardy estimates that the urban subsidies increased significantly between 1978 and 1982, and in 1982 the value of urban subsidies per worker actually exceeded the average wage.[18] One reason for the increase in urban subsidies was that the 1979 price increases for grains and vegetable oils were not passed on to consumers for cereals and vegetable oils. In addition, to offset retail price increases for pork a per capita subsidy of approximately 7 yuan per month was paid.

It is one of the ironies that in an economic system that claims to be egalitarian, numerous and large subsidies have gone not to the least well off but to the best off. This is exactly the case for subsidies that go primarily to workers in state enterprises, including governmental units. Even within urban areas most such subsidies do not go to workers engaged in private activity. The per capita subsidies for urban families working for state enterprises in 1982 were about 300 yuan; the subsidies to peasants were 10 yuan. This is a shocking discrepancy unrelated to either equity or productivity.

The increase in production of agricultural products of 52% between 1978 and 1984 was the result of much greater productivity, as well as an increase in inputs or resources (see table 2). If we accept the official statistics, the increase could be about equally divided between more inputs (primarily labor and fertilizer) and a higher level of productivity. The recent increase in productivity is in sharp contrast to the period of the communes when productivity actually declined, an almost unparalleled development and a sharp condemnation of the communes and Mao's agricultural policies. As expected, the agricultural output growth rate slowed down in both 1985 and 1986 with only a 3% and 3.5% increase, respectively. It is reasonable to assume that the rapid productivity gains of about 3% per year have

Table 2. Indexes of Output, Inputs, and Total Factor Productivity of Agriculture, People's Republic of China, 1952 and 1977–86 (1978 = 100)

	Farm Output	Farm Inputs	Total Factor Productivity
1952*	(43)	(40)	(109)
1977	92	95	97
1978	100	100	100
1979	108	103	105
1980	109	107	102
1981	115	111	104
1982	128	116	110
1983	138	121	114
1984	152	125	122

Source: Anthony M. Tang, *An Analytical and Empirical Investigation of Agriculture in Mainland China, 1952–1980* (Taipei: Chung-hoa Institution for Economic Research, 1984), table 3, pp. 73–83; table 4, pp. 84–86; table 5, pp. 87–91.

Note: The official Chinese measure of the gross value of agricultural output includes the value of output by brigade and village run industries. The measure of farm output in this table excludes the output of these enterprises since we do not have adequate input measures. If the output of these enterprises were included, the index of agricultural output for 1984 would be 167. It should be noted that handicraft and industrial output of households is included in the output measure in the table. The farm input indexes are extensions of Tang's estimates in tables 4 and 5. The prices used as weights are those of 1952 prices for current inputs and capital. The aggregate weights are 50% for labor, 25% for land, 10% for farm capital, and 15% for current inputs. The index of farm output in all years but 1952 is in terms of 1970 prices. The link between 1952 and 1978 is based on output measured in 1952 prices. Compared to all other difficulties with the data, the error introduced by using somewhat different prices weights for output and inputs is probably quite small.

*The data for 1952 are not directly comparable to those in the rest of the table. The indexes are taken directly from Tang's table 7, based on the newly released Chinese data, made available as his book was nearing completion. The link was made with his data for 1952 and 1978. His output measure includes the output of the village and brigade run industries. Such enterprises were unimportant in 1952, but in 1978 accounted for 11.6% of the gross agricultural output value. If the output of these enterprises is excluded in 1978, the 1952 output index would be 49 and the index of total factor productivity 122.

come to an end and that future output growth will depend much more on increases in inputs.

Unfinished Business

The remarkable success of the agricultural reforms should not leave the impression that there are no serious problems on the horizon. Unfortunately for the farm people of China such is not the case. Some unfinished business has been inherited from the past, such as the grossly inadequate rural road system. But other problems were created by the reforms themselves.

It is not too much of an exaggeration to say that the Chinese government has abrogated many responsibilities for carrying out appropriate governmental functions in rural areas. Under Mao's commune system government seemed to be omnipresent, intervening in almost all aspects of peoples' lives. But with the abolition of the communes, the new county and township governments seem to have neither the authority nor the inclination to provide minimal desirable levels of certain governmental functions. It seems that there has been a shift from one extreme to another—from too much governmental authority to too little exercise of such authority. The abolition of the communes, which was part of a concerted and desirable effort to separate political and economic authority, left a governmental vacuum in rural areas. Thus a number of functions that can only be effectively carried out by government and by units that have the authority to tax either are not being carried out or are being seriously neglected.

With the abolition of the communes, a major source of resources to finance medical and hospital facilities, schools, maintenance of irrigation and flood control facilities or the expansion of processing and marketing facilities was lost. Consequently there is some evidence that the amount, availability, and quality of health care has declined in rural communities. It has been suggested that infant mortality has increased since 1980, though the evidence for this assertion is not very strong. There are complaints that irrigation canals and facilities are not being adequately maintained. Individual farm families with less than two acres of cropland cannot accept the responsibility for maintaining canal irrigation. Maintenance of such canals requires either a governmental unit or some association that has the authority to collect fees and make the necessary expenditures. The communes and/or brigades could perform these functions. Apparently in some rural areas no responsible institution has been created to carry out the required maintenance of irrigation facilities.

Earlier it was noted that when the reforms were first announced, state investment in agriculture was to increase from 11% of total state investment to more than 18%.[19] Even 18% of total state investment going to agriculture and rural areas reflects an enormous urban bias in China's policies. But instead of agriculture's investment share increasing, it was cut in half by 1984 when agricultural investment was a mere 5% of total state investment. Prior to 1980, a large fraction of state investment in agriculture was allocated to water conservancy and irrigation investment—a very appropriate governmental function. Between 1976–79 and 1982 the amount of such investment declined by 45% and even with an increase in 1983 remained a third below the prereform level. It is not obvious who the state planners expected to offset these declines.

Let us now turn to specific items of unfinished business, in addition to the problems referred to earlier. Space permits only a limited number of examples.

As economic growth occurs, farming provides a declining share of total employment opportunities in the economy. This is a continuing, long-run problem in addition to sharp reduction in the number of farm workers required to perform farm work with existing technology resulting from the abolition of the communes. In other countries, farm people find alternatives to the decline in farm job opportunities by either migrating to cities or towns or engaging in nonfarm work while retaining their farm residence. In China migrating to cities, especially the large cities, is severely restricted, one might say almost prohibited. Chinese officials talk about developing new moderate-sized cities in rural areas or rapidly increasing the population of towns and small cities. It is recognized that it may be necessary to find nonfarm employment for 100 million farm people within the next two decades. Where is the capital required to create these jobs to come from? So far as I can tell, the capital is to come primarily from the rural areas with very little state investment planned. This is in sharp contrast to the frenzied construction program that one sees in Beijing and most other large cities.

There are serious implications to this neglect. The investment for nonfarm jobs must come almost entirely from local savings. For farm people who live near large cities and have relatively high incomes, this expectation is a realistic one. But for the majority of less favorably located farm families who do not have easy access to urban markets and with relatively low incomes, such an expectation is most unrealistic. Local financial resources will not be sufficient to provide the necessary capital. Consequently, area and regional income inequality—which is already very substantial—will grow over time.

According to measures of income inequality published in *Beijing Review*, there has been an increase in income inequality in the countryside since 1978.[20] In urban areas income inequality has declined. These differences may well help explain why rural reforms have been so successful in increasing production and urban reforms have been largely ineffective. But the increased inequality of income in rural areas does not mean that most low income people have failed to achieve some improvement in their real incomes. The lowest third of rural income recipients were clearly better off in 1984 than in 1978, though available data make it impossible to say by how much. In 1978, 35% of the peasants had per capita incomes of less than 80 yuan per year. In 1984, only 16% had less than 150 yuan per capita. Thus even if prices almost doubled, which is almost certainly too large an estimate, the real income of the many poor peasants increased by a significant amount. True, millions are still very poor, but less poor than a few years ago.

There is danger in the growing inequality in rural incomes. Egalitarianism is still a strong ideological position in China. Increasing income inequality may be used as an argument either to withdraw many of the rural reforms or to halt further experimentation. This will be true even though the increased inequality, especially if it is largely due in significant part to greater regional inequality, results from distortions and market failure in the rest of the economy rather than from the rural reforms themselves. I refer to such matters as the strict limits on migration from the countryside to the city. Inadequate transportation of farm products, monopolistic barriers to the movement of farm products from one area to another, failure to develop nonfarm employment opportunities in rural areas that are disadvantaged by their location, long-term neglect of processing and storage facilities for farm products act to increase the disparity between villages located near to cities compared to those located some distance away.

Limited efforts are being made to alleviate some of these problems. China has approximately 600,000 kilometers of rural roads compared to 5.2 million kilometers in the United States.[21] Only a fifth of the Chinese rural roads are asphalted. Between 1985 and 1987, construction of 96,000 kilometers of new rural roads is planned. The scheme for building the rural roads is described in a document issued by the Central Committee of the Communist party of China on January 1, 1985, as follows: "Continue to build highways by adopting the methods of 'hiring semi-voluntary local workers with low pay in grains or clothes' or 'running by the local people with state subsidies.'" The following was also stated: "In areas where the financial condition is comparatively well developed, the State advocates to build highways by collecting funds from the society and adopting the principle of 'those who invest shall derive benefit from it.' In the mountainous areas and those areas with difficulties, the fund for building highways shall be financed by the local region and the labour be provided by the peasants. The State will provide a certain amount of grains, cotton, clothes to be used as investment for building the highways and provide a portion of materials such as drill rods, dynamites, etc."[22]

The investment policies that are being applied to road building in rural areas differ greatly from the policies applied to construction of office buildings and apartments in Beijing and most other large cities in China. While it is good that the Chinese government recognizes the need to build roads in rural China, the differences in the financing of investments in rural roads and in construction in cities clearly reflects the very strong urban bias in official policy. It is perhaps worth noting that much of the road building in the rural areas of the American Midwest and the South before 1930 was by the use of "semivoluntary" labor—farm people could pay their poll tax by road building work.

Reforms Are Not Completed

The process of rural reform in China has shown a great deal of flexibility, trial and error, and pragmatism. In the 1970s there was a saying, which was a source of some political consternation of the times: "I don't care if the cat is red or black as long as it catches mice." Many experiments have been conducted, some with official sanction and some without. But if something has worked, it often became a new policy that was sanctioned.

There is evidence that the rural reforms are still incomplete. Let me give an example from a village near Shenyang in northeastern China that I visited in 1985. Officially there is a limit on the number of workers that can be hired by a family—the limit is of the order of seven. However, there are many cases in which this limit is exceeded in the operation of industrial facilities, such as a textile plant. In such cases the building and machinery are owned collectively and those who operate it are paying for the facilities; thus this is defined as a collective and not private enterprise, and there is no limit on the number of employees even though they are being paid wages in the same manner as in private enterprises anywhere. In 1983 I visited a village in which three families were operating a textile mill that had more than 100 employees and this seemed not to be a matter of concern to any one.

What was new in the 1985 experience was that in a township that had 650 farm workers, there were just 27 families who were farm operators. Ten of the families provided all of their own labor. The remaining 17 families, designated as Big Families, employed the remaining 625 workers. The Big Families paid the village for the use of land and the agricultural taxes. The two payments came to 102 yuan for one-sixth of an acre or, at the 1985 exchange rate, about US$230 per acre.

I asked to talk to one of the Big Families and the request was immediately met. In this family there were three adult workers, including the man who met with us. His family operated 7.5 acres of land, of which a little more than 5 acres was under plastic for growing vegetables, and had 52 hired workers. The plastic, which lasts only one year, cost 3,600 yuan per acre or about 20,000 yuan (about US$7,400) for this one family. The Big Family was committed to pay the workers at least the average income the workers had received in the previous two years. The family must pay the wages regardless of the crop, but the risks in growing vegetables under plastic were probably not too great. There was an agreement that if the operation were profitable, 80% of the net profits was to be paid to the workers and 20% to the Big Family. It was not clear if this arrangement was compulsory for the Big Family or not.

But this was not the whole story; in fact, it was only about half the story. In October 1984, the same family had taken over the operation of a factory that produced auto locks. The factory had been established the previous April by the production team to provide employment for workers that were no longer required for farm work after the responsibility system was introduced. The family employed 56 persons in the factory. There were 778 families with 1,600 workers, including the agricultural workers, in the township; thus this one family employed 108 workers or nearly 7% of all the workers in the village. The Big Family that we visited did not have the largest number of farm workers; one family had 80 farm workers.

This form of organization of agriculture does not have official approval, but it is well known to governmental officials in Beijing that it exists. It has not received approval because it does not seem politically productive to call attention to what many would call a deviation from socialism. But if it results in increased productivity and higher incomes, it may well be sanctioned at some future date. Since the land and means of production (buildings and machines) are collectively owned, the system can still be said to be socialistic.

Urban and Industrial Reforms

As is evident, I believe that rural reforms have been remarkably successful on virtually all counts. The food supply of China, at least for the next decade, now seems assured. The shift from being a net agricultural importer in the early 1980s to a significant net exporter by the mid-1980s can only be described as remarkable. This is particularly true since there is considerable evidence of a substantial buildup of stocks of cotton and grain. China may have achieved the enviable status of having even larger unwanted stocks of cotton than the United States. And the incomes of most farm people have increased, and some have increased very substantially. True, there has been an increase in inequality, but if one is to believe the available official data, there has been a significant increase in real incomes of the poorest third of the rural population.

As noted, there may not have been a significant reduction in the urban-rural income differential since 1978. I mention this because it is relevant to the prospects for urban and industrial reforms. There appears to have been great unwillingness to make any changes that would reduce the position of relative privilege that has been bestowed upon urban workers and staff over the past 4 decades. And this is almost certainly a major reason why urban and industrial reforms have been of relatively little significance, especially as compared to rural reforms. But it is not the only reason, as I will note.

I have little to say about urban and industrial reforms. One reason is that either little reform has occurred or, when there have been significant reforms, most reforms have misfired because the reform was incomplete, and the results that occurred had serious negative consequences. Another reason is that the urban and industrial reforms have been less transparent than the rural reforms and thus much more difficult to evaluate, either prospectively or retrospectively.

But I shouldn't be too negative. It is important to note that the numerous reform failures have not resulted in an unwillingness to keep trying new approaches. At the present time there are many reforms being tried; it is clearly too early to judge their effectiveness. I must admit I am relatively pessimistic about the effectiveness of reforms that rely on shifting decision making and financial responsibility to the enterprise level until there is a fundamental reform of the price system. How much will be gained by permitting firms to go bankrupt if the prices they face for inputs and outputs are such that no matter how efficient they were in the use of their resources they could not cover their costs? Or how much is gained by permitting enterprises to retain large profits if the prices they face are such that large profits are related hardly at all to efficient resource use? Many relative and absolute prices have remained unchanged for years and years. Where there have been significant technological improvements, such as probably have occurred in consumer electronics, profits may say very little about productivity. These criticisms are well known to Chinese economists yet there is great reluctance, as in most countries, to readjust prices that are under state and political control. As a result, the current price structure is inimical to a rational use of resources in a system in which many decisions are decentralized to the enterprise level. Planned control by the center of inputs and output may well be a superior nth best solution to decentralized decision making with an inappropriate price structure.

But it should not be forgotten that modest reforms can be important in improving the quality of life. In particular, the policy changes that have permitted (and in some cases actually encouraged) private enterprises in urban areas have made urban life much more satisfactory. Private repair services, restaurants, small shops, public markets for the sale of an enormous range of goods and services by state enterprises, collectives and individuals, and markets where rural people sell their agricultural and handicraft products in urban areas have all made a great difference to urban people. In most cities, major department stores are now open during evening hours—some as late as 10 P.M.—in contrast to the former practice of pushing all the customers out the doors at 5 P.M. Given the underinvestment in retail space that occurred between 1949 and 1980, a change in hours of operation of retail shops has important positive implications for the ordinary person. These and similar changes require little or nothing in the way of new investment or in reforms of a more system-wide nature. But such reforms are to be applauded.

Private economic activities and employment have continued to grow in urban areas, starting from the miniscule level of 150,000 in 1978 and increasing to 1,130,000 in 1981, to 2,310,000 in 1983, and to 4,500,000 at the end of 1985.[23] While the growth rate starting from a tiny base has been a large one (62% annually), such private or individual employment represented little more than 3.5% of total urban employment in 1985.

There is some anecdotal evidence that the preference for increased individual employment varies significantly from city to city and that in this matter, as in others, local governments retain considerable authority over the pace at which many reforms occur.

The reform of urban income policies merits very serious attention. One objective of reforms in both rural and urban areas has been to improve the connection between effort and reward. It is not obvious that there has been a significant improvement in that regard since 1978 in urban areas. As noted earlier, subsidies for employees in state enterprises increased from 78% of average wages in 1978 to more than 100% in 1982. Thus at least half of the income of an urban worker is wholly unrelated to work performance; the share of income unrelated to effort and productivity has increased during a period where there was an emphasis on reform of incentives.

One of the first reforms of urban income policies should be the elimination of food subsidies so that market prices can have a more effective role in influencing the allocation of agricultural resources. There is no reason, other than their political influence, why urban people should have access to cheap food while such access is denied to farm people who are on the average much poorer than urban residents. Most other subsidies should be gradually reduced if not entirely eliminated, including housing subsidies for urban people. Farm people are forced to bear the full cost of their housing while urban people bear a minor part of the cost.

A closely related aspect of urban income policies is the ability of an enterprise to discipline its work force for poor performance, including malingering, absenteeism, and generally poor quality work, and to reward individuals for good performance. While it has been claimed for a number of years that enterprises could fire employees for poor performance, there is little evidence that any significant use has been made of this means to minimize the negative consequences of the "iron rice bowl" or the lifetime contract. Nor does there appear to have been effective use of wage bonuses or the authority that has been delegated to some enterprises to adjust their wages to encourage improved performance by significantly differentiating payments to reflect individual differences in productivity. Bonuses usually are constant percentages of base wages and salaries and thus do not reward superior performance or productivity.

The reform of prices and urban income policies represent only the beginning of what is required to make the most effective use of China's human resources. China has the most stringent barriers to migration and mobility that exist anywhere today, paralleling what must have prevailed during the feudal period in Europe. Not only is it difficult to move from one city to another, from one village to another, and especially from one countryside to the city, it is also nearly as difficult to transfer from one enterprise to another in the same city. Enterprises have strict control over such transfers, which in many cases results in arbitrary and abusive exercise of authority.

In this brief discussion of urban and industrial reforms I have neglected two major reforms—money and credit and "the opening to the outside world." Both reforms remain incomplete. In particular, it appears that there has been a lack of coordination between the control of domestic money and credit and of foreign exchange. This lack of coordination was made evident by the great loss of foreign reserves in 1984 and 1985. But the task of adjusting the monetary and credit systems to the changes in the planning process and the introduction of markets in the rural areas has been an enormous one, and it is understandable that perfection has not been achieved. The management of foreign exchange is difficult enough when starting with extensive experience with a functioning monetary and credit system; to try to allow a considerable degree of local decision making with respect to both the use of foreign exchange and of domestic credit simultaneously at least speaks to the seriousness with which the reform efforts are being pursued.

Concluding Comments

The agricultural reforms have been very successful. The successes have been achieved at minimum cost to the central government. Farm production has increased by half and the number of workers required in agriculture has declined significantly. Nonfarm sources of employment in rural areas have increased to absorb many of the workers no longer required in agriculture. An appropriate criticism of the reforms is that in abolishing the political and governmental functions of the communes, little has been created to replace the abolished functions. Thus there appears to have been deterioration in health services, in education, and in a number of other functions normally carried out by local governments. There are complaints that certain collective production-related functions, such as maintenance of irrigation canals and facilities, are not being properly maintained. One of the major shortcomings of rural life in China is the very limited road network. The central government expects the local communities to pay most of the cost of building roads, yet the government makes enormous expenditures in improving office and living conditions in the major cities. To repeat, there is an enormous urban bias in the allocation of governmental expenditures.

Further moves toward a market economy in agriculture will require major changes in urban areas—the removal of price controls on grains and vegetable oils, the elimination of food price subsidies, abolishing restrictions on the movement of agricultural products across province boundaries.

I have emphasized the reforms that have greatly increased the role of the market in rural areas. Urban and industrial reforms have been attempted but have been far less successful than the rural reforms. Numerous urban reforms have been tried and later abandoned because enterprises were given incentives that resulted in undesirable outcomes. Any reforms that are to result in significant decentralization of decision making and increasing the role of market forces wait on major price and wage reforms. This is now officially recognized, but it is also recognized that there are major resistances to such price reforms. Perhaps we can understand the reluctance to undertake such reforms when we remember how difficult it is and how long it takes to eliminate market price interventions in our own country. Some examples that come to mind are price controls on energy and airline and truck deregulation or agricultural price supports that encourage additional farm output when we already produce more than we know what to do with.

Notes

1. Xu Dixin et al., *China's Search for Economic Growth: The Chinese Economy since 1949* (Beijing: New World Press, 1982), p. 4.

2. Ibid., pp. 9–10.

3. B. Ashton, K. Hill, A. Piazza, and R. Zeitz, "Famine in China," *Population and Development Review* 10, no. 4 (December 1984): 619.

4. Xu et al., pp. 11–12.

5. Ibid., p. 12.

6. Ibid., p. 13.

7. Xue Muqiao, *China's Socialist Economy* (Beijing: Foreign Language Press, 1981).

8. Ibid., p. 99.

9. Ibid., p. 101.

10. *People's Daily*, January 20, 1979.

11. *People's Daily*, May 14, 1980.

12. World Bank, *World Development Report, 1984* (Washington, D.C.: World Bank, 1984).

13. World Bank, *World Development Report, 1985* (Washington, D.C.: World Bank, 1985).

14. I am not certain that the commune has been abolished everywhere. The commune or a reasonable facsimile of it may continue to exist near three of the largest cities—Beijing, Tianjin and Shanghai. In 1984 just 10% of the incomes of peasant households came from collectives. However, in Beijing 66% came from collectives, in Shanghai 52%, and in Tianjin 39% (*China Agricultural Yearbook* [Beijing: China Agricultural Publishing House, 1985]). These three municipalities

or provinces have 1.4% of China's rural population and accounted for 13.5% of total peasant household incomes from collectives. There are nonagricultural collectives in rural areas and presumably much of the national collective income comes from such organizations.

15. U.S. Department of Agriculture, Economic Research Service, *China: Outlook and Situation Report* no. RS-85-8 (Washington, D.C.: Government Printing Office, July 1985).

16. Ibid.

17. Ma Hong, *New Strategy for China's Economy* (Beijing: New World Press, 1983), p. 46.

18. Nicholas Lardy, "Runaway Subsidies," *China Business Review* 10, no. 6 (November/December 1983): 21–24.

19. Nicholas Lardy, "Prospects and Some Policy Problems of Agricultural Development in China," *American Journal of Agricultural Economics* 68, no. 2 (May 1986): 452.

20. *Beijing Review* 28, no. 29 (July 22, 1985): 22.

21. U.S. Department of Agriculture (n. 15 above).

22. State Council, People's Republic of China, *1985 Document No. 1* (Beijing: State Council, January 1, 1985), p. 4.

23. *Statistical Yearbook of China 1986;* and *Beijing Review* 29, no. 19 (July 22, 1986): 19.

12

Why Is It So Difficult to Replace a Failed Economic System: The Former USSR

Nearly every day we are told of the latest conflict that has unfolded in Moscow or Kiev that impinges on the transition from the former Soviet system to a market economy. Everywhere there seems to be conflict and disagreement concerning how the transition to a new economy should be made. Increasingly there are those who are so disoriented by the disorder they see about them they wish for the return to the more orderly old days. If there wasn't much freedom then, there was relatively little crime against persons and you were quite certain that with some patience and effort adequate food and clothing would be available. And to Gorbachev's great credit, in the last years of the USSR, the fear of government had nearly vanished.

My answers to the question "Why is it so difficult and taking so long?" consists of three parts: (1) the state of the Soviet system in terms of its productivity was much worse than either most outsiders or insiders understood it to be; (2) while the system reform is underway, it is necessary to create new nations, new constitutions, new laws, and radically new institutions required by a market economy; and (3) in Russia, given the enormity of the task, a great deal has been accomplished since December 1991 when the USSR collapsed. Unfortunately, much more needs to be done before a functioning market economy will emerge and the Russian people will realize the benefits that a market economy can and will bring.

To many it seems that the process of reform has been underway for a long time and I suppose if we were living with the disarray that has accompanied the transition, it would seem like a long time. But it should be remembered that the USSR was disbanded only eighteen months ago and many actions had to be taken with little time for preparation. In November

University of Chicago, Office for Agricultural Economics Research, Paper no. 93:8 (May 1993).

and December 1991 the Soviet system had very nearly ceased to function; store shelves were empty of most ordinary items of food or other daily necessities. In that great nation, aspirin was no longer available. When something was available in the stores, people snapped it up whether they wanted to consume it or not since their only protection against inflation in a system that lacked modern monetary and financial instruments was the holding of physical goods.

Why the Soviet System Collapsed

There are many reasons why the Soviet system collapsed in 1991. There were a number of political reasons, of which I will note only those related to the failure to hold the budget deficit within reasonable bounds and the suppressed and open inflation generated by this failure. These were perhaps the immediate causes that established the timing of the failure though the effect of the failed coup on Gorbachev's capacity to govern cannot be ignored. But even without the coup, the inflationary pressures and the accompanying disintegration of the economy would have made the end inevitable though at a somewhat later date.

The demise of the Soviet economy did not occur overnight. Nor did the efforts to reform the economy start with Secretary Gorbachev. There was a lively debate on the need to reform the Soviet economy in the mid-1960s. This debate, lead by the ideas of L. Kantorovich and E. G. Liberman, was based on the need to follow or mimic market principles, if not to actually rely upon markets. For a variety of reasons, including opposition by the planners and bureaucrats, instead of carrying through with significant reform, under Brezhnev there was what Gertrude Schroeder (1979), my good friend and keen observer of the Soviet economy, aptly labeled "The Soviet Economy on a Treadmill of 'Reforms'" that changed nothing fundamental about the organization of the Soviet economic system but did create uncertainty and confusion due the repeated efforts to tinker through decrees and administrative changes.

The slowdown in the growth of the Soviet economy became evident in the early 1970s. This slowdown was politically tolerable due to an external windfall, namely, the large increases in the price of oil during the 1970s, and the expansion of oil and natural gas exports made possible by the construction of pipelines to Western Europe. Fuel exports for hard currencies increased from $493 million in 1970 to $15.1 billion in 1980 (Cooper 1982, p. 462).

The windfall from petroleum exports permitted increasing imports of consumer goods or raw materials that were converted into consumer goods. Imports of agricultural products increased from $2.3 billion in 1970 to $17.3 billion in 1980 (OECD 1991, pp. 187–88). Large imports of agricultural products continued through the 1980s for a total of $179 billion for the decade (OECD 1991, p. 188). There were substantial imports of other consumer goods as well. The increased earnings from the oil exports covered up the slowdown in the domestic production of consumer goods and services. The windfall from the higher prices of oil and natural gas during the 1970s and the 1980s was not used primarily to increase the productivity of the economy but instead was consumed.

In 1970 the USSR had no significant hard currency debt; by 1980 the debt was $18 billion (Zoeter 1982, p. 490). By the mid-1980s the debt had increased moderately to $28 billion and when the USSR was dissolved was on the order of $75 billion.

One of the editors of *Ogonyok*, a national periodical that was a strong supporter of Gorbachev, said at a meeting I attended in 1989 that the reason Gorbachev came to power was that finally the national leaders recognized that unless there was major reform, the Soviet Union would soon become a Third World country. During the 1980s there was increasing recognition in the USSR that the country was losing—had lost—the economic race with Western Europe and the United States. But it was probably the economic successes of Singapore and Hong Kong and, especially, South Korea that drove home the point. By the mid-1980s the estimates of per-capita real incomes based on purchasing power parities indicated that the average income in South Korea would soon surpass the USSR level.

There has been considerable dispute concerning the performance level of the Soviet economy during the 1980s. Some Soviet scholars claimed that per-capita gross national product (GNP) was less than 15 percent of the U.S. level. I find the recent appraisal of relative consumption levels by Abram Bergson (1991) persuasive. He estimates that as of 1985 USSR per-capita consumption was about 29 percent of that in the U.S. (p. 31). Bergson's estimate of per-capita consumption in the USSR is based on a quite remarkable comparative study of USSR-U.S. consumption prepared by Gertrude Schroeder and Imogene Edwards for the Central Intelligence Agency. The study was published by the Joint Economic Committee of the U.S. Congress in 1981. One part of the study involved purchasing more than 150 products in the USSR and bringing them to the United States for comparison in terms of quality to similar U.S. products.

The USSR per-capita consumption was estimated to be above Turkey's but below Portugal's and approximately equal to Mexico's. Because the USSR devoted a larger percentage of its GNP to investment and to defense than the U.S., its per-capita GNP might have been as much as 35–40 percent of that of the U.S. But whatever the exact figures were, it was obvious that the USSR had not overtaken the U.S. as Khrushchev claimed it would.

By the mid-1980s it had become evident that in terms of technology the USSR, as well as the other socialist economies of Central or Eastern Europe, was falling farther and farther behind the market economies. In particular, the USSR had permitted the computer information revolution to pass it by. By 1990 the USSR had no more computers than Thailand, a country with less than one-fifth the population and a significantly lower per-capita income.

But while the lagging position of the USSR in the economic competition may have triggered the recognition that reform of the Soviet system was essential and thus had a major role in bringing Gorbachev to power, this was not the primary reason for the demise of the USSR. In my opinion, the end of the USSR was due to Gorbachev's failure to undertake the necessary reforms of the Soviet economic system and to maintain a macroeconomic balance. There is not space to outline his failures as an economic reformer, though one might note that one reason for the success of the Chinese reforms was that they were begun while the economic system was functioning reasonably well and while there was macroeconomic balance. Gorbachev's failure to institute appropriate reforms when he came to power condemned the USSR to forever forgo this opportunity. While the Soviet economy could hardly have been described as a robust one in 1985, most goods and services were generally available and in reasonable quantities. While Gorbachev made great improvements in the political and civil situation—perhaps his greatest achievement was to remove fear as a major determinant of people's actions and speech—his economic reform efforts were either without effect or made matters worse.

When the history of the downfall of the Soviet empire is written, it may be said that the real cause was they ate too much meat and drank too much milk. Let me explain. One of the great achievements claimed for the Soviet system by Communist Party ideologists was that it had been successful in keeping the *nominal* prices of housing, household energy, urban transportation, and food (especially bread, meat, and milk) constant for long periods of time. Until recently the prices of the first three had been the same since the 1920s, the price of bread from at least the mid-1950s and meat and milk since 1962. Holding these prices constant was considered to constitute a social contract between the Communist Party and the people.

Carrying out the commitment to hold the prices of meat and milk at their 1962 nominal levels had enormous consequences for international trade and the Soviet budget. It became evident by 1965 that the demand for meat and milk was growing faster than supply. The decision was made to increase prices paid to the farms for the products while holding retail prices constant and to make up the difference by budget subsidies. The subsidies were about 3 billion rubles in 1965, increasing to 12 billion in 1970 and to 25 billion in 1980 (World Bank 1992, table 7.11). The 1982 food program resulted in a further huge increase in the subsidy bill, increasing it to 58 billion in 1985, and to about 100 billion in 1990. How much was 100 billion rubles in 1990? It was 14 percent of net material product and more than 10 percent of national income. It represented 20 percent of the budget of the government of the USSR—this is a consolidated budget for all governmental units. Imagine: one-fifth of the budget went for price subsidies primarily for the meat and milk consumed by the urban population; the rural population had little access to subsidized meat and milk.

The rapid growth of meat and milk subsidies after 1982 was associated with the increase in the deficit in the Soviet budget from an average of about 17 billion rubles for 1981–85 to 83 billion in 1988 or 10 per cent of GNP (Ofer 1989, p. 155). Andrei Markov, professor of economics at Moscow State University, has estimated that the government budget deficit might have been as much as 31 percent of GNP in 1991 (1992). There is no doubt the budget deficit grew enormously under Gorbachev. The close parallel between the growth of the food price subsidies and the deficit is striking. Gorbachev at one time, circa 1987, proposed reducing the subsidies by increasing the meat and milk prices, but when industrial workers opposed the increases he backed off. Little did he realize that this decision would be an important factor in the dissolution of the USSR and that he had sealed his own political fate.

Why was the size of the budget deficit so important? Until early 1991 the official Soviet policy was to maintain controls over the prices of nearly all goods moving through the state retail network. But the budget deficits were leading to inflationary pressure—the people received incomes that exceeded the value of the goods and services available for sale at the fixed prices. This was a classic case of suppressed inflation. Eventually it led to empty shelves for nearly all consumer goods—in early 1989 the store shelves of electronic products in Leningrad and Moscow were emptied in a matter of weeks. This was not an isolated example. An analysis of the availability of 211 groups of food products as of 1988 found that only 23 were to be obtained without difficulty (FBIS-SOV-89-019, 31 Jan. 1989, p. 78). In April 1991 the state store prices of meat, milk, and bread were increased by 200 to 300 percent. But it was too late. Demand still exceeded supply and meat and milk were available only sporadically in Moscow, a generally favored city, in November and December 1991, based on my personal observation.

The other consequence of the fixed nominal prices for meat and milk was the huge increase in the imports of feed and other agricultural products referred to above. It was not possible to meet the growing demand for food products at the fixed prices relying solely on domestic agricultural resources. The government was faced with a number of unattractive alternatives—increasing retail prices, introducing formal rationing at the consumer level, permitting the queues to grow and grow or expand imports of livestock products or feed to supplement the domestic supply. It first chose to import feed but by the mid-1970s also found it necessary to become a large importer of meat. As noted above a large fraction of the value of petroleum exports was used to pay for feed and food rather than to increase the productivity of the economy.

The price subsidies contributed to the downfall of the USSR through the growth of the budget deficit and the inflationary forces that resulted. By the end of 1991 the failure to limit the price subsidies and control the budget deficit resulted in a near collapse of the economic system. The distortion in trade due to the effort to meet the demand for meat and milk was enormous. Resources that could have gone to modernizing the economy were not available.

One final comment about the meat and milk subsidies. In 1989 Soviet consumers paid less than half of the cost of bringing meat and milk to the retail store (World Bank 1992). Gorbachev once noted that the price of meat was so low that the cost of an average person's annual consumption of meat was less than the price of a pair of women's boots. I thought he was exaggerating but he wasn't. The cost of a year's consumption of meat at official prices was about 120 rubles; in 1989 the prices for women's boots in the state stores was 120 rubles and up.

So Much Needs to Be Done

In one essay, even a relatively lengthy one, it is impossible to describe all the laws and institutions that the successor republics to the USSR must create if democracy and a market economy are to be created and flourish. The constitution of Russia was written while the USSR existed; a new constitution suited to a democratic society with a market economy has yet to be adopted. The legislature of Russia was elected before the USSR collapsed and still has three more years. Many members—perhaps most—were elected through the support of the Communist Party; the party is now outlawed but this doesn't mean that all or most of the members have forsaken its objectives.

One necessary foundation for an efficient market economy is that property rights be well defined and that there be a system for protecting those rights and for adjudicating any conflicts that may arise. In Russia property rights remain very ambiguous. An individual, whether a citizen or a foreigner, wishing to acquire a piece of real property, say, a building, can seldom be certain who now owns it. One way to enhance the power of the bureaucracy under communism was to obscure many things, including property rights. The more obscure matters were, the greater the potential for bureaucratic initiative, authority, and arbitrary behavior. The more clearly rights are defined, the less the scope for the exercise of arbitrary power. Rather little has been done to clarify the ownership of existing property or to create the framework for the establishment of property rights as privatization occurs though in recent months significant progress has been made.

The private ownership of land is still prohibited by the Russian constitution. Limited forms of ownership rights have been authorized by law or decree but there does not exist an unrestricted right to sell farm land. The land can be passed on by inheritance. But lacking the right to sell land has several negative consequences. First, a land market will not develop. Second, in the absence of a land market it will be very difficult to create a rural credit system that will serve the needs of farmers. Land mortgages will not emerge and even production credit will be limited because land is not available as collateral for a loan. Third, many families will be reluctant to establish their own farms if they cannot be secure in the rights to the land they are to farm.

But even if property could be bought and sold there does not exist an independent judiciary to resolve conflicts among individuals or to prevent the illegal exercise of power by a governmental unit. An independent judiciary is a concept that is largely foreign to the people of Russia. It was not a concept or idea that met with favor with the czars nor, obviously, did the communists embrace it.

But creating an independent judiciary is only a beginning for a market economy. There must be a commercial code of law developed or adopted. It must be illegal to break a contract that is voluntarily entered into; the rights of those who sign such contracts must be defined before they can be enforced. There is a whole body of law and practice that underpins market economies that have taken centuries to evolve. We should not be too surprised that they have not emerged literally overnight in the former Soviet Union.

Basic Agreement on the New Framework Has Not Been Achieved

There is not much point in belaboring the enormous number of steps that remain to be taken without facing perhaps the most important point of all, namely, that there does not exist a broad agreement on how the new economic system should be organized. While one seems safe in saying that there is no possibility of returning to the political-economic system as it existed in the USSR, there is a lack of consensus on what should replace that system. Equally clear, even if one assumes it is agreed that the objective is to create a market system, there is lack of agreement on how to go from where they are to where they want to be.

Let us first consider the issue of the speed of the transition—was the Big Bang or shock therapy—the best way to handle the transition to a market economy? There are those inside and outside Russia who answer in the negative, that there should have first been created a safety net for all those who were to be adversely affected by the transition. This included those on pensions and those who would become unemployed, to mention but two large groups. While the safety net was being created, gradual steps in preparation for the market should be undertaken. One of those steps would be privatization of significant sectors of the economy, especially retail and wholesale trade. Those who hold such a position have strongly opposed the price liberalization that took place in Russia at the beginning of 1992.

The sharp rise in prices and the large fall in real income that occurred in 1992 has been blamed on the failure to adequately prepare for the price liberalization. This position ignores several important things. First, there was an attempt to rather slowly liberalize but not decontrol prices of most foods and ordinary consumer goods in April 1991. The prices of most foods were increased by 200 to 300 percent in an effort to mop up the purchasing power generated by the budget deficits and to reduce the burden of the milk and meat subsidies. It was successful in neither regard. Second, a number of economic reform efforts were undertaken by Gorbachev in an effort to revitalize the economy. But none achieved their objectives. Third, the state-store system for supplying food and ordinary consumer goods to the population had collapsed in the last months of 1991, much as the USSR itself had collapsed as a government and country. The shelves in all stores were empty most of the time. This doesn't mean that the stores had nothing to sell but that the time cost of queuing and searching had become an enormous burden, especially for women. The price increases in April were simply inadequate to equate supply and demand; instead, the discrepancy between demand and supply steadily grew throughout 1991.

Fourth, there was no possibility in the last months of 1991 that either the government of the USSR or the Russian Republic could enact the legislation required for privatization or the creation of a safety net. Following the coup attempt the USSR government was paralyzed and the Russian Parliament was unwilling then as now to support wholeheartedly those measures needed for a transition to a market economy. Fifth, it was not possible to continue with fixed prices in the state stores given the inflationary pressures that existed at the end of 1991. With the partial price liberalization on 1 January 1992, retail prices rose approximately tenfold. Even if prices had been doubled or quadrupled in November in an effort to soften the impacts of price changes, the store shelves would have been empty still.

Prices were liberalized in Russia at the beginning of 1992 because there was no other feasible alternative. Some action was required; the situation as it evolved in November and December 1991 could not continue. What must be said is that price liberalization by itself accomplished relatively little, though what it did accomplish was very important. It put goods—true, in limited variety and still of very poor quality—back in the stores. Money now became significant; it could buy something, most things that had formerly been available in the Soviet economy. What it did not accomplish was a sudden burst in production and increased overall availability of consumer goods. In fact, prior to January 1992 the state enterprises had not hoarded or stored large quantities of products in anticipation of the price increases, as one would have expected "normal enterprises" to have done, as was the case in West Germany in 1948. The hoarding, and I use that term in a positive and not in a pejorative sense, was done by ordinary people and not by state enterprises. Why not? For the same reason that there has been little or no output response to the price liberalization, namely, that neither the managers nor workers of these enterprises obtain any benefit from responding to opportunities for increased profitability. Neither their pay nor their jobs depend on lowering costs or producing more or in adopting an output mix that responds to consumer demand. Only when the enterprises are privatized or as state enterprises they can no longer depend upon subsidies or loans will there be the kinds of output responses that are normal in a market economy.

While the supplies of food in the stores increased after 1 January 1992 it must be said that most who favored the price liberalization were disappointed with the modest supply response to the freeing of most prices. Put simply, the state enterprises that still controlled the food sector from the farm to the retail store continued to operate as they always had. They priced by formula; but if that price were too high—that is, if supply exceeded demand—they had no incentive to change the price or to try to reduce their costs. They were neither rewarded nor punished, as the case might be, for their behavior. Thus meat would remain in the meat case until it spoiled rather than lower its price. There is no evidence that most state enterprises have significantly changed their behavior during the year. And because their losses are covered by loans that are never expected to be repaid, there is no reason for them to change.

It is very important that the state enterprises be privatized or be required to operate under a hard budget restraint. In other words, they must take in enough to pay their bills or go bankrupt. Unless this happens it is highly probable that real per-capita incomes will continue to fall.

Could Reform Have Been Slower?

There are those who argue that the transition to a market economy would be much less painful if the pace of change—of reform—had been much slower. There are several reasons why this conclusion is incorrect. The first is that the former system had collapsed by the end of 1991 and was no longer able to function as it had in the past; immediate changes such as price liberalization were required if the economy were to function at all. The second is that there had been efforts, as noted earlier, to reform the Soviet economy for at least a quarter century and with no avail. And, third, there is the experience of Hungary that started its reform to create a socialist market economy in 1968, nearly a quarter century ago.

The Hungarian experience clearly justifies the conclusion that a *socialist* market economy cannot be created out of a centrally planned or directed socialist economy. Briefly, in 1968 Hungary abolished detailed central planning—its *Gosplan* was abolished—and instituted the New Economic Mechanism. While there was a period of recentralization in the 1970s— the old guard won a temporary ascendancy—significant market reforms were reinstituted in the 1980s. The casual visitor to Hungary during the 1980s might well have concluded that the reforms were successful. Private enterprise was permitted in numerous service fields, such as repairs, restaurants, boutiques, and other retail stores. It was easy to find an excellent meal in terms of both food and service at low cost and shopping areas in Budapest and elsewhere were bright and attractive.

While these changes clearly improved the life of the ordinary citizen, by the end of the 1980s over 90 percent of the gross domestic product (GDP) was still produced in the state or cooperative sector (Hare 1991). Real income growth during the 1980s was slow—about 1 percent per capita annually. Real wages were low for a well-educated population and most industrial products were of poor quality. The publicly owned housing stock was of low quality, poorly maintained, and badly deteriorated. The external hard currency debt reached 50 percent of GDP and thus some of the appearance of plenty in the shops had been created by spending borrowed funds. The reform effort was not successful in creating a market economy or in significantly increasing the productivity of the centrally directed economic system. Instead, Hungary fell farther and farther behind Austria in terms of real per-capita incomes and consumption.

A comparison of the economic welfare of the citizens of Kiev and Moscow as of March–April 1993 provides a final reason for rejecting the conclusion that slower reform efforts would have minimized the hardship. The Ukraine has done relatively little to reform its economic system; there has been very little privatization and most of the controls over production and marketing have been retained. While consumption levels may have been similar in late 1991, a recent visit to the two capitals reveals that per-capita real consumption in Kiev was substantially below that of Moscow. And the end of the decline in production in the Ukraine is not in sight while in Russia there is some evidence that the slide in output has ended or nearly so.

Progress Has Been Made

We should not be overly pessimistic in spite of the rather grim situation that now prevails. The former republics of the USSR and other countries in the former Soviet bloc are now going through a transition that has never been experienced before. When we consider how long it takes us to correct our policy mistakes we should display a lot more patience and understanding of what has been accomplished given the magnitude of the tasks involved and the time that will be required to complete the transitions.

We should not forget what has already been accomplished in Russia over the past two years. Until quite recently people were imprisoned for buying and selling—for engaging in speculation. Today in Russia all forms of private trading that are not specifically prohibited by law, such as selling drugs or unsafe products, is sanctioned. True, such trade has been legalized by decree rather than by changing the law. Almost all prices have been decontrolled. International trade has been significantly liberalized though the scarcity (high price) of foreign currency has limited the role of imports as competition for domestic producers. Efforts are being made to implement an antimonopoly law. Privatization of housing and state enterprises are being pursued, including the introduction of a voucher system, and progress is apparent.

Through the decontrol of prices and other measures, one of the major achievements is that money now matters—it now provides command over goods and services. Before one's command over goods and services depended upon who you were, whom you knew, or how much time you were able and willing to devote to search and standing in line. Many premium goods were not generally available since they were sold in special stores available only to the elite. There is a political achievement that should not be overlooked, even though at the moment this achievement represents a major impediment to the transition. In Russia there is now a separation of executive and legislative powers, something that has hardly ever existed before in Russian history. True, much of what Yeltsin has done has been by decree rather than through enforcing laws passed by the parliament. But it has been the parliament that has given the authority to rule by decree for a limited time.

Much of the current disorder in economic matters in Russia results from conflicts over the operation of the Central Bank. The Central Bank is not an independent agency but is legally responsible to the parliament. Pressure from the parliamentary opponents of Yeltsin and his economic programs has resulted in an extraordinary outpouring of money and credit. Credit has been provided in enormous quantities to prevent enterprises from going bankrupt. Virtually none of the credit has been used to help firms restructure or for any other constructive purpose, for that matter. Thus the flow of credit has done two very negative things: delay adjustments that must occur and create an inflationary situation bordering on hyperinflation. This effort to countermand the Big Bang ends by creating great apprehension among those who are harmed by inflation.

Imagine if you will the state of mind of those millions on pensions who had looked forward to a life of reasonable ease and enough income to cover the costs of housing, food, and modest amounts of other necessities. This was to be their reward for their dedication to, support of, and sacrifice for the Soviet system. They were confident that their pension would be forthcoming and that they could draw upon their modest savings. During the past two years they have seen their cash savings wiped out by inflation and the purchasing power of their pensions fluctuate widely and around a declining trend. With prices increasing by 20 percent or more per month, they worry that their pensions will not keep pace. Unfortunately, experience indicates that their concern is a valid one. They feel they have been cheated, and in a real sense they have been. For people who are young, vigorous, and risk takers the current transition provides opportunities for economic gain that could not have been imagined a few years ago. But for the elderly and probably for many of the unskilled, the present represents a frightening experience and there is no reasonable hope that the near future will be less so.

It is difficult to imagine that any significant progress can be made toward the restructuring required for a market economy until the enormous macroeconomic imbalances that exist are reduced to manageable levels. A monthly inflation rate of 21 percent compounds into an annual rate of 1,000 percent. In an economy in which the ordinary citizen does not have access to any financial instrument other than cash, it is difficult to see how any constructive changes in the real economy can occur until inflation is brought under control. It can only be hoped that the political conflicts that are contributing to the untenable inflationary conditions will be resolved before the suffering of the Russian people becomes unbearable and all hope is lost.

Concluding Comments

The years that lie ahead will be filled with difficulties and great uncertainty for the people of the former USSR. But it must be remembered that the economic (and political) systems that existed prior to 1992 were no longer viable. These systems had condemned their people to a level of living that was no more than half what it could have been, if we can judge by the evidence provided by economic developments in West and East Germany and by the fates of two members of the Austro-Hungarian empire, Czechoslovakia and Hungary, compared to a third, Austria. And the systems achieved what they did through varying degrees of personal oppression and enormous environmental disruption. It was not a matter of *whether* the previous system would collapse, but *when*. And it seems best that it was now rather than later.

References

Bergson, Abram. 1991. "The USSR Before the Fall: How Poor and Why." *J. Econ. Persp.* 5 (Fall): 11–28.

Cooper, William H. 1982. "Soviet-Western Trade." *Soviet Economy in the 1980s: Problems and Prospects*, part 2. Joint Economic Committee Print, 97th Congress, 2d session, pp. 454–78.

Hare, Paul G. 1991. "Hungary: In Transition to a Market Economy," *J. Econ. Persp.* 5 (Fall): 195–202.

Markov, Andrei R. 1992. "Russian Policy of Economic Reform and Creation of Markets." Paper presented at Conference on Institutional Economics and the Transition to a Market Economy in Russia, University of Wisconsin, 9 October.

Ofer, Gur. 1989. "Budget Deficit, Market Disequilibrium and Soviet Economic Reforms," *Soviet Economy* 5 (2): 107–61.

Organization for Economic Cooperation and Development. 1991. *The Soviet Agro-Food System and Agricultural Trade*. Paris: OECD.

Schroeder, Gertrude. 1979. "The Soviet Economy on a Treadmill of 'Reforms.'" *The Soviet Economy in a Time of Change*. Joint Economic Committee Print, 96th Congress, 1st session, pp. 312–40.

World Bank. 1992. *Food and Agricultural Policy Reforms in the Former USSR: An Agenda for the Transition*. Washington, D.C.: World Bank.

Zoeter, Joan Parpart. 1982. "U.S.S.R.: Hard Currency Trade and Payments." *Soviet Economy in the 1980's: Problems and Prospects*, part 2. Joint Economic Committee Print, 97th Congress, 2d session, pp. 479–506.

PART FIVE

Developing Countries and World Agriculture

13

Agriculture and Foreign Economic Policy

Although the origins of the English Corn Laws are obscure, it is clear that in a rudimentary form this peculiar group of economic regulations existed at the beginning of the fifteenth century. Over the next four centuries the regulations became increasingly complex, until the combined effects of Cobden and his Manchester liberals and the potato famine led to their abolition in 1846. Contrary to general belief, the Corn Laws were not simply a set of import duties but also consisted of export bounties, variable import levies, and export and import prohibitions, with special preferences for using British ships.

During this period of British history, and also after the repeal of the Corn Laws until World War II, agricultural and trade policy were closely linked though there was more to agricultural policy than the regulation of trade. The objectives of the Corn Laws have a modern ring. Their aims were to stabilize the price of corn for the "benefit" of both producer and consumer; reduce the dependence upon foreign supplies of a critical food; achieve adequate land rent, since rent was viewed as a good measure of national prosperity; and maintain a large and prosperous rural population.

Corn Laws All Over Again

The Common Agricultural Policy of the European Economic Community could well be called the European Corn Laws. The variable import levy was introduced into the Corn Laws in 1670. But the parallel with the Corn Laws is even closer. When direct payments and subsidies to farmers are withdrawn, the Common Agricultural Policy will be basically a trade and agricultural policy. The agricultural price and income goals will be sought

Reprinted by permission from the *Journal of Farm Economics* 46, no. 5 (December 1964): 915–29.

through manipulation of international trade by variable import duties and export bounties.

To an increasing extent, especially during the last decade, the United States has come to rely upon international trade as an adjunct of agricultural policy. Import limitations and export subsidies were a necessary adjunct of farm price programs beginning in 1933, but with the inception of P.L. 480 export subsidies have become a major and positive aspect of our agricultural policies. Limitations on imports have begun to play an even more important role in recent years, especially with the "voluntary" limitations on certain dairy products and beef and veal that were negotiated with New Zealand, Australia and Ireland in 1963 and 1964. The Sugar Act of 1962 introduced the variable import levy into our agricultural policy. We have long used variable export subsidies—the size of the subsidy representing the difference between domestic prices and whatever was needed to export the desired volume.

On July 1, 1964 the British imposed minimum import prices on grains. An offer of grain at a lower price results in the imposition of a duty to bring the landed cost of the grain to the specified minimum level. Both the EEC and British minimum import prices are an open invitation for sellers to collude to raise the cost of grain.

The industrial countries seem to have come almost full circle for the English Corn Laws are now the pattern for agricultural and trade policy. Thus, there are important lessons for students of current agricultural policy to be drawn from English history prior to abolition of these laws in 1846. This is particularly true when we consider the possibilities of success or failure at the current trade negotiations in Geneva. The U.S. position is especially similar to that of the British following the War of 1812 up to the abolition of the Corn Laws in 1846.

The repeal of the Corn Laws was a part of a general movement toward free trade, just as the negotiations at Geneva now underway are part of an effort to achieve freer trade. Prior to the repeal of the Corn Laws, as now, the difficulties of modifying barriers to trade in agricultural products were important considerations in general trade relations among nations. This was made clear by J. S. Nicholson in his lectures on the Corn Laws when he discussed efforts to reduce barriers to trade.[1]

> At the same time, both at home and abroad, the Corn Laws were the most prominent and noticeable part of the protective system. When foreign nations were invited to reduce their tariffs, they always pointed to our Corn Laws. A remarkable instance occurs in connection with the framing of the tariff of the United States in 1824. Our Minister at Washington, in that year, wrote to Mr. Canning: "Had no restriction existed on the importation of

foreign—i.e., American—grain into Europe, and especially into *Great Britain*, there is little doubt that the tariff (that is, of the U.S.A.) would never have passed through either house of Congress, since the great agricultural states, and Pennsylvania especially, the main mover of the question, would have been indifferent, if not opposed, to its enactment."

In the same despatch, it is said that the retention of the Corn Laws by Britain led the Americans to suspect the real intention of any removal of other restrictions as in the modifying of the Navigation Acts. . . . They suspected this country of insidious designs, with the view of afterwards taking advantage of the concessions obtained from other countries.

How similar these views and concerns are to those of today! The view exists in some quarters that the current U.S. interest in reducing the barriers to our agricultural exports is largely a self-serving one and that any concessions we might make on agricultural products will be negated subsequently if the increase in our imports becomes embarrassing. Actions that have been taken under Section 22 to limit imports, our recent responses to increases in imports of certain dairy products and of beef through negotiations of so-called voluntary agreements, or recent congressional consideration of beef import quotas that would violate our GATT obligations serve as a basis for the view. Thus just as we suspected the British almost a century and a half ago of "insidious designs," our behavior in negating reductions that we have made in tariff duties leads others to suspect us.

There is also a growing concern in the United States of the long-run effects of the Common Agricultural Policy of the EEC. This concern is due to the adverse effects of that policy upon our agricultural exports, for example, the effects of the variable levy imposed upon poultry meat; our government cried foul and exercised its right to retaliate under GATT regulations. But exercising the rights did not prevent a sharp decline in our poultry exports to the EEC.

It is feared that high grain prices and consequent high prices for livestock products will both increase EEC production of grain and reduce EEC consumption of livestock products, which in turn implies a reduction of imports and perhaps even some export surpluses in the EEC.

The EEC has rejected the use of deficiency payments to meet price support commitments. Thus the difference between a support or intervention price and world prices must be maintained by an import restriction— the variable levy—or, if there is an export surplus, by an export subsidy. Thus, in effect, the EEC price support effectively determines the EEC trade policy for those agricultural commodities produced in the EEC. The United States finds that it must negotiate with the EEC about the Community's price support levels. In turn, the U.S. must be willing to negotiate with respect to our domestic farm programs.

Secretary Freeman, in presenting the negotiating plan of the United States for agricultural trade for the then forthcoming GATT meeting in Geneva, summed up the objectives and hopes of that plan:[2]

> Finally, it is a plan for trade liberalization, and that is what the trade negotiations are all about.
> It would require tariff cuts. . . .
> It would assure markets to efficient producers and would require some limit to the measures encouraging inefficient production.
> It would expose the trading practices and the domestic farm policies of the Free World to the test of the high principles under which the trade negotiations were launched.
> It would inject new strength and vigor into the world's established commercial trading system, a system which has served man well through the ages and which continues to offer best promise for effective and rewarding distribution of his production.

I consider the above to be an extremely important statement. It represents a new direction; it is the clearest official statement that I have seen which says that it is wrong for us as a nation to continue to develop our agricultural programs without concern for their effects upon others. As the Secretary said on an earlier occasion,[3] "No matter how much it complicates our problems, agricultural policy must be considered in terms of the needs of all our people, of every segment of our economy, and of the position and responsibility of this Nation as a leader of the free world."

If the present trade negotiations are to have a successful conclusion, the close interconnections between agricultural and trade policies require that the negotiations concern themselves with price support levels, subsidies, and other methods that tend to encourage output increases. The export subsidies, the quantitative import restrictions and the variable levies are only the obvious requirements of the domestic farm policies. These interferences with trade are not desired for themselves, but only as means for achieving the objectives of domestic agricultural policies.

If present policies are maintained, the EEC will expand its high cost agriculture and will burden its population with high food prices; the United Kingdom will be saddled with a huge drain upon its treasury to finance the deficiency payments and production grants unless it, too, transfers these costs to consumers; and the U.S. will continue to incur farm program costs of upwards of 6 billion dollars. And the farm problems of these countries will be no nearer solution in 1970 than today.

But my main concern here is not the problems that the rich industrial nations of the West have made for themselves. They have what they well deserve! I am concerned about *the consequences of the agricultural policies of the industrial nations for the less developed areas of the world.* The industrial countries no doubt can easily afford their present farm programs from a purely domestic viewpoint. But is this true when all of the international implications are taken into account? I do not promise to answer this question, but I know that it is a critically relevant question which requires analysis and which, in terms of policy choices, calls for an open mind.

The Web of Trade

It should not be necessary here to show the importance of the expansion of agricultural production and the growth of agricultural exports as significant contributors to the growth of per capita incomes in the underdeveloped areas. In most of the less developed countries agriculture provides employment for half or more of the labor force and for three-fourths or more of total foreign exchange earnings. More farm output is required to feed a growing population and to feed it better; more farm exports are required to provide some of the capital and other requisites for modernization.

The primary link between the agricultural policies of the industrial nations and the development of agriculture in low income countries is the value, origin and composition of international trade in agricultural products. The major industrial nations influence the value of agricultural exports from the less developed countries by restricting imports and expanding exports. Western Europe reduces the value of exports of the less developed areas by expanding indigenous agricultural output. If present policies are maintained in Western Europe, that area may become an important exporter of some farm products for which it has been a traditional importer. The United States also reduces the value of such exports by the use of export subsidies and by expanding domestic production in the face of less costly imports.

The less developed countries suffer from two important disabilities because of the composition of their export trade. The first is that the major components of that trade—food, fibers and beverages—have relatively low income elasticities which means that the growth in demand in the industrial countries for products of the type they produce is primarily a function of population growth and only secondarily a function of increases in per capita income. The second difficulty is that the industrial countries are engaged in protecting or subsidizing their own producers of similar products. The first disability should not be confused with the second. At best, the growth in demand for the exports of the less developed areas will be much slower than the growth in real national income of the major industrial countries, unless there is a substitution of imports for domestic production in the industrial countries.[4] When this difficulty is added to the second—the practice of most industrial nations of subsidizing a number of competing products—the outlook for future growth of exports is indeed bleak.

It is argued by some that since most of the lesser developed countries are semitropical or tropical and most industrial nations are in the temperate zone, there is relatively little competition between the agricultures of the two areas. While it is true that a wide variety of tropical products are admitted into industrial countries under favorable terms and that tariff rates on many of these products are zero or nearly so, there are some tropical products that face relatively high internal excises—coffee in West Germany and France, for example. And while this point is not directly within the province of this paper, it should be noted that generally only the raw tropical products are found on the free list. Most industrial nations, including the U.S. and the EEC, have what appear to be moderate duties on the first processed product or products (oil and oil meal from an oilseed) but have what in fact are duties that are a very high fraction of the value added during such first processing. A 10 percent tariff on the first processed product, with a zero duty on the raw material, can result in protection of the first processing of 50 to 150 percent. It is virtually impossible for the underdeveloped countries to develop certain processing industries that could compete for international markets.

But let me return to the competition between the tropics and the temperate zone. Some of it is very direct—sugar, tobacco, cotton, rice, fats (butter and lard versus vegetable oils), many fruits and nuts. In recent years, for example, approximately 40 percent of the world's output of sugar has been grown in temperate areas subject to all manner of protection and subsidization. But the less direct forms of competition—one food grain for another (e.g., wheat for rice) and one food relative to all others—must in the aggregate be of substantial importance. If the output of temperate zone food products increases as a result of the policies of the industrial countries, the demand for tropical zone products in the industrial countries will almost surely decline.

Development and Trade

As already noted, the economic growth of the developing countries is dependent upon agricultural exports. This statement is an oversimplification; what the developing countries require is a substantial increase in imports. An expansion of imports can be financed in only two ways—by an increase in exports, or by loans, grants and aid from the industrial nations. The expansion of agricultural exports is emphasized because the present composition of the exports of the less developed countries is so largely agricultural, and especially so when areas that have large petroleum exports are excluded.

While loans and grants to the less developed areas are likely to increase in value over time, it is difficult to imagine that loans and grants can ever provide for most of the total import requirements of the poor countries. Only about 16 percent of total imports of the developing countries were provided by all forms of private and governmental loans and grants in 1960–1961.[5] If the capital flow from rich to poor countries in 1960–61 had conformed to the standard that is now being put forward as an objective— namely, 1 percent of the gross domestic product of the advanced countries—it would have financed about 21 percent of the total imports of the poor countries.[6] Thus the magnitude of the capital flow actually achieved in 1960–61 was rather close to what some consider to be a reasonable upper limit to the amount of transfers. While it is not possible to predict exactly how much the imports of the less developed countries would need to increase over the next decade or two if they were to achieve rapid economic growth, projections made by the Economic Commission for Europe indicate the probable order of magnitude of the required increase. If the less developed areas achieve an annual growth rate of per capita income of 3 percent between 1957–59 and 1980, the commission's projection is that their imports would increase from $22.5 to $60.0 billion.[7] If net capital inflow stays at a constant fraction (approximately 16 percent) of imports, exports would have to increase by more than $30 billion. Put another way, this projection implies a required increase in the exports of the developing countries in two decades that is greater than their exports as of 1957–59.

The share of agricultural exports in the total exports of the less developed countries will decline if per capita national output increases. However, a large increase in the absolute value of agricultural exports must be achieved. But if the industrial countries maintain their present domestic agricultural policy course, is this in the realm of what is possible?

Agricultural Protection in Industrial Countries

Agricultural protection of a significant magnitude is almost universal among the industrial countries. This protection takes numerous forms— higher prices, deficiency payments, production grants and input subsidies. Because of the many forms that protection takes, it is difficult to measure.

Gavin McCrone has made an estimate of the amount of one type of protection for Western Europe for 1955–56.[8] His estimate is the difference in the value of output measured at the prices received by farmers and the value of output at import prices of the same year. He did not consider any of the many other forms of subsidization. I have slightly modified his estimates and have made similar estimates, on a consistent basis, for 1961–62.[9] The degree of protection, indicated by the percentage excess of output valued at national prices over output valued at import prices, was the following:[10]

	1955–56	1961–62
France	24	17
West Germany	22	39
United Kingdom	33	29
Italy	19	25
Sweden	26	41
Norway	20	43
Netherlands	5	14
Belgium	6	13
Denmark	3	0

By the limited measure used, the degree of protection decreased significantly only in France during the six-year period, though there was also a small decrease in the United Kingdom. Major increase occurred in West Germany, Sweden and Norway. Tiny Denmark had virtually no protection in either year.

How does the United States compare with Western Europe? I have made estimates similar to those made by McCrone and they show that the difference in value to output valued at national prices and import or commercial export prices amounted to 16 percent of the latter in 1961–62. This estimate does not include the direct payments to farmers which were about 6 percent of the value of farm production for sale and home use. However, similar payments (except the deficiency payments in the United Kingdom) were not included in the estimates of protection for Western Europe.

The amount of protection afforded agriculture in the industrial countries has increased significantly in the last three decades, and the ratio of farm output to domestic use of farm products has risen both in the United States and Western Europe. FAO estimates indicate that Western Europe imported 31 percent of its food supply before World War II; by the late 50's the area imported only 25 percent of its food; projections made before the implementation of the Common Agricultural Policy of the EEC and other recent protectionist measures in Western Europe indicated that by the mid-60's, 22 percent or less of the total food supply would be imported.[11] During the past decade, U.S. imports of agricultural products have remained approximately stable in value terms, while exports of farm products have increased by more than 60 percent and total agricultural exports which at the beginning of the decade were no larger than imports now exceed total agricultural imports by roughly 60 percent.

An admittedly extreme example of agricultural protectionism in the industrial countries is that of sugar. U.S. sugar production has long been highly protected, first by tariffs and since 1934 by import quotas and direct subsidy payments to farmers. Since the first sugar act, the quotas allocated to continental U.S. have more than doubled. In the late 20's, U.S. producing areas (excluding the Philippines) supplied only 37 percent of U.S. sugar consumption; during 1956–60, 51 percent. Under the Sugar Act of 1962, the share of the sugar quota reserved to U.S. producers was increased from 53 to 60 percent. Sixty-five percent of any increase in total quotas due to increased U.S. consumption is now allocated to U.S. producers.

The record of the EEC with respect to sugar is about the same as ours. Before World War II, the present EEC imported about a quarter of their sugar even though all domestic sugar was heavily protected.[12] Between the late 30's and 1957–58 sugar imports declined 68 percent while internal use increased about 60 percent and local production doubled.

While sugar appears to be an extreme case, it is extreme only because the degree of protection has been so great and not because protection is atypical. Moreover, it is exceedingly relevant to this analysis because sugar is a commodity that can be grown throughout much of the underdeveloped world—and sugar would be produced to a much greater extent in that part of the world than it now is if the major markets were not so protected and so precarious.

Of course, not all the agricultural policies of the industrial countries work to the disadvantage of the underdeveloped countries. For more than a quarter century, the U.S. has set a floor under the world price of cotton and tobacco. There was a marked response on the part of producers in other regions to the relatively profitable prices achieved by the efforts at output restriction and stock accumulations in the United States. In the case of cotton, some of "the bloom has been taken off the rose" in recent years by export subsidies equal to roughly a quarter of the U.S. domestic price. However, there is evidence that U.S. output of cotton would not be much if any below the present level if there were no farm program for cotton. In both tobacco and cotton the U.S. has lost its position of dominance in international trade. In the late 20's the U.S. exported about two-thirds of the cotton moving in international trade; now its share is little more than two-fifths. The loss in the case of tobacco is not so dramatic—from in excess of 40 percent in the late 20's to about 30 percent recently.

The United States, alone among the industrial countries, has some programs designed to limit the output of certain agricultural products and to some degree, total agricultural output. Although the official view of the Department of Agriculture is that the United States is effectively limiting the output of farm products, I believe, as I have argued elsewhere, that the net effect of all federal farm programs—irrigation, reclamation, soil conservation, conservation payments, high support prices, imported labor, soil banks and acreage allotments—has been to increase farm production.[13]

It is important to emphasize that we have not been able to convince our major trading partners that our farm programs, including both output restrictions and inducements, has had any effect other than to increase output and depress world prices. One of the important rationalizations of the EEC variable levy system is that it is required to protect EEC farmers from subsidized farm product exports from the U.S. This argument was used during the poultry dispute, with little justification in my opinion, but this does not mean that their position is without general validity. The following two quotations are from material that came to my desk in the same week— in one case the authorship is French and in the other it is American:[14]

> The variable levy applied by the EEC to agricultural imports from nonmember countries is aimed at making up the difference between the abnormally low world prices at which products are usually sold and prices within the Community. It is therefore nothing more than the counterpart of the export subsidies used by these nonmember countries; these subsidies also vary in relation to world prices so that products may be marketed below domestic prices. While this variable levy presents an effective counter to the disturbing effects of dumping foreign surpluses, it is not intended to prevent the entry of products that meet a need on the Community market, either because of their quality or the services provided. . . . if market prices plus the certificate result in wheat being priced higher than world prices, appropriate adjustments will be made by way of export subsidies so we will continue to be competitive.

So long as these two positions on the output effects of U.S. farm programs prevail, there is little possibility of reaching mutually acceptable solutions to the major agricultural trade problems. It would appear to be in the interest of the United States to ask for an independent evaluation of the output effects of our farm programs, just as we asked for an evaluation of the trade impacts of the EEC variable levy on poultry.

Trade Really Matters

In 1960–62, world exports of major agricultural products totaled approximately 22 billion, of which $12 billion was exported by the less developed regions.[15] According to USDA estimates, the total value of agricultural output in the developing countries was approximately $53 billion in 1958; by coincidence, the value of agricultural output in Western Europe and the United States was also estimated to be $53 billion.[16] Thus if the industrial countries displace a billion dollars of the agricultural exports of the less developed countries by decreasing imports or by subsidizing exports, the amount is large relative to either the value of LDC's agricultural exports or the value of their total farm output.

During the 50's the annual rate of growth of the quantity of exports from the less developed countries was 3.6 percent compared to 6.9 percent for the industrial countries.[17] But what is even more important is that much of the growth in exports occurred by 1956. In value terms the exports from the nonindustrial countries to the industrial countries increased by 33 percent between 1950 and 1956; from 1956 through 1962 the increase was only 14 percent.[18]

Some of the reduction in import growth of the industrial countries of the products of the underdeveloped countries may have been due to a decline in the rate of national income growth during the 50's and to overcoming the effects of World War II on agricultural production in Western Europe. But some role must surely be assigned to the rising tide of agricultural protectionism in Western Europe after the end of the Korean War and the large increase in surplus disposal by the United States, especially under P.L. 480 beginning in 1954 and increasing in later years.

The significance of the increase in surplus disposal resulting from P.L. 480 can be seen by comparing it with the change in the value of world exports of the main agricultural products among the major regions between 1952–53 and 1960–62.[19] The increase in the interregional trade in agricultural products was $3,283 million during the period. This increase may be partitioned as follows: (1) the increase in the exports of the less developed countries was only $815 million; (2) the increase in commercial exports of the industrial countries was $1,097 million, and (3) the increase in the value of products made available as economic aid or surplus disposal by the United States was $1,371 million. Thus the increase in the value of shipments by the U.S. under programs such as P.L. 480 was much greater than the increase in exports of agricultural products from the less developed regions and somewhat greater than the increase in commercial sales by industrial countries, even though a substantial fraction of the latter increase was due to the use of export subsidies. How much substitution there has been between P.L. 480 shipments and commercial exports of all countries, including the U.S., is subject to dispute, but it is highly probable that there has been a significant degree of substitution.

The protection given agriculture in the industrial countries restricts exports from the less developed countries in one or both of two ways—increasing output and reducing consumption. Only in the United Kingdom have the consumption effects been largely eliminated through the use of deficiency payments. If the increase in prices received by producers averages 20 to 25 percent and if the increase in prices paid by consumers averages 15 to 20 percent, even very low elasticities of supply and demand will result in substantial contraction in the demand for exports of the less developed regions. If supply elasticities are as low as 0.15 and demand elasticities 0.2, the effect on the value of agricultural imports of the industrial countries would be as much as $3.5–4.5 billion. Some of the increased import demand would be met by developed countries such as Canada and Australia. But if the increase in LDC exports were to be only half that indicated or about $2 billion, the importance is very great when compared with total LDC exports of agricultural products of approximately $12 billion.

Trade and Aid and Agricultural Policy

For a number of years, the slogan "Trade, not aid" had a certain attractiveness in some circles in the United States. If current trends in agricultural policy in the industrial countries persist, the situation in the years ahead, as viewed by the less developed countries, may be best described as "Aid, not trade." Western Europe will import less and less and the United States will export more and more.

One of the most alarming prospects facing the less developed areas is that the United States soon may lose its near monopoly on the provision of food aid. This is a dim prospect, not because the new participants will be less concerned about the export implications of their activities than the U.S. has been, but because it will mean that the total flow of such aid will increase, more and more markets will disappear and relative farm prices will be further depressed in the world markets and in the recipient countries. If farm output in the EEC continues to expand relative to EEC consumption, the EEC will have no acceptable alternative but to inaugurate its own version of P.L. 480.[20]

An expansion in the volume of food aid will not offset the adverse effects of a decline in export earnings. While it is comforting to the rich to believe that their bounty is being used to prevent malnutrition among the poor and, perhaps on occasion, to prevent actual starvation, it is a delusion to assume that such aid will be a major factor contributing to a rise in real per capita incomes in the recipient countries.

If, as a result of the agricultural policies of the industrial nations, the less developed countries are not able to finance the imports that will contribute to rapid economic growth, the alternatives are to follow policies of autarchy or for the industrial countries to substantially increase the flow of aid, grants or loans. If export earnings are demonstrably decreased by the farm policies of the industrial countries, the claim of the less developed countries for increased aid or other capital flows will be a strong one. Thus it is not at all unlikely that a reduction of export earnings will be offset, at least in part, by an increase in aid and loans.

If present agricultural policies in the industrial countries are continued, the costs to consumers and taxpayers will surely increase as they have for the past decade. When the cost of additional foreign assistance necessitated by the consequences of these policies is added to the direct costs, the case for searching for alternative methods of meeting the income objectives of the agricultural policies of the industrial countries becomes even stronger.

The very high cost of the current farm programs—to consumers and to taxpayers—is due to a substantial degree to the measures that result in increased output. A very high fraction of the total costs of the farm programs is required to pay for additional inputs attracted to or retained in agriculture as a result of production subsidies or high product prices. I refer not so much to labor or land, though these are involved, but more to capital inputs, fertilizer and other current purchased inputs. The U.S. farm programs, with an emphasis upon acreage limitations, appear to be no more effective in creating additional income for farm labor and management than the programs used in Western Europe.

Almost exactly two decades ago I argued that while price policy could be used to improve the efficiency with which resources were used within agriculture, price policy was not an effective tool for meeting important income goals.[21] Price policy can make little or no contribution to the elimination of poverty in agriculture, to the reduction of the dispersion of income within agriculture or to the achievement of returns for farm labor equal to the returns to comparable labor in the rest of the economy. The most that price policy can do is to contribute to greater stability of farm income. But most industrial countries have placed major or sole reliance upon price policy as the means for increasing farm incomes.

The adverse consequences for the less developed countries of the agricultural policies of the industrial countries are primarily due to the efforts by the latter to achieve income objectives by the use of commodity price policy. Only as the industrial countries turn to direct approaches for meeting their farm income objectives—education, training, aids to mobility, health, employment services, and income grants not associated with production or productivity—will it be possible to eliminate the existing inconsistencies between their agricultural and foreign economic policies.

If we are guided by consideration for the interests of others and if we are imaginative in devising ways for meeting the legitimate income objectives of the farm populations of the industrial countries, we can improve upon the present highly unsatisfactory state of international trade and create a basis for international trade in which the less developed countries will have a real opportunity to compete effectively for markets. If, for lack of imagination and consideration for the interests of others, we continue along our present paths, we must accept responsibility for the adverse consequences of our actions upon others who are immeasurably less fortunate than we are.

Notes

The University of Chicago Office of Agr. Econ. Res. Paper No. 6422.

I wish to call attention to the following publications, which include material that is directly relevant to the contents of this paper: Department of Economic and Social Affairs, UN, *World Economic Survey 1962, I. The Developing Countries in World Trade*, New York, 1963; Economic Commission for Europe, UN, *Economic Survey of Europe in 1960*, Geneva, 1961; and Lawrence W. Witt, *Towards TransAtlantic Agricultural Policy Bargaining*, Natl. Planning Assoc., Washington, 1964. The first draft of this paper and Witt's paper were prepared at about the same time and have much in common.

1. J. S. Nicholson, *The History of the English Corn Laws*, London, 1904, pp. 129–130.

2. From a speech given by Secretary of Agriculture Orville L. Freeman to the annual meeting of the Rice Millers Association, Houston, Texas, January 31, 1964.

3. From a speech given to the Agricultural Policy Forum, Chicago Board of Trade, Chicago, December 12, 1963.

4. Most of the exports of the less developed regions are to industrial countries—in 1962 approximately 75 percent.

5. UN, *World Economic Survey 1962, I. The Developing Countries in World Trade*, p. 115.

6. Based on *ibid.*, p. 119.

7. Economic Commission for Europe, UN, *Economic Survey of Europe in 1960*, Geneva, 1961, Chap. 5, p. 6. The estimates and projections exclude the major petroleum exporters.

8. Gavin McCrone, *The Economics of Subsidizing Agriculture*, London, 1962. Rather similar estimates for the major agricultural products may be found in ECE, *Economic Survey of Europe in 1960*, Chap. 3.

9. The modifications included a change in the basis for measuring the degree of protection—from output valued at national prices to output valued at import prices—and the inclusion of poultry meat in the calculations.

10. The estimates of agricultural protection are not subject to a simple interpretation; in all of the countries, other sectors of the economy are protected in varying degrees and in ways that are not reflected by calculations of average tariff rates. Furthermore, the specific percentages should be considered as general approximations of the degree of protection since many difficult problems of comparison are involved. These include differences in quality of products imported or exported and local production, exchange rate complications and location and stage of processing effects upon prices.

11. "Trends in European Agriculture and Their Implications for Other Regions," *Monthly Bul. of Agr. Econ. and Stat.*, Vol. 9, November 1960, pp. 1–8.

12. The Netherlands was an exception until recently.

13. "Efficiency and Welfare Implications of United States Agricultural Policy," *J. Farm Econ.*, Vol. 45, May 1963, pp. 331–342.

14. *France and Agriculture*, no author, distributed by French Information Service, New York, December 1963, p. 48, and speech by Secretary Orville L. Freeman in Wenatchee, Washington, June 11, 1964.

15. FAO, *The State of Food and Agriculture 1963*, Rome, 1963, p. 50. The data exclude the trade of eastern Europe, USSR and Mainland China. The data also exclude intraregional trade.

16. USDA, *The World Food Budget, 1962 and 1966*, Foreign Agr. Econ. Rpt. No. 4, Rev., January 1962, p. 9. Values based on 1958 wholesale or export prices or major exporting countries.

17. UN, *World Economic Survey 1962, I. The Developing Countries in World Trade*, Chap. 1, p. 1.

18. GATT, *International Trade, 1962*. The country classifications differ slightly in the GATT and UN studies.

19. FAO, *State of Food and Agriculture 1963*, p. 51. Data exclude the communist areas and the changes in value of exports are for broad regional groups, not the change in the value of trade of each country.

20. See *France and Agriculture*, pp. 50–55; and Pierre Uri, *Partnership for Progress, a Program for Transatlantic Action*, New York, 1963, pp. 27–40, esp. pp. 33–40.

21. "Contribution of Price Policy to the Income and Resource Problems in Agriculture," *J. Farm Econ.*, Vol. 26, November 1944, pp. 654–664.

14

World Agriculture, Commodity Policy, and Price Variability

The primary emphasis in this paper will be upon governmental agricultural commodity policies and their effects upon price variability. It is the commodity policies of the governments of the world that provide the links between what occurs in one part of the world and in the rest of the world's food and agricultural systems.

When one discusses price variability or price stability in today's world, one must be quite specific in indicating the context. The market for most farm products is so fragmented as a result of governmental regulations and interferences with trade across national boundaries that there is often little relationship between the behavior of a particular price series, such as prices received by farmers for grain, over time in different countries. Not only are there substantial differences in prices for approximately the same product at a moment of time, but there are major changes in the differences over time. The differential changes reflect primarily the effects of governmental policies, though to some small degree variations in the costs of transportation can affect the difference in prices between two points in space.

There is an obvious point, which I must admit eluded me in several abortive efforts to prepare this paper, that I feel is worth making. If governments are interested in price stability for agricultural commodities, their primary interest is in stability of prices within their own countries. This is not to say that governments have no interest in the stability of prices at which farm products are traded among nations, but past behavior of most governments and even a cursory examination of policies and programs designed to stabilize prices indicate that there is far less concern with the

Reprinted by permission from the *American Journal of Agricultural Economics* 57, no. 5 (December 1975): 823–28.

stability of prices outside than inside national boundaries. This is hardly a surprising conclusion.

In fact, the concern of most governments with internal price stability, with little or no regard for external effects, is comparable to the primary concern of governments with internal resource adjustments in agriculture. The agricultural and trade policies that were followed in recent years by most industrial nations to minimize their own need to adjust forced other nations to undergo relatively larger adjustments than would have been needed if all nations had participated on a more equal basis in the required resource adjustments.

There has been little recognition of the extent to which one nation or region achieves price stability at the expense of instability to others. This has not been an important issue in international negotiations or in trade negotiations. Where price stability has been considered an issue, it is in terms of arrangements that would limit fluctuations in international prices through commodity agreements or buffer stocks. The effects of national policies on price instability elsewhere have received almost no attention in such discussions.

The causes of international price instability have generally been attributed to supply fluctuations due largely to output variations resulting from natural phenomena, the breakdown of buffer stock arrangements, or fluctuations in demand over the course of business cycles. Instability has also been attributed to cobweb-like phenomena for tree crops or sugar where the time lag between investment and production can lead to alternating periods of high and low levels of production. These causes are real; there can be no doubt about them. But what can be doubted is whether these causes are the primary ones, at least for the very wide variations in international prices of most farm products such as we have seen in the past three years or perhaps even during the great depression.

National Price Stabilization

Market price stabilization requires that either the demand or supply functions be very elastic. Practically, for a given geographic area relatively little can be done to make demand functions highly elastic. Thus, programs designed to achieve market price stability must work through modifications of the supply function. The supply function for a given geographic area can be made very elastic in one of two main ways—by managing exports and/or imports and by storage.[1] Obviously the two techniques can be combined, as they have been in the United States and Canada for most of the past three decades.

The different methods of achieving an elastic supply curve for a given geographic area have very different effects upon prices in international markets. The control of imports and/or exports to stabilize internal prices increases the variability of prices elsewhere in the world. If internal prices are fully stabilized by controlling the flow of trade, this means that the price elasticity of demand for imports or the price elasticity of supply for exports, whichever is relevant, is zero. None of the variations in world supply and demand is absorbed by a country or region following such a system. All of the price effects of variations in supply or demand thus must be absorbed by others.

The effects of such policies of national price stabilization through the control of trade can perhaps be visualized best through a hypothetical example. Assume that half of the world's consumption of grain occurs within economies that stabilize internal prices through the control of trade. There is an autonomous shock that reduces the world's output of grain by 4%, and the only stocks that exist are working stocks. Assume further that the short-run price elasticity of demand for grain for the world is -0.1. The effects of the national price stabilization schemes are to require prices in the part of the world that normally consumes half of the world's grain to reduce their use by 8%. If the price elasticity of demand were -0.1 in this part of the world, the increase in price from a world production shortfall of 4%, assuming stable demand, would be 80% (approximately). If there were no national price stabilization schemes through the control of trade, the increase in price for the world would be 40% (approximately). Thus, half the world following such schemes doubles the price swings for the rest of the world unless there are stocks to absorb the shortfall in production.

Price Stabilization through Storage

If prices are stabilized through accumulation and de-accumulation of stocks, demand and production variability would be absorbed through changes in stocks. At some cost, prices could be stabilized within a specified price range—not with certainty unless the cost approached infinity but with a very high probability of success.

In fact, during the 1960s for wheat and the feed grains, the world came close to having a storage system that stabilized the international prices of these grains to a remarkable degree. It was a policy operated primarily by the United States and Canada with a late assist from Australia. The primary objective of the storage policies was not price stability; the storage function was largely an inadvertent outgrowth of efforts to increase prices and returns for the grains. In fact, the storage role was not only inadvertent but was also largely unwanted.

One of the major factors in the substantial modification of the U.S. farm programs during the early 1960s was the political concern over the high costs of storing the grain (and cotton) that could not be disposed of at the price support levels then prevailing. Similarly, the revisions in our farm programs that came in the late 1960s and early 1970s were motivated by the same considerations—the fear that stocks would increase to levels that could not be politically sustained. This was the view not only in the United States but also in Australia and Canada. The three governments took steps to drastically reduce the production of wheat and, in fact, accomplished this end. In the process, stocks of wheat held by the major grain exporters were substantially reduced from mid-1970 to mid-1972 by almost 20 million tons or by one-third (Johnson, p. 55).[2] The reduction in the stocks of wheat and the unwillingness of the United States to accumulate large quantities of feed grains occurred even though the absolute level of grain stocks in the exporting countries was significantly lower than in the early 1960s. In mid-1960 and mid-1961, the grain stocks of the major exporters represented about 15% of world grain production. In mid-1970, such stocks equalled 10% of world production. Even so, the three major grain exporters desired to reduce stocks further and did so.

As noted earlier, the storage and pricing policies of the major exporters achieved substantial stability of the export prices of grain during the 1960s (Johnson, pp. 54–55). For the crop years 1960–71, wheat prices were held within a range of $59 to $65 per metric ton in eleven of the twelve years; in one year (1969–70) the annual average price was $53. Corn prices were nearly as stable being held within a range of $47 to $57 per ton except for 1970, the year of the corn blight. Even in that year, the annual average export price was $61.

The price stability during the 1960s was achieved during a period of significant variability in world grain production. In fact, the absolute shortfall of world grain production below trend during 1961–62 through 1965–66 was greater than during 1971–72 through 1974–75, 72 million tons compared to 36 million tons. Even if 1970–71 is added to the later period to include the effects of the corn blight on U.S. and world production, the shortfall for the period in the 1970s was 62 million tons. The shortfall of production below trend in the 1970s, relative to trend production and consumption, was at most two-thirds as large as during the first part of the 1960s (Johnson, p. 51).

Why, then, was the behavior of the prices in the international markets so different between the two periods? One reason was that the major exporters had held their stock levels to a lower level in the 1970s than in the 1960s. There is absolutely no evidence that, except for India, any country in the world made any effort to increase stocks as an offset to the declines in North America and Australia (UN, FAO 1974, p. 7). Thus, the change in storage policy of the major exporters appeared to be acceptable to the major importers. If there was any anxiety, it did not find expression in increased stocks.

Price Policies and International Instability

But I believe that a second reason was far more important as an explanation of the different price behavior in the 1970s as compared with the 1960s than the lower level of grain stocks in the later period. This reason was that a much larger percentage of the world's grain production and consumption in the 1970s than in the 1960s occurred within the framework of policies to achieve internal price stability through the control of imports and/or exports. It was not so much that basic policies had changed as it was that either the ability or the will to pursue price stabilization policies more effectively had changed.

For example, the basic features of the announced agricultural and food price policies of the Soviet Union were the same in 1972 as in 1963. Prices paid to producers were fixed, and prices at which farm products were sold as farm inputs or to consumers were also fixed and stable. The difference between 1963 and 1972 was that a much greater effort was made in the later year to make the prices effective prices, to more nearly equate supply to demand at those prices. In the earlier period, substantial shortfalls of supply relative to demand were tolerated; in the later period, serious efforts were made to eliminate or minimize the shortfalls. Thus, after the poor crop of 1963, the Soviet Union imported only about one-third of the grain production shortfall; the same relationship held following the poor 1965 crop. But in 1972–73, net grain imports exceeded the production shortfall relative to the previous year by approximately enough to maintain use at the trend level for 1972–73 (Johnson, p. 28).

Similar changes in the effectiveness of implementing price stabilization policies occurred in the European Community and, probably, in China. It is generally ignored that China has imported more grain, on the average, during the past three years than during the very difficult years in the early 1960s, or that since 1969–70 China has had larger aggregate net imports of grain than the Soviet Union (USDA, p. 24).[3] It appears that the countries of Eastern Europe and Western Europe also have effectively implemented policies to stabilize prices and use (around a rising trend) in recent years.

I earlier used a hypothetical example in which it was assumed that half of the world's grain use occurred within the framework of national price stabilization achieved primarily by control of trade. The level of one-half was not chosen arbitrarily. Approximately one-half of the world's grain use in recent years has occurred in the Soviet Union, the rest of Europe, and China (USDA, p. 24). These regions of the world increased their share of world grain use from 49% in 1969–70 through 1971–72 to 52% in 1974–75. In fact, the absolute increase in grain use of 68 million tons in these areas in 1974–75, compared to the earlier period, almost equalled the increase in world grain use of 73 million tons; the rest of the world increased grain use by only 5 million tons.

It would be an interesting exercise to determine how much the increase in the average price of grain received by farmers increased in the world between, say, 1971 and 1973 and 1974. A farmer in the United States would refer to an increase of approximately 175% in nominal prices, though perhaps 75% in real prices.[4] I have made a rough guess for the world as a whole, and it is little more than a guess. It is that the real price of grain received by the world's farmers increased by no more than 40% between 1971 and 1974. In the European Community it appears that the real grain price actually declined over this period (Johnson, p. 34).

If a nation or region is successful in achieving price stability, prices do not serve the function of influencing either consumption or production when the world's demand-supply balance has changed. Thus, as noted before, all of the adjustment to the variability of supply and demand must be made elsewhere in the world. In the recent period these adjustments fell primarily upon two groups of countries: the major grain exporters and the low income, developing countries that imported grain.

There were, of course, other factors that increased world prices of grain. One was the devaluation of the Canadian and American dollar. The dollar prices of grain could have been increased by such devaluations by perhaps 15%; with that increase the real price of grain to the major importers would have remained unchanged. There was obviously some speculative overreaction to the situation that developed in 1973 and 1974. However, it is not at all obvious that the major speculators consisted of evil individuals that frequent the grain pits of the Chicago Board of Trade. Governments or governmental purchasing agents may well have been far more important, though this is only an impression that I cannot document. Another factor was that the major exporters held to low export prices for grains for too long during the summer of 1972. Pricing policies that had worked reasonably well for more than a decade were simply inappropriate in the situation that arose.

The radical interference with the operation of the market due to the U.S. wheat export subsidy resulted in maintaining the export price of wheat at too low a level. Without the export subsidy, market prices would have much more promptly reflected the impact of the enormous grain exports contracted for in 1972. No one knows, outside of a few individuals in Moscow, how much impact substantially higher grain prices would have had on the amount of Soviet imports. Given the level of purchases already made in 1975 at significantly higher real prices than in 1972, it is not clear that higher prices in 1972 would have had a significant impact on their imports. This may sound as though their behavior was irrational. However, imported grain at $140 to $150 per ton is in the range of the Soviet average procurement price and significantly below marginal procurement prices.[5] It could be true that in the range of grain prices of $75 to $150 per ton, their import demand was very inelastic. I do not know that this is the case, but I would not be surprised if it were.

Reserves and International Price Stability

The conventional argument for a reserve is to offset uncontrolled variations in supply. This argument may be valid for an individual country that does not engage in international trade. It is not the valid explanation for the holding of substantial stocks in excess of working stocks for the world as a whole. Yagil Danin, Daniel Sumner, and I have estimated the optimal grain reserves for the world for 1948–1973 if there were free trade in grains (p. 27).

The criterion for optimal grain reserves was that the expected increase in price would equal the expected increase in marginal cost of storage. Storage costs were estimated to be $7.50 per ton and a real rate of interest of 5% was assumed. Given the probability distribution of world grain production, based on actual variability of grain production for a period of approximately twenty-five years, we found that in only one year out of five carry-over stocks would be expected to be positive, and in only one year out of twenty would such stocks exceed 10 million tons. This was for a level of world grain production of approximately 1.2 billion tons. While if we had taken into account demand variability—the demand function was assumed constant except for a trend coefficient—carry-over levels would have been increased by a few million tons. However, we assumed a rather low price elasticity of demand (−0.1), and this probably resulted in an overestimate of carry-over levels.

Thus, for the world as a whole, grain production variability is not large enough to make it profitable to hold large reserves. What may make it profitable to hold substantial reserves are the governmental policies designed to achieve a high degree of price stability for individual countries or regional groupings such as the European Community. These policies result in significant year-to-year variability in the excess demand and supply functions for grain by these countries or regions. In the absence of reserves, such variations in the demand for imports or the supply of exports result in variations in the international prices of grain.

Would it be profitable for someone—governments or private traders—to hold carry-over stocks in response to largely policy-induced variations in import demand and the production variability in the major exporting countries? The answer to that question is clearly in the affirmative. Before the massive direct and modern governmental intervention in the markets for farm products, which can be dated from about 1930, the private market did hold substantial carry-over stocks of grain, especially wheat. Stated approximately for wheat, in the United States about half of annual production deviations, either positive or negative, were offset by variations in carry-over and most of the remainder by variations in exports from 1896 through 1927 (Working, p. 173).

During the first part of this century, substantial interferences with the trade in grain existed, but the interferences consisted of specific tariff duties. In many countries, especially in Western Europe, the tariffs were highly protective, but imports were determined primarily by market phenomenon, not by a bureaucrat or a legislature. Thus, it is possible that the current governmental policies have introduced such a greater degree of uncertainty into the international grain market that the private trade would be less effective in minimizing price fluctuations than it was a half century ago.

Quite frankly, we do not know whether it would be in the interest of the governments of the major exporters to jointly or singly adopt a carry-over policy for the grains not as a price support measure but as an investment. I hope that research that I am just now beginning, supported by the National Science Foundation, will provide at least a partial answer. An attempt will be made to determine the probability distributions of import demand functions for wheat and the feed grains. If this can be done, it should be possible to determine what the carry-over levels for the United States or for the major exporters should be for any given total supply at the beginning of a year. One assumption that will be made is that the expected marginal return from the investment in carry-over stocks should equal the expected marginal costs.

Some may argue that this approach will result in relatively small levels of carry-overs, certainly much smaller than held by the major exporters in the early 1960s and probably lower than was held in 1972. If true, and I do not know if this will be the case, who should pay for the losses incurred in holding larger stocks than implied by the optimal inventory rule? Should it be producers in the exporting countries in return for greater price stability? Should it be the taxpayers in the major exporting countries? Or should it be the taxpayers in the importing countries and consumers generally who should pay?

A persuasive case has been made that it is consumers who gain from a reserve policy (UN, FAO 1975, p. 7). The case depends, to a considerable degree, on the assumption that the price elasticity of demand becomes smaller absolutely as price increases. If this assumption is correct, then shortfalls in supplies such as were witnessed in 1973 and 1974 result in very large transfers of income from consumers to producers. Consumers thus might find it in their interest to subsidize the holding of stocks in a greater amount than would be called for by the optimal storage or profitability rule.

If the case for consumer benefits is valid, then it is probably not in the interest of grain producers to subsidize or to encourage the holding of stocks larger than indicated by the optimal carry-over rule. However, it is possible that the exporters may find it necessary to hold fairly substantial reserves as a means of inducing importers to hold their degree of self-sufficiency in check or to actually decrease it (Johnson, p. 58).

Concluding Comments

The world need not have a period of price instability for major storable farm products such as it has witnessed since 1972 and is likely to have over the next year or more. If there were substantial liberalization of trade in farm products, price instability would be significantly reduced for internationally traded products. Trade liberalization would permit private traders and marketing firms, whether publicly or privately owned, to engage in price- and supply-stabilizing reserves. There would remain considerable price instability, but the wide swings of recent and near future years almost certainly would be avoided.

Realistically, there is little hope of enough trade liberalization over the next decade to make a significant contribution to international price stability. It is not only Western Europe and Japan that would have to modify domestic agricultural policies but also the Soviet Union and China.

Given the numerous and uncoordinated national efforts to achieve internal price stability, the only feasible approach for achieving price stability in the international markets is through the creation of commodity reserves. Probably the only significant possibility of establishing a reserve policy that could be sustained and would contribute a significant degree of stability to international markets without destroying the capacity of the price system to influence the allocation of resources and consumption decisions in a reasonably efficient manner would be through the cooperative efforts of the three major grain exporters. But if such a cooperative effort attempted to hold price changes within very narrow limits, such as 25%, the effort would fail due to the unacceptably large costs that would be involved.

Price stability has economic and social values. However, with national agricultural policies as they are in countries that consume one-half of the world's grain, the costs of achieving a substantial degree of price stability in international markets will be large. It is a truism that the price stability objective must be related to a level of costs that is acceptable to those who will bear those costs.

Notes

The preparation of this paper was partially supported by a grant from the National Science Foundation to the University of Chicago. The author is solely responsible for the views expressed.

1. It could be argued that storage is a means of making the demand function highly elastic. When stocks are being increased, it is clearly appropriate to speak in terms of the demand function. However, since stocks can be decreased as well as increased and it is the supply available for consumption that adjusts rather than prices and consumption, I have considered a buffer stocks operation as a means of making the supply function for a given time period highly elastic. The underlying effects are the same, of course, whether one views a buffer stock operation as either a demand or supply phenomenon.

There are some other methods of making the supply somewhat more elastic than it would otherwise be, such as marketing limitations or acreage controls or destruction of part of the output. Price discrimination, as in fluid milk markets, can be used to make the supply to one segment of the market highly elastic by reducing the elasticity of supply to other segments of the market, but the methods discussed in the text are the major ones with relatively broad applicability.

2. The tons used in this paper are metric tons.

3. However, Chinese imports have not exhibited the erratic behavior exemplified by the trade of the Soviet Union. Chinese grain imports do not appear to have been significantly influenced by the real price of grain.

4. The estimated changes in prices do not include the direct payments received by U.S. farmers. If these were included in the returns for 1971, the increase in returns for the later years would be significantly less than 175%. The data refer to crop years.

5. In this calculation, the official exchange rate of 1 ruble = $1.45 is used.

References

Danin, Yagil, Daniel Sumner, and D. Gale Johnson. *Determination of Optimal Grain Carryovers.* Office of Agricultural Economic Research, University of Chicago, Paper No. 74:12, Mar. 1975.

Johnson, D. Gale. *World Food Problems and Prospects.* Washington: American Enterprise Institute, 1975.

United Nations, Food and Agriculture Organization. *Food Reserve Policies for World Food Security: A Consultant Study of Alternative Approaches.* ESC:CSP/75/2. Rome, Jan. 1975.

————. *World Food Security: Draft Evaluation of World Cereals Stock Situation* (Preliminary Draft). CCP:GR 74/11. Rome, July 1974.

U.S. Department of Agriculture. *World Agricultural Situation*. ERS WAS-7, June 1975.

Working, Holbrook. "Disposition of American Wheat since 1896." *Wheat Studies of the Food Research Institute*. 4 (1927–28):135–80.

15

An Optimization Approach to Grain
Reserves for Developing Countries

D. GALE JOHNSON AND DANIEL SUMNER

The increased interest in grain reserves, in the last few years on the part of decision-makers and scholars, has been met by a growing body of research designed to consider various aspects of food carryovers. This paper represents an attempt to specify quantitatively optimal grain carryovers for various countries and regions of the world. We use the concept of optimality in a precisely defined sense and make other particular assumptions which allow the application of a calculation methodology which is developed below to recent data. We deal with those year to year carryovers which might be held against the uncertainties of price caused by variable supplies (or potentially demands). We do not consider the problem of working-stocks nor of holdings over a yearly cycle. Our problem is the costs and gains of quantity arbitrage over time horizons of greater than a single year.

The basic unit of analysis is a single country or region (which may be as large as the whole world) which is then treated as a separate market area. Each region has a stochastic process which represents grain production at each point over the relevant time horizon. This gives information concerning expected supplies for each year and the potential variability of these supplies. The demand facing a region's supply is assumed to follow a deterministic path over time with international trade built into the relevant demand concept to keep this demand equal to long run trend supply at the base price. Thus the long run path of the relative price of grain is deterministic and constant. The price elasticity of demand is an important parameter in the calculation of optimal carryover because it determines the marginal value of moving supplies (over time) from high supply situations to periods of low supply. Storage costs and discounting by the real rate of

Reprinted from *Analyses of Grain Reserves: A Proceedings*, compiled by David J. Eaton and W. Scott Steele (U.S. Department of Agriculture, Economic Research Service Report no. 634, August 1976), pp. 56–76.

interest convert the gross positive returns from storage into a net welfare calculation.

Our calculation method, which is developed below, is based directly on the seminal work of Robert L. Gustafson.[1] After setting up the formal procedure to be applied, we discuss the empirical specification of the problem and of each parameter. We have aggregated all cereal crops together using FAO production data. Further we have used the FAO regionalizations because of data convenience and because the empirical results of this paper are meant as representative and not as specific policy proposals. Discussion of the demand and cost specifications of the model is followed by an explanation of the use of time series on grain production to derive forecasted probability distributions for future grain supplies which enter the carry-over determination. Our results from applying the model as specified are summarized in Tables 1 to 5 below. A potential international approach to stabilization based on an insurance concept is discussed quantitatively and its impacts on carryovers of countries and regions are presented.

It should be noted explicitly that this analysis does not imply any particular governmental role in the holding of grain carryovers. Our procedures yield an optimal quantity which is equal to that which would be held in a free market situation by private firms under the conditions specified and given that other barriers to such stocks did not exist. The issue of location of stocks within a country or region is also not considered in this paper and would involve spatial optimization based on transport costs, etc. Further we do not discuss specific operations of grain reserves either publicly or privately held.

While our analysis might be used in developing policy proposals both nationally or in an international situation, our purposes here are only to demonstrate the determination of carryover levels under specified conditions.

Basic Model and Empirical Specification

Formal Model

The next few pages outline the basic formal model and calculation method which was applied to various countries and regions. Begin by specifying a finite end point T periods (years) in the future, where T is large enough so that effects of this finiteness are negligible. Cereal production in year t, denoted X_t is a random variable with probability density $f_t(X_t)$. Supply for year t is

$$S_t = X_t + C_{t-1},$$

the sum of current production and carryover from the previous year denoted C_{t-1}. Consumption in year t is

$$Y_t = S_t - C_t,$$

and the relevant welfare measure in period t is a function solely of consumption in that year;

$$W_t = W_t(Y_t).$$

Storage costs of holding supply from t to the next period are

$$G_t = G_t(C_t)$$

and the discount factor for intertemporal exchange is denoted by δ_t.

The problem in a given year is to maximize the expected value of the discounted sum of welfare over the horizon. That is in period ℓ select C_ℓ, $0 \le C_\ell \le S$ to maximize:

$$E = \int_{-\infty}^{\infty} \int \cdots \int_{-\infty}^{\infty} \sum_{t=\ell}^{T} (\delta_t (W_t - G_t) f(X_\ell, X_{\ell+1}, \ldots, X_T)) dX_\ell \cdots dX_T.$$

We employ the assumption of independence of the production distributions across years, so that

$$f(X_\ell, \ldots, X_T) = f_\ell(X_\ell) \cdot f_{\ell+1}(X_{\ell+1}) \cdot, \ldots, \cdot f_T(X_T).$$

This simplification imposes potentially important restrictions, is empirically testable on past history and will be discussed below with the presentation of our empirical implementation.

For solution, the idea is to work back from the "final" period to the present using knowledge of the realized values of the stochastic supply function to solve the maximization problem. In the "final" period no "future" production is relevant. For the year T, with S_T given, we want to maximize the net welfare,

$$W_T(S_T - C_T) - G_T(C_T).$$

But from the point of view of year $T-1$, S_T is a random variable because X_T is not yet realized. Letting V_T be the solution to the period T maximization problem for a given supply, we may write the expected value of V_T as

$$EV_T = \int_{-\infty}^{\infty} V_T(X_T + C_{T-1}) f_T(X_T) dX_T.$$

After integration over X_T this is a function of the carryover from the previous period C_{T-1}. So in year $T-1$ optimal carryover is found as, that C_{T-1} which maximizes

$$W_{T-1}(X_{T-1} + C_{T-2} - C_{T-1}) - G_{T-1}(C_{T-1}) + \delta_T (EV_T(C_{T-1})).$$

This simply says that current *loss* in welfare in $T-1$ arising from positive carryover plus the storage costs are balanced against the expected *gain* from additional supply in year T. The solution values of this problem and the optimal carryover may be written as functions of $T-1$ supply; $V_{T-1}(S_{T-1})$ and $C_{T-1}(S_{T-1})$ respectively. In $T-2$, X_{T-1} is still unrealized so S_{T-1} is a random variable. The expectation of the solution value of the problem for $T-1$ may be written as a function of C_{T-2},

$$EV_{T-1}(C_{T-2}) = \int_{-\infty}^{\infty} V_{T-1}(X_{T-1} + C_{T-2})f_{T-1}(X_{T-1})dX_{T-1}.$$

Generally, then, for period $T-k$, the problem is to find $0 \le C_{T-k} \le S_{T-k}$ to maximize,

$$W_{T-k}(S_{T-k} - C_{T-k}) - G(C_{T-k}) + {}_{T-k}(EV_{T-k+1}(C_{T-k})).$$

The final term subsumes the value of the carryover to the next period given the probability distributions of production for all relevant future periods. (The number of relevant periods obviously depends on the discount factor δ.) So selecting an optimal C_{T-k} requires that S_{T-k} be given or alternatively any potential S_{T-k} may have an optimal C_{T-k} associated with it. Our results reported below are a consequence of this methodology being carried out empirically.

Empirical Specification

The solution method calls for a finite horizon after which the value of storage is zero. We set $T = 1990$ which is 15 years after the year of any of the results we report and is far enough in the future that no finite end point affects are felt on the optimal carryover results. With the horizon set each of the other components of the model may be specified. A constant elasticity demand function with a constant exponential rate of growth over time was the basis of the consumption side of the model. This is represented by:

$$Y_t = \exp(\gamma t) \cdot Ap_t^{\eta},$$

where Y_t is the relevant grain demand in a region and P_t is the price of grain in year t. The parameters are: the growth rate γ, the constant term α, and the price elasticity η.[2] This demand function yields our welfare measure which is simply an index of the area under the demand curve:

$$W_t(Y_t) = \int_{\varepsilon}^{Y_t} (A\exp(\gamma t))^{-\eta} Y_t^{1/\eta} \, dY_t$$

where ε is a small positive number entered for computational convenience. Carrying out the integration:

$$W_t(Y_t) = (A\exp(\gamma t))^{-1/\eta} \cdot \frac{1}{1+1/\eta} \, (Y_t^{(1+1/\eta)} - \varepsilon^{(1+1/\eta)}).$$

The model has been applied to various individual countries and regional aggregates which, within the context of the model, are then held, in a specific sense, separate from the international market in cereals. Given a particular country or region we allowed for free market movement of grain within the unit but assumed that trade with the outside world was *not* available to offset year to year variation in own supply. In each year the "normal" levels of imports or exports of grain are committed so that the "relevant" demand for a region's carryover calculation is that demand (including net "normal" international trade) that may be expected to be met by domestic supply. It is assumed that long range adjustments in domestic demand and supply in trade will be made to keep this equilibrium in the "normal" levels. Thus, the function of regional carryovers is only to meet the short term fluctuations. If the world as a whole were the only relevant market unit for grain only the relative fluctuation in aggregate world supply would be relevant for carryover calculation. However, given policy restrictions and other barriers, it may be appropriate to estimate optimal carryovers for smaller market aggregates.[3]

To operationalize these ideas we calculated A from

$$A = Y_m P_m^{-\eta}$$

where $t = $ m was taken as a convenient base year. P_m is an approximation of grain price for the region[4] and Y_m is set equal to the trend level of regional grain production in year m. Thus Y_m includes what would have been "normal" exports or imports in that year added to or subtracted from domestic demand. In order to keep this condition throughout the horizon γ was set equal to the growth rate in production which was estimated for the supply side of the model from past grain production data as described below.

Our measures of η, the price elasticity of demand, were approximated after considering regional estimates made by Rojko et al. [3] of the USDA. After calculating the appropriate weighted aggregates of these individual own and cross elasticities for wheat, rice and coarse grains, we arrived at an estimate of own price elasticity of all grains.[5] These elasticities ranged between -0.10 and -0.30 but since it may be true that demand becomes more inelastic at higher prices we chose to use the lower absolute value elasticity in most of our work. Optimal carryovers were quite sensitive to changes of η over this range as was to be expected. A 5% drop in supply implies a 50% rise in price at $\eta = -0.1$ and only a 15% rise in price at $\eta = -0.3$. The carryovers reported below based on $\eta = 0.1$ may then be overestimates of the optimal carryover if demand were truly more elastic.

A simple cost of storage function of the form,

$$G_t = MC_t$$

was applied, where M is the average and marginal yearly cost per ton of storing grain including in and out charges and physical loss. For simplicity we used $M = \$7.50$ for all regions and this was held constant over time. Long range real rate of interest was held constant for all regions at 5% leading to the discontent factor, $\delta_t = \left(\dfrac{1}{1.05}\right)^t$.

The assumption that the demand function relevant to carryover calculations follows a deterministic trend equal to that predicted for production implies that the probability distribution of domestic prices over the whole horizon will be generated only by the probability distribution of production. World trade price or long run resource costs of grain are thus held to be deterministic and constant. Further, because the adjustments are made sequentially to keep expected excess demand equal to zero at base price, no long term variance of the prediction of production is relevant for carryover decisions. That is, the potential variability in production that is appropriately used to specify the $f_t(X_t)$ over the horizon includes only variance in the production distribution itself, not the variance embodied in the estimates of parameters used to forecast production. Insofar as those estimates deviate from true parameters the long term adjustments on the demand side correct for the error.

Time series of grain production from the FAO were used to develop the forecasts of the probability distributions $f_t(X_t)$ for each year t of the carryover horizon. In order to convert information about grain production in the past into useful information about the future a simple log linear trend was estimated. For each country or region

$$ln\ X_i = \alpha + \beta i + U_i, i = i_1, \ldots, i_n$$

was fit by OLS, where X_i is production in each year i for which the FAO data was available (in most cases i runs from 1948 to 1973). The estimates of α, β ($\hat{\alpha}$, $\hat{\beta}$) and $\hat{U}_i = ln\ X_i - (\hat{\alpha} + \hat{\beta}i)$ were used to generate a probability distribution for each of the years $t = m, \ldots, T$ of the carryover horizon. For each year, each production quantity

$$\hat{X}_{ti} = exp\ (\hat{\alpha} + \hat{\beta}t + \hat{U}_i)\ i = i_1, \ldots, i_n$$

was assigned equal probability. For example with $n = 26$ and $t = 1975$, $f_{75}(X_{75})$ is a multinominal distribution with each point assigned a probability of approximately 0.0385.[6] So that while the theory above was framed with continuous distributions our computation used discrete probability distributions.

Because of the log linear specification of production trends with an additive disturbance term, the variability in production is assumed to be a constant proportion of trend production. That is, homoskedasticity of the disturbance over the fitted period extended to the carryover horizon implies that the variability from trend production enters multiplicatively. Thus as expected production grows at rate $\hat{\beta}$ over the time the absolute variation also grows at that rate. For example the probability of a 5% shortfall below trend production will be equal in each year while the probability of, say, a 5 ton shortfall will increase over time as expected production rises. This specification of production variability is very important to the implied optimal carryover results. Assuming constant proportional variability implies that, given constant elasticity demand, the incentive to storage will remain constant over time. Since equal proportional deviations from trend supply generate equal price changes a falling proportional variability, as mean production grew over time, would imply falling carryover amounts. In fitting the model to available data we found no evidence to suggest a rejection of the assumption of homoskedasticity in the U_i. Obviously it is an empirical issue and the predicted trend in variability of production depends on what factors are behind the expected growth in mean production levels. Increases in yield per acre due to new varieties and additional fertilizer may have different implications for production variability from increases in the area planted to cereal crops.

Both our computational procedure and our estimation of the production time series have imposed independence of production probability distributions over time. In the regression the classical assumption of zero covariances between any U_iU_j, $i \neq j$ was made. When applied to the carryover horizon this assumption implies that the probability of, say, a 5 percent shortfall in year t is unaffected by the realized production in year $t-1$. Alternatively positive autocorrelation would mean that above trend production in $t-1$ would make above trend production more likely in t and likewise shortfalls would tend to follow one another.[7] Obviously a storage policy should take such information into account. A further implication of non-independence of the probability distributions over time would be that the best forecasted production in year t would depend on the realization of the production level in the previous period or periods. In this case the forecasted trend production would not follow any smooth path but would include "predicted" fluctuations.

Fluctuation in the forecasted production may result from either a simple trend model with non-independence or from a more complicated forecasting model which includes other "explanatory" variables. Given this fluctuation, its implication for domestic price response must be considered. If the base price demand remained on a simple exponential trend determined by long run factors then the "expected" deviations from trend production would produce price movements and hence would affect optimal storage. Alternatively if international trade were responsive to predicted production fluctuations then the relevant demand, inclusive of trade, might follow the path of predicted supply and thus meet the expected variations to hold domestic base price constant. In such a model only unexpected supply variability would generate a carryover response with international trade adjusting to expected fluctuation. Choosing between these potential specifications is again an empirical issue and in fact turns on geography, institution and policy. In applying forecasting models to the carryover decision the assumption of the time lag for trade adjustments is vital. For example in some situations international trade agreements may offset yearly changes in acres planted so that only yield variability would enter the probability distributions used in the storage model. As is discussed above our applications are implicitly based on a year long lag in trade adjustments to supply and our predicted production follows a smooth trend so full variability in production (but not prediction error) enters into the dispersion $f_t(X_t)$.

Results

Results of applying the method and empirical specifications described in the first section to selected countries and regions of the world are presented and discussed in this section. The model was applied under alternative assumptions to illustrate various aspects of the optimal carryover analysis. Our focus is on developing countries and regions though some results for developed regions may be of interest and are included below. This section is in two parts. The first deals with direct applications of the optimal carryover model, while the second presents and discusses a proposal for an international insurance program for cereals and analyzes its impact on optimal carryovers. We conclude with a brief summary of the method and results of this paper and suggestions for continuing the research.

Optimal Carryovers under Specified Conditions

The model generates an optimal carryover amount for each year of the horizon given a supply level for that year. Supply in a year t is the carryover from the previous year plus current year's production and hence is stochastic prior to year t. At any time prior to a given year t a probability distribution on optimal carryovers for that year may be computed based on the probability distributions of production in the years leading up to year t and some given beginning year supply level. We have computed such a probability distribution of optimal carryovers for 1975 based on the specification described in the first section and conditioned on a given supply equal to trend production for the year 1970. Thus given this 1970 supply as a starting point the probability distributions in each ensuing year determined the distributions of supplies for the next year up to 1975. For 1975 we then have the combination of the probability distributions of 1974 carryover and 1975 production which in turn determine a probability distribution of optimal carryovers for 1975. The results of such an experiment for various countries and regions are summarized in Table 1. It should be stressed that the input into these calculations is such that it applies to calculating the likelihood of levels of carryovers in future years and is not based on the best information which would actually be available for carryover decision in that year. We would have just as easily chosen 1980 as our example year for the results reported in Table 1. Below in Table 2 we illustrate an alternative experiment which uses more directly the information for years just prior to the decision year.

Table 1 shows three points on the cumulative probability distribution of 1975 optimal carryovers for some selected countries and regions based on the experiment described in the previous paragraph. The quantities which would cover the lower 0.50, 0.75 and 0.95 portion of the cumulative probability distributions are displayed in the respective columns.[8] Thus for India optimal carryover would have been 6.5 million tons or less with 0.50 probability; 9.5 million tons or less with 0.75 probability and the probability is 0.95 that carryover would have been 13.5 million tons or less. For purposes of comparison 1975 trend production levels for each entry are also listed.

Several points should be discussed concerning the results of Table 1. First, note that there are considerable differences across countries and regions in their optimal carryovers relative to their trend production levels. These differences are mainly due to differences in the variability of production. For example, at the 0.75 probability level Africa has a carryover of less than 4 percent of trend production while the Near East the indicated carryover is 9.5 percent of trend production. From the time series analysis of production trends Africa had a variance of residuals of 0.23×10^{-2} while that of the Near East was $.68 \times 10^{-2}$. The extremes are even greater for the individual countries of the Far East region. The effect of the variability of production over time is related to the implications of aggregation on optimal carryovers. The larger the aggregate the more scope for interspatial vs. intertemporal arbitrage to offset production variability.

In our model, within each unit of analysis a single market prevails so that only supply variability of the whole region causes price fluctuation. When the covariance of production across regions is not perfect the fluctuations in production will be offsetting and hence lead to less optimal carryover. The results of such offsetting variations may be noted from Table 1. In the top portion of that table the Far East region has been broken down into five individual countries plus Bangladesh—Pakistan, and all

Table 1. Optimal Carryovers for Selected Countries and Regions, 1975
(million tons)

Country or Region	1975 Trend Production	Cumulative Probability Levels		
		0.50	0.75	0.95
A. Demand elasticity η = −.10				
Burma	6	0.3	0.7	1.2
India	100	6.5	9.5	13.5
Indonesia	16	1.6	2.9	4.4
Pakistan-Bangladesh	23	1.4	2.4	4.2
Philippines	6	0.1	0.2	0.3
Thailand	13	3.5	4.7	6.2
Other Far East	19	1.4	2.1	3.1
Africa	46	1.5	3.0	5.0
Far East	184	3.0	7.5	12.5
Latin America	78	2.5	5.0	8.5
Near East	48	2.5	4.5	8.5
All Developing Regions	353	2.5	7.5	15.0
Europe	231	1.3	5.5	9.5
North America	270	10.0	18.0	33.0
Oceania	18	8.0	10.5	15.4
USSR	199	28.0	41.0	49.0
World	1,304	0.0	2.0	18.0
B. Demand elasticity η = −.20				
India		2.0	4.0	7.5
Africa		0.0	0.5	2.5
Far East		0.0	1.0	7.0
Developing Regions		0.0	1.0	7.0
North America		1.5	8.5	22.0
USSR		13.0	24.0	37.0
World		0.0	0.0	7.0

other countries. Note that with this disaggregation India alone has as large an optimal carryover as the whole Far East region when treated as a single unit. The impact of aggregation may also be seen by noting the rows for All Developing Regions and the Whole World. Because covariance between optimal carryovers is not unity, the probability distribution of the sum of carryovers in each region is not simply the sum of the carryovers under each probability level. However we can easily compare the expected values of optimal carryovers under disaggregation compared to aggregation. The expected value of the sum of the optimal carryovers of the 7 countries and groups which make up the Far East is the sum of the expected values of the carryovers for these countries, which is 16.7 million

Table 2. Effects of Carryover Program on Available Supply Based on Actual Production, India and Africa, 1968–1975 (million tons)

Year	Actual Production	Optimal Carryover	Available Supply	Trend
		India		
1968	81.6	3.0	78.6	80.7
1969	85.1	5.0	83.1	83.2
1970	91.7	10.0	86.7	85.8
1971	90.2	10.0	90.2	88.5
1972	86.6	5.5	91.1	91.2
1973	95.4	7.0	93.9	94.0
1974	86.7	1.0	92.7	96.9
1975		at probability level		99.9

$$\text{Carryover} < \frac{0.5}{4.0} \quad \frac{0.75}{4.5} \quad \frac{0.95}{6.5}$$

Year	Actual Production	Optimal Carryover	Available Supply	Trend
		Africa		
1967		(1.0)		
1968	40.5	2.5	39.0	38.2
1969	41.3	3.0	40.8	39.3
1970	40.2	2.0	41.2	40.4
1971	40.9	1.0	41.9	41.5
1972	44.4	2.0	43.4	42.7
1973	37.1	0.0	39.1	43.9
1974	42.2	0.0	42.2	45.1
1975		at probability level		46.4

$$\text{Carryover} < \frac{0.5}{1.0} \quad \frac{0.75}{1.5} \quad \frac{0.95}{3.0}$$

tons. The expected value of carryover for the Far East is 4.5 million tons. For the four developing regions the expected value of the sum of the carryovers is 13.7 million tons compared to 5.0 million tons for the Developing Regions. This impact of regionalization on the probabilities of optimal carryovers demonstrates the value of reducing international barriers to quick and free movement of cereals in a world market. Some impediments to short term trade are geographical and technical but others are institutional and a function of policy. These regionalization examples also show the importance of very careful attention to the appropriate specification of region trade considerations when actual applications for policy are made using this framework.

Also included in Table 1 are results of applying the model to some of the FAO defined developed regions. These results are informative and show that especially for the USSR and Oceania carryovers become quite sizable relative to trend production. In both of these regions grain production is much more variable than in any other major producing areas. The application to the developed regions and to the world as a whole is more preliminary than the analysis of the developing regions because some of our maintained assumptions may be more at variance with reality for these regions. Specifically, the assumptions of: (1) non-stochastic demand, (2) no variability offsetting trade, and (3) deterministic and stable long run world market grain prices, cause more problems for these regions that include major exporters. Regions like North America and Oceania which export a significant part of their grain production face a world market for grain which may violate more drastically than for importers our assumption of non-stochastic demand. To some degree the demands they face depend on stochastic elements in production in other countries. Further exporters may have lower costs of short run changes in their trade arrangement, thus a large crop might be sold in part to new buyers and would drive down the price less than is indicated by our model. Finally for these regions and especially for the world as a whole, the long run world market price cannot be considered deterministic. The price on traded grain is stochastic and must be estimated. Thus the choice between carryovers or the use of future production should include the fact that even at mean production the future base price cannot be known with certainty. For the world as a whole especially there is no possibility in adjusting trends in production and demand by imports or exports as some stable world market price. This means that estimation error in determination of production trends is a relevant component of the variability of future production which generates an incentive to carryovers. Fortunately our regression for world grain production trends fit very well and such variance from prediction error is not large.

For comparison with the results of Table 1 the calculations reported there were repeated for some of the countries and regions using a price elasticity of demand of $\eta = -.2$. Table 1.B summarizes the significant effects of the greater elasticity in reducing the optimal carryover quantities. Since our knowledge of the appropriate price elasticities is weak we can only suggest that these results are illustrative. The USDA estimates of Royko et al. lend credence to the belief that assumptions in the range of -0.1 or -0.2 are not far off for most countries and regions.

In Table 2 we report on a somewhat different application of the model. The optimal carryover quantities for each of the years 1968 through 1974 were calculated based on actual production in those years. In order to more closely approximate actual conditions the estimates of the potential variability of production were based on residuals prior to each decision year. Thus for 1968 on the values of $\hat{U}_i$ from 1948 through 1967 were used. Results are presented for India and Africa. For India it was assumed that no carryovers were held for 1967. Both 1965 and 1966 were very bad years in India and 1967 had below trend grain output. For Africa 1967 carryovers were assumed to be one million tons which is also reasonable since 1967 was just above trend production and followed a bad crop in 1966.

Table 2 shows how a carryover program would have affected the year to year variations of supplies of grain from domestic production. For comparison the smooth exponential trend of production is also listed. In India the relatively good harvests would have caused a gradual build up of stocks until 1972; 1972 and 1974 would have seen reductions in stocks. In Africa production variability from trend was slight except for 1973 which was a disastrous year for grain production. Available reserves would have only partly made up for this unexpected shortfall. For India over the seven years 41.5 ton-years would have been held under the model we have specified. Africa would have held 10.5 ton-years.

The results from calculating optimal carryover for 1968 through 1974 given actual production levels were also used to calculate the probability distribution on optimal carryovers for 1975 conditional on the optimal 1974 carryover from Table 2. These probabilities differ from those of Table 1 because in that application supply in 1975 was based on 1970 supplies with uncertainty about the following years. For Table 2 the 1974 carryover is taken as given so the 1975 probability distribution of supply is simply the distribution for 1975 production plus the deterministic 1974 carryovers. It is interesting to compare these Table 2 results with those for India and Africa in Table 1. Differences are the result of less uncertainty embodied in the Table 2 experiment and also that for India 1974 carryovers were relatively high while for Africa they were low.

An International Cereal Insurance Program and Its Carryover Implications
The results presented in Table 1 demonstrate the value of reducing barriers to international flows of grain at short notice in terms of savings in carryover costs. We shall now consider a proposal of an international insurance program for providing security against grain production variability. We have made some calculations concerning an insurance arrangement which would guarantee that all shortfalls in grain production, in a country, below some specified percentage of the expected production in each year would be made up by insurance payments. Naturally in any actual situation the cutoff point for payments as well as the payments schedule and premiums would be a matter of negotiation. In our examples each region is insured fully against a grain harvest below 94 percent of their trend production. The difference between the actual harvest and the 94 percent level would be delivered by the insurer.

In Table 3 we show the insurance payments which would have been made from 1954 through 1973 under three grouping of countries and re-

Table 3. Insurance Payments to Selected Groups of Countries and Regions, 1954–1973 (million tons)

Year	Developing Regions F.E. Disaggregated as in Table 1	Developing Regions No Disaggregate of F.E.	Disaggregated F.E. Developing + Developed Regions
1954	1.0	0.0	17.8
1955	1.5	0.0	1.5
1956	0.0	0.0	1.9
1957	5.0	0.2	9.8
1958	2.2	0.0	2.2
1959	1.3	0.0	2.1
1960	1.8	1.3	1.8
1961	0.0	0.0	7.5
1962	0.2	0.0	.2
1963	0.8	0.0	30.0
1964	0.0	0.0	11.6
1965	6.8	3.4	29.3
1966	9.4	5.5	9.4
1967	1.2	0.0	4.7
1968	0.0	0.0	0.0
1969	0.5	0.0	.5
1970	1.0	1.0	20.5
1971	0.0	0.0	0.0
1972	5.5	3.2	19.5
1973	8.7	8.7	8.7

gions. These payments are computed as simply the sum of the payments to each country or region for each year. That is:

$$P_t = \sum_{k=1}^{K} P_{kt}, \text{ where } P_{kt} = .94 \, \hat{X}_{kt} - X_{kt} \text{ iff} > 0 \text{ otherwise } P_{kt} = 0,$$

where t refers to years, k countries or regions, P refers to payments, $\hat{X}_{kt}$ is expected production and X_{kt} is actual product for a given region k in year t.

When only the developing regions are considered the insurance payments would have been quite small, the largest payment coming in 1966 when India had such a disastrous grain harvest. Both 1972, when the Far East and Latin America had poor crops and 1973, when it was Africa and the Near East which had below trend production, were major payment years. When the developed regions are added to the insurance coverage the payment amounts rise dramatically. This is mostly the effect of the USSR which is a large grain producing area and has a history of periodic huge shortfalls in their grain production. The largest USSR shortfall over our data period was 1963 when the crop was only 75 percent of the trend value and would have generated an insurance payment of 29.2 million tons.

These payment totals in the past can be used to produce a probability distribution of payment quantities in future years. By using the percentage payments for each country or region in the past to calculate the quantities they represent a percentage of, say, 1975 trend production, and then summing across regions a probability distribution for payments to the four developing regions in 1975 was derived. This was computed such that each quantity, $Q_t =$

$$\sum_k (.94 - \frac{X_{kt}}{\hat{X}_{kt}}) \, \hat{X}_{k75} \text{ iff } > 0 \text{ otherwise } Q_t = 0$$

was assigned equal probabilities of $1/T$ in a multinominal distribution, where $\hat{X}_{k75}$ is 1975 expected production, where T is the number of years of data. This distribution showed a probability of .65 that zero payments would be made, .80 that payments would be below 2.5 million tons, .95 that they would be below 7.5 million tons with the uppermost payment possibility in this distribution being 9.2 million tons.

Note that no specific international *reserve* is implied by this insurance scheme. Given some particular operating condition an optimal carryover amount could be calculated for the insurers on the basis of the stochastic nature of the demands they face. If either the agency had some access to a short notice world market in cereals or if some of the major producing countries or regions took the insurance agency position then optimal carryovers for the agency could be calculated using the models presented in this paper.

An advantage of the insurance approach to world co-operation is that it would not seriously interfere with the incentives of participating countries to encourage growth of production over time nor to operate their own carryover programs. Since the insurance payments would only be made in the case of very bad harvests relative to a long run trend, a country would not find much encouragement to reduce its production growth. Further, all stabilization above the insurance point would be left to the individual country or region. The insurance participants carryover decisions would be made on the basis of an altered probability distribution of supplies.

The area on the production probability distribution below the critical percentage would be shifted to exactly the critical insurance payment point. In our example all production possibilities below .94 of trend would be set equal to .94 $\hat{X}_t$. With the chance of major disasters eliminated by an outside source, optimal carryovers will obviously be less than with no such insurance.

Table 4 shows the results of the same experiment as reported in Table 1 but with the insurance policy in effect. Note the significant savings in the carryovers held by most countries and regions when these quantities are compared with those of Table 1. No insurance policy effects for the world

Table 4. Optimal Carryovers for Selected Countries and Regions, 1975 Insurance Policy in Effect (million tons)

Country or Region	Probability Levels for Carryovers		
	0.5	0.75	0.95
A. Demand elasticity, $\eta = -.10$			
Burma	0.1	0.3	0.7
India	1.5	3.5	7.5
Indonesia	0.7	1.5	2.9
Pakistan-Bangladesh	0.3	1.5	2.7
Philippines	0.0	0.1	0.3
Thailand	0.6	1.2	2.1
Other Far East	0.3	0.7	1.5
Africa	0.0	1.0	3.0
Far East	2.0	5.0	10.0
Latin America	0.5	2.5	5.5
Near East	0.5	1.5	5.0
All Developing Regions	2.0	6.0	14.0
Europe	1.0	4.5	8.0
North America	3.0	10.0	28.0
Oceania	1.0	2.5	6.0
USSR	6.0	15.0	32.0
B. Demand elasticity, $\eta = -.20$			
India	0.0	1.0	3.5
Africa	0.0	0.5	2.0
Far East	0.0	0.0	5.0
Developing Regions	0.0	1.0	7.0
North America	0.0	4.5	16.5
USSR	1.0	8.5	23.5

as a whole were calculated because of the lack of an "outside" agency to act as the insurer.

The experiments for India and Africa with the hypothetical operation of an optimal carryover program over time was also repeated with an insurance policy in effect. These results are reported in Table 5. For India over the seven year period insurance payments would have been made only in 1974 but the security of potential insurance would have reduced carryovers to 12.0 million ton years, a savings of 29.5 million ton years or around $400 million in storage costs and interest. For Africa payments would have been made in 1973 and 1974 which would have had very significant effect on available domestic supply, especially in 1973. There would have been no savings in carryovers for Africa. This follows because prior to the disaster of 1973 African grain production had only had one

Table 5. Effects of Carryover Program on Available Supply Based on Actual Production, Insurance Policy in Effect, India and Africa, 1968–1975 (million tons)

Year	Actual Production	Optimal Carryover	Insurance Payment	Available Supply
		India		
1968	81.6	0.5	0.0	81.1
1969	85.1	1.5	0.0	84.1
1970	91.7	5.0	0.0	88.2
1971	90.2	4.0	0.0	91.2
1972	86.6	0.0	0.0	90.2
1973	95.4	1.0	0.0	94.4
1974	86.7	0.0	4.3	92.0
1975		at probability level		

$$\text{Carryover} < \frac{.5}{1.0} \quad \frac{.75}{2.0} \quad \frac{.95}{4.0}$$

Year	Actual Production	Optimal Carryover	Insurance Payment	Available Supply
		Africa		
1968	40.5	2.5	0.0	39.5
1969	41.3	3.0	0.0	40.8
1970	40.2	2.0	0.0	41.2
1971	40.9	1.0	0.0	41.9
1972	44.4	2.0	0.0	43.4
1973	37.1	0.0	4.2	43.3
1974	42.2	0.0	1.2	43.4
1975		at probability level		

$$\text{Carryover} < \frac{0.5}{0.5} \quad \frac{0.75}{1.0} \quad \frac{0.95}{2.0}$$

very slight shortfall of more than 6 percent, thus the calculations based on the insurance were essentially the same as those without insurance prior to 1973.

Concluding Comments

This paper has developed and illustrated the application of a specific procedure for calculation of optimal grain carryovers. We have incorporated information on production variability, demand elasticities and storage costs with restrictive assumptions of non-stochastic demand and simplified trade barriers in a mathematical optimization model. Our results, mainly for developing regions of the world, show the major impact of variability in grain production on carryovers. This emphasizes the role of policy barriers and other restrictions to international trade which allows for offsetting production variability across regions on the optimal reserves. We illustrated the potential impacts of a carryover program over time using India and Africa as examples. An insurance approach to world co-operation on food security was analyzed and found to have several useful characteristics and a major effect of reducing individual regions optimal carryovers while improving supply stability.

There is obviously ample opportunity in continuing this line of approach to world grain reserves, both in theoretic improvements and applications. Some of the most useful generalizations of this model might include the incorporation of stochastic demands, non-independent production probability distributions over time and non-constant elastic demand curves. The development of the insurance policy concept involves a thorough consideration of the interaction between the optimal behavior of the insurance agency and the insured regions. Empirically, better estimates of demand elasticities, production distribution and especially appropriate trade barriers and regionalization are important before conclusive recommendations should be forthcoming.

Notes

The preparation of this paper was supported by grants from the National Science Foundation and the Rockefeller Foundation. We wish to recognize the contribution of Yagil Danin to earlier phases of the work. The authors are solely responsible for the views expressed.

1. The intellectual lineage of this dynamic programming approach to grain carryovers goes back through Gustafson [2], to the inventory models of the early 1950's.

2. This assumption of constant elasticity form of demand might plausibly be altered to test the effect of more complicated demand functions. For example, one might argue for functions in which the price elasticity for grain was very small (in absolute value) for low consumption levels, reflecting the approach to widespread hunger in poor countries, but which had larger elasticity at higher consumption.

3. The recent famines in Africa suggest that in some cases transportation costs of moving grain within a country or region at short notice might dictate very small geographical units for carryover analysis or alternatively may suggest a particular locational distribution of storage throughout a region. Further, by adjusting the extent and composition of regions one may approximate the effects of wider scope of free trade on the analysis. Our units of aggregation range from individual small countries to the entire world. However, with the numerous governmental interferences in trade in grains the full problem of adequately incorporating international trade is extremely complex and has not yet been solved.

4. These prices were based roughly on FAO Trade Year Book data, and were in the range of $100 per metric ton for most regions. Region differences are mainly a reflection of different compositions of the all-cereal aggregate of wheat, rice and coarse grains. Since price enters only to set levels, the carryovers were relatively insensitive to variations in price in the range of $20 or $30.

5. The own price elasticity for all grain G with respect to its relative price Pg was computed as:

$$\eta_{g,Pg} = \frac{W}{G}\,(\eta_{W,Pw} + \eta_{W,Pr} + \eta_{W,Pc}) + \frac{R}{G}\,(\eta_{R,Pw} + \eta_{R,Pr} + \eta_{Rm\,Pc})$$

$$+ \frac{C}{G}\,(\eta_{C,Pw} + \eta_{C,Pr} + \eta_{C,Pc})$$

when W, R and C refer to wheat, rice and corn, and Pw, Pr and Pc are their respective prices. Each $\eta_{i,j}$ is the elasticity of cereal type i with respect to the price of j.

6. An alternative specification would have been to use a regression estimate of the variance of production together with an assumption of the form of the probability distribution of production to generate distributions with this estimated variance over the carryover horizon. For example, if U_i were assumed normally distributed this would imply log normal production and the estimated variance on the U_i would yield a particular predicted dispersion of the X_t at any year in the horizon.

7. Some analysis of the independence of the production time series over the data period was conducted using runs tests and Box-Jenkins [1] time series techniques. So called ARIMA models were fit to the available data and it was found that for most countries and regions the independence assumption roughly held. Further investigation of both the appropriate auto-covariance structure of grain production time series and of implication of nonindependence on optimal storage is certainly in order.

8. The data for the estimation of the probability distributions of production come from FAO Production Yearbooks [4]. Our definitions of countries and regions are thus based on these FAO breakdowns. Each of the developing regions refers to non-centrally planned developing countries in a given area. For details see any FAO Yearbook or other publication. With respect to the developed regions we broke with the FAO regionalization to use the FAO defined continental Europe rather than just the developed market region. Our all grain aggregate follows the FAO data except that we have converted rice paddy to the usable grain by a 0.65 conversion factor. Further information on our time series analysis on grain production and the specification of other inputs into our calculations can be made available on request.

References

[1] Box, G. E. P., and G. M. Jenkins. *Time Series Analysis, Forecasting and Control.* San Francisco, Calif., Holden-Day, Inc., 1970.
[2] Gustafson, Robert L. *Carryover Levels for Grains: A Method for Determining Amounts That Are Optimal under Specified Conditions.* USDA, Technical Bulletin No. 1178, 1958.
[3] Rojko, Anthony S., F. S. Urban, and J. S. Naive. *World Demand Prospects for Grain in 1980 with Emphasis on Trade by the Less Developed Countries.* For. Agr. Econ. Report No. 75, U.S. Dept. Agr., Econ. Res. Serv., Dec. 1971.
[4] United Nations, FAO. *Production Yearbook, 1972.* Rome, 1973 (and other years).

16

World Agriculture in Disarray Revisited

Almost fifteen years after *World Agriculture in Disarray* (Johnson 1973) was completed, I return to the topic to ask whether there have been any fundamental changes in the basic conditions affecting the production, consumption and trade of farm products and the employment and remuneration of resources engaged in agriculture. Unfortunately it takes little more than casual observation to permit one to conclude that the disarray has not diminished in the intervening period. It is apparent that the disarray has, in fact, deepened. But before I document the changes that have made the disarray more prevalent and pervasive, I want to review some of the other aspects of *World Agriculture in Disarray* that are of some analytical significance.

The basic conception of *Disarray* was that the numerous product market and trade interventions pursued by national governments had little or no effect on the welfare of farm people. This conclusion was derived from a reasonably simple application of standard neoclassical economics to conditions affecting agricultural activities in the industrialised countries.

The underlying ideas in *Disarray* were simple ones. The starting point was that economic growth requires significant adjustments in agriculture and by farm people if farm people are to share the benefits of economic growth in their economy. These adjustments impose substantial costs on farmers in terms of the need to change resource combinations, often with striking rapidity. In particular, the most difficult adjustments are required of labour.

Why Agriculture Declines

As economic growth occurs, defined as an increase in real per person incomes, agriculture's relative importance in the economy declines. The de-

Reprinted by permission from the *Australian Journal of Agricultural Economics* 31, no. 2 (August 1987): 142–53.

cline results from certain simple but absolutely fundamental relationships that are an essential aspect of a growing economy. Agriculture has the unfortunate fate of producing goods that are necessities, which means that the income elasticities of demand for farm products are less than one. Since the income elasticity of demand for all products and services in an economy is one, this means that the income elasticity of demand for non-farm goods and services is greater than one. These relationships mean that as real per person incomes increase, the demand for farm products grows more slowly than the demand for all other products and services. In fact, the disparity in the rates at which the demands for farm and non-farm products increases becomes greater as per person incomes rise from relatively low levels, such as $200 per person, to levels of $2500 or more. At low levels of income the increase in demand associated with a given increase in real per person income for non-farm goods may be two to three times that for farm products. As per person incomes increase, the income elasticity of demand for farm goods gradually declines from perhaps 0.6–0.7 to 0.2–0.1. Thus, at incomes of $5000 or more, the growth in demand for non-farm goods will be ten times more than the growth in demand for farm goods.

While the classical economists were concerned that demand, largely driven by population growth, would outstrip the growth of food supply, it has been evident for at least a century that productivity change in agriculture can be at a rate comparable to that which occurs in the rest of the economy. The increase in supply of farm products at constant real prices may not be at the same rate as the growth in the supply of non-farm goods for the reason already alluded to, namely the much slower growth in demand for food than for non-food goods. Consequently, with agricultural productivity increasing at about the same rate as non-agricultural productivity, agriculture's fate is to shrink. This is not a recent historical phenomenon. The industrial revolution brought with it the seeds of change in labour productivity in agriculture and in the capacity to cultivate lands that could not have sustained a significant farm population just a century earlier.

It has been argued that the conclusion that agriculture must decline need hold only for a closed economy, either a particular nation or the world as a whole. If the comparative advantage of agriculture is what I call absolute—only farm products are exported and only non-farm products are imported—could it not be that in a small open economy agriculture's relative importance might not decline? Empirical experience, of course, indicates that decline occurs even with these restrictive conditions very largely met. One needs only trace the experiences of Australia, New Zealand and Iowa as small open economies that now employ no more than a tenth of their labour force in farming compared with 80 per cent or more within the past century.

Agriculture and farm people create many of the conditions for the decline of agriculture in a growing and open economy. When farm incomes increase, farm people change their consumption patterns away from food toward non-farm goods and services, just as people do generally. Many of the goods and services for which demand increases are non-tradables—ranging from roads and local transportation to beauty parlours, barber shops, retail services and movie theatres. Labour transfers out of agriculture to provide such services. Even if the shift in consumption were toward imported manufactured products, such products require a significant component of non-tradable services to be available for consumption.

But perhaps the most striking contribution of agriculture to its own decline comes from a whole host of production decisions. Productivity change in agriculture has required the use of new inputs, almost all having a significant non-farm component. One needs only to illustrate—tractors replacing horses; combines replacing human labour, the scythe and the threshing floor; hybrid seeds replacing seeds from local production with an on-farm labour input; and herbicides replacing farm labour. But this is not all. As labour of farm people increases in value, they find it advantageous to support specialisation in the processing of their output—using the creamery instead of making butter at home, purchasing processed grain in place of using their labour to mill the grain, or buying meat rather than slaughtering their own livestock.

These changes are apparent in the rapid growth of the importance of current operating expenses as a percentage of the value of gross farm output from 1950 through to the mid-1960s. During that period, in northwest Europe current operating expenses, in constant prices, grew at twice the rate of gross output in constant prices. And current operating expenses did not include payments for labour or investment (neither current investment nor depreciation).

As a consequence, in a small open economy or in the world as a whole, agriculture will decline in a relative sense because of the way farmers and farm people respond to economic growth. A decline in the real prices of farm products is not required for this conclusion, though obviously if real prices decline, the fall in agriculture's relative importance will be even more rapid.

I have given so much stress to the fact that economic growth requires agriculture to decline because most national agricultural price interventions are motivated by an effort to offset the effects of the required adjustment process. As economists we apparently have done a very poor job of educating our politicians about the inevitability of the decline of agriculture and of explaining that government intervention, if it exists, should be designed to facilitate the adjustment rather than to impede it.

Output Prices and Welfare of Farmers

Put very simply, the basic conclusion of *Disarray* was that product demand is of little importance in determining the returns to mobile farm resources, namely management, labour and capital. Product demand is a factor in determining the return to land or rent. But it is supply conditions that primarily determine the returns to labour and capital. Demand can be important in influencing the level of employment, but when the supplies of farm labour and capital are very elastic, as they are in the long run, wage rates and capital returns are determined primarily by the supply functions for labour and capital.

The case for intervention in product markets, through price supports, through subsidies related to output or by control of imports or exports to achieve a domestic price objective, is based on two erroneous assumptions. The first is that resource returns are determined by demand conditions and the second is that the economy is static. As I argued in *Disarray*, an output price increase has a once-and-for-all effect on resource returns. In the case of labour, this is not enough. If farm people are to share in economic growth, the returns to, or wage of, labour must increase continuously so that farm workers do not fall behind non-farm workers in terms of income.

Farm Resource Returns Depend on Alternatives

If the level of product prices is relatively unimportant in determining the returns to farm labour and capital, what is important in that determination? The answer is a simple one, yet the answer is almost universally ignored in the design of farm price and income policies. In the industrialised countries the returns to labour and capital are determined primarily by the returns to comparable labour and capital in the rest of the economy. This is not to say that such has always been the case or that there cannot be short-run variations in the returns to farm labour that are not matched by similar variations in the economy as a whole.

But if you look at the differences in the returns to farm labour across countries it becomes obvious that the levels of output prices explain very little of such differences. The difference in returns to farm labour that exists between France and Portugal, for example, cannot be explained by output price differences. In fact, in recent years output prices in Portugal have been higher than in France. The differences are, in fact, explained by the general levels of returns to labour in the two economies.

Simple models of farm labour supply and demand may be used to show that output prices have a very modest effect on the returns to farm labour compared with other changes that affect the returns. The other much more powerful factors are the growth in non-farm earnings and increases in the level of schooling, that is, increased human capital. Using the available empirical estimates for the United States, it was shown that one year's growth in non-farm wages of just 2 per cent had as much effect on the returns to farm labour as an output price increase of 10 per cent. In a growing economy, the 2 per cent growth in non-farm labour earnings can be repeated indefinitely. In fact, for most industrialised economies for most of the period since the Second World War the growth in real labour earnings has exceeded 2 per cent a year. A real output price increase of 10 per cent clearly cannot be repeated, at least not more than once or twice.

The increase in schooling has two effects on the returns to farm labour. The direct effect is that more schooling increases the marginal product of labour engaged in agriculture; the indirect effect is that more schooling increases the rate of migration out of agriculture. Taken together, an increase in schooling of farm workers of 10 per cent resulted in increased returns to farm labour of approximately 10 per cent. For the time period of the study, an increase in schooling of 10 per cent meant an increase of approximately one year in the number of years of schooling for farm youth.

In the industrialised economies there have been substantial improvements in the education of rural people over the past half century. However, neither these improvements nor the improvements in transportation and communication have ever been considered as critical components of farm policy in any country. Yet these changes, together with the growth in real non-farm labour earnings, have been the primary source of the substantial growth in the real incomes of farm families in the industrialised economies over the past four decades. In fact, over this period real farm output prices have declined in almost all industrialised economies. But due to productivity improvements and the greater integration of farm people into the rest of the economy, the real incomes of farm families have generally doubled (or more) over the past four decades.

I know of no more striking proof of the futility of using higher product prices to increase the return to farm labour than the conclusion reached by the Commission of the European Communities (1984) in its 1983 report on the agricultural situation. It was concluded:

> During the period from 1964/65 to 1976/77, regional disparities in agricultural incomes (as measured by gross value-added per agricultural worker) increased in the Community. The ratio between the regions with the highest agricultural incomes and those with the lowest rose from 5:1 to 6:1.
>
> Generally speaking, the regions with above-average levels of agricultural income are to be found in a favourable general eco-

nomic context; the converse is true of regions with a low level of agricultural income.

Later in the same report, the Commission referred more directly to the issue of factors determining the incomes of farm families:

> ... in 1975 it was found that about one quarter of farm holders had a second gainful activity. In this context, it should be borne in mind that the availability of a second gainful activity to part-time holders varies significantly from region to region—for example, while the majority of part-time farm holders in Southern Germany have a second income earned outside their farm, in the Mezzogiorno (Southern Italy) few such opportunities exist and most farm holders are under-employed.

But the Commission stopped short of addressing the full implications of non-farm income opportunities for the withdrawal of labour from agriculture and the consequent effects on the earnings of those who remain in agriculture. Given the dynamic effects of economic growth on agriculture, farm people can share in the gains of economic growth only as the supply of workers to agriculture responds to the rising level of real labour earnings in the rest of the economy. Due to the slow growth of demand for farm products, the growth of productivity in agriculture at a rapid rate, and the substitution of farm inputs for labour due to the increase in real earnings in farming, labour use in agriculture declines. It declines both through partial labour withdrawal (part-time farming) and through migration from the countryside to non-farm areas.

If high output prices have little effect on the real earnings for farm work, do they have a significant effect on the level of employment in agriculture? A few years ago I looked at this issue for the European Community (of nine countries), North America and Japan (Johnson 1982). Annual rates of decline in farm employment were calculated for 1955–60, 1960–70 and 1970–79 and compared with wheat and feed grains prices in 1970 and 1979. For this period of a quarter of a century there did not exist any relationship between the level of prices and the rate of decline in farm employment. I do not imply that the level of output prices had no effect on farm employment, but it seems clear that other factors were far more important than output prices in determining the rate of employment decline. These other factors include the discrepancy in farm and non-farm labour earnings at the beginning of the period, the rate of increase in real non-farm earnings, the rate of growth in non-farm employment, the regional dispersion of non-farm employment, and the improvement in the rural infrastructure, such as roads, buses and communications.

One reason that output price levels did not appear to be related to changes in farm employment was probably that in these countries in the mid-1950s there were different degrees of long-run disequilibrium in the returns to farm and non-farm labour. It seems quite clear that, at least in Japan and North America, farm labour returns increased relative to non-farm labour returns, at least until the early 1980s. Thus the actual changes in farm employment may well have been different from those that would have occurred had there been no increase in the farm wage relative to the non-farm wage. Obviously these data merit a more systematic statistical analysis than I undertook.

High product prices are not enough for farmers to share in the fruits of economic growth. Increasing product prices, in the short run, can increase the returns to farm labour by increasing the demand for labour while the supply elasticity is very low. But the increase in the returns to labour would be a once-and-for-all increase. This is not enough. What the policy makers who assume that high output prices will have a favourable long-run effect on farm incomes fail to recognise is that farm incomes are pursuing a moving target in a growing economy. The moving target is the result of the growth in real non-farm labour earnings. Since farm employment declines absolutely in growing industrialised economies, a once-and-for-all increase in labour earnings is soon eroded by even a very modest fall in the rate of decline in farm employment.

On the whole, I believe that the analysis and conclusions presented in *Disarray* have stood the test of time quite well. Obviously I am not an unbiased observer, but I have seen nothing in the unfolding of farm policies and developments in agriculture in the industrialised countries that contradicts any significant element in *Disarray*. In fact, I think it can be said that there is now increasing recognition among policy makers that manipulation of output prices is not an effective way of improving the lot of farm people. Unfortunately, there does not seem to be any agreement among the policy makers in Western Europe and North America about what the appropriate policy alternatives are.

Important Omissions

Let me now turn to what I consider to be two most important shortcomings of *Disarray*. In terms of the emphasis of the book on the industrialised countries, a major shortcoming was my failure to recognise the full impact of macroeconomic variables on farming, farm people and the sectors closely related to agriculture. In particular, I completely ignored the role of exchange rates and monetary and fiscal policies in influencing decisions with major consequences to agriculture. The significance of macroeconomic policies for the agricultural sector is discussed in the following section.

The other major shortcoming was the failure to consider the farm price policies of the developing countries. Much of the motivation for *Disarray* was to make clear the adverse effects that the farm price and trade policies of the industrialised countries had on the agricultural sectors of the developing countries. Unfortunately I did not realise at the time that, in all too many developing countries, the negative consequences of the industrial-iscd country policies were exacerbated by the strong urban bias of economic policies in many, if not most, developing countries. Import substitution policies, price ceilings or low procurement prices, and use of trade in agricultural products to hold domestic farm prices below international market prices were clearly adverse to the development of agriculture in the developing countries. I did view the consequences of food aid as an adverse factor in the development of agriculture in the developing economies and argued for increased investment in research but I did not consider the full implications for agriculture of the strong urban bias of developing country policies.

Economic Instability and Agriculture

It is now all too clear that changes in economy-wide variables have major effects on the economic state of agriculture. These effects can be such as to increase greatly the variability and uncertainty faced by farm people. It is not only national macroeconomic policies that are important. The increased importance of international trade, combined with the integration of capital markets, has meant that many of the changes in macroeconomic variables within a country originate from outside the country and are thus beyond the control of that country.

In an important article published in 1974, G. Edward Schuh brought to our attention the role of exchange rates in properly understanding the economic circumstances of agriculture. In that article he showed the impact of the overvaluation of the US dollar on the prices received by farmers, the resource adjustments that had been required and efforts of government to offset the adverse consequences. Thus to some degree the government payments made to farmers offset the negative impact of the overvaluation, which acted as an export tax on traded farm products. He argued that the dollar was overvalued throughout the 1950s and 1960s, and that the overvaluation increased as time passed and reached an unsustainable level resulting in the 1971 devaluation of the dollar.

But the overvaluation of the dollar that existed during the late 1960s and early 1970s was relatively mild in terms of its effects on agriculture compared with the effects of other macroeconomic variables from 1973 until the present. This is not the place to detail the many changes in variables that had major short-run and, in some cases, long-run consequences for agriculture in the United States as well as in other industrialised countries. But a brief summary seems appropriate.

In the United States, as in Australia, for most years during the 1970s real interest rates were negative and the rapid increases in farm land prices reflected the effect of such low interest rates as well as the expectations concerning the level of farm prices. After 1973 the recycling of the petro-dollars resulted in such a large excess supply of the dollar that the exchange value of the dollar had fallen substantially by the late 1970s. One major effect of the worldwide inflation plus the devaluation of the US dollar was the sharp increase in world trade in farm products and the sharp increase in the US share of that market for several key products, especially grains.

With the change in US monetary policy in the late 1970s, interest rates began to increase, first in nominal terms and then in real terms. The exchange value of the dollar increased in the early 1980s and, with the decline in US and world inflation, the process that started in the mid-1970s was reversed—export growth slowed and the US share of world trade declined. The sharp swings in farm incomes were due primarily to the effects of macroeconomic policies and only to a limited degree to the effects of farm price and income policies in the United States. During the 1980s the US farm price policies exacerbated the adverse consequences of the change in the macroeconomic environment. The 1981 farm bill set the farm price supports too high and the United States once again became a residual supplier in world markets. However, even if the price supports had been substantially lower, though the United States would have retained more of world markets, it would have done so at very low prices.

During the 1970s the macroeconomic setting was far more important than the farm price and income policies in all the industrialised economies, with the possible exception of Japan, where the farm policies were quite effective in isolating agriculture from the swings in the macroeconomic variables.

Policy makers in the European Community have taken pride in achieving quite stable output prices, particularly during the 1970s and early 1980s. Between 1975–76 and 1983–84, the lowest index of real output prices (1980–81 = 100) was 98.4 and the highest was 109.3. Estimates of net value added per labour unit employed in agriculture in the EC countries appear to be reasonably stable for 1974–83 (Commission of the European Communities 1984, p. 39). But value added, even net value added, includes interest costs as a positive component. Interest costs increased dramatically in some of the EC countries in the unstable macroeconomic environment of the period. Interest costs per hectare in both the Netherlands and the United Kingdom increased from about 160 ECU in 1979 to nearly 350 ECU just two years later (Commission of the European Communities 1984, p. 44). Danish agriculture has long operated with a major dependence on credit. In Denmark, the relative share of interest payments in the gross value added (which is net value added plus depreciation) increased from 23 per cent in 1975 to 48 per cent in 1981 (OECD 1983, p. 42). It is striking that in 1981 interest costs were 58 per cent of net value added in Denmark.

However, relatively stable output prices have not been translated into stable net incomes of farm operators. In Great Britain, for example, real net farm operator income fell by more than 50 per cent from 1977–78 to 1981–82. From 1982 to 1983 the change in real net value added in Germany was a fall of 22 per cent and in Denmark a fall of 18 per cent.

The share of interest payments in gross value added doubled in Canada and the United States between 1975 and 1980, and in the latter year equalled 19 per cent in Canada (Cloutier and MacMillan 1986) and 21 per cent in the United States. In the United States total interest payments increased from US$8000m in 1978 to US$21 000m just four years later. The importance of this increase is indicated by the fact that in 1978 net farm operator income was US$25 000m and in 1982 was US$23 000m. By 1982 interest payments were almost as large as net farm operator income—the bankers and other creditors got as much from farming as did the people who did the work, so to speak.

The changes in interest costs were so great in a number of countries that the stabilising effects of direct price and income policies were significantly offset. There were, of course, other macroeconomic variables that negated some of the desired effects of national agricultural policies. In the United States, the increase in the foreign exchange value of the dollar, combined with high fixed price supports, resulted in a loss of export markets and the accumulation of stocks. The latter, in turn, induced the government to attempt to manage supply through acreage diversion. The acreage diversions then called forth responses from competing producers and further loss of export markets occurred.

Disarray Has Increased

One must conclude that the disarray in world agriculture is now greater than it was fifteen years ago. In real terms the cost of government intervention in agriculture and food imposed on consumers and taxpayers has clearly grown since the early 1970s. We have a variety of estimates that indicate a significant increase in rates of protection of agriculture over the past fifteen years. The nominal rate of protection of Japanese agriculture increased from about 100 per cent to 150 per cent from 1980 to 1982

(Tyers and Anderson 1984) and has certainly increased since then. Protection levels in the European Community are currently much greater than fifteen years ago for grains and in recent years significant protection has been introduced for oilseeds where none existed in the early 1970s (BAE 1985). There can be no doubt about the increase in protection in the United States in recent years. Taking all factors into account, the United States had relatively low rates of protection for grains, cotton and livestock products, except dairy products, until 1981.

If protection levels are to be measured as the differences between domestic prices plus producer subsidies and world equilibrium prices if there were free trade, the recent sharp decline in the international market prices of several important farm products results in exaggeration of the current levels of protection. The recent declines in international prices can be attributed primarily to a significant policy change by the United States. In terms of the interests of US farmers and farmers elsewhere in the world, the 1981 US farm legislation was a disaster of major proportions. The target prices, which provided the incentives to produce, and the price supports were set too high. At the time the bill was passed, the conventional wisdom in Washington was that world food demand was going to grow more rapidly than food supply. While it is now hard to believe that such views could have been held at that time, such was the case.

The legislation had a number of undesirable consequences. One was that instead of assisting in the necessary transfer of resources out of agriculture, it encouraged resources to remain in farming. Given the level of the target prices, farmers were misled concerning their long-term prospects. A second negative effect was that the high support prices for grains and cotton made the United States once again the world's residual supplier. Competing exporters, including Australia, were quick to take over markets at prices a little below what US sellers could offer. The sharp increase in US stocks of grains and cotton led to the 1983 payment-in-kind program, under which nearly a third of the cropping land normally devoted to grains and cotton was idled. The US officials apparently believed that they could manage world supplies, but the responses of farmers elsewhere in the world showed that such a view represented little more than self-delusion.

Export markets continued to disappear and stocks continued to increase. The 1985 farm legislation represented an effort to regain export market share by sharply reducing the levels at which market prices were to be supported. The impact was soon felt in international markets, with the prices of rice and cotton declining by about half, the price of corn by nearly a third and the price of wheat by about a quarter. But the declines in market prices were not accompanied by any reduction in production incentives, since the target prices were left unchanged for 1986 and 1987 and were then to be decreased slightly for the subsequent three years. It is particularly alarming that even with the sharp declines in prices, world grain stocks may increase by as much as 50 Mt or by 15 per cent, during the 1986–87 crop year.

The effect of the 1985 legislation is to make all or nearly all of the large US stocks of grain and cotton available for sale. This explains most of the large drop in international prices. But since world stocks will not be reduced during the 1986–87 crop year, it is unlikely that we have seen the end of the price declines.

I have given this detail about US policy changes and their consequences to make a number of points. One is that US farm programs contributed significantly to excess production capacity in agriculture in the industrialised countries. They did so domestically by providing inappropriate signals through the high target prices and did so internationally by holding prices higher than would equilibrate world supply and demand. The second is that to the best of my knowledge there was no effort by government officials in other industrialised countries to warn their farmers that the prices they were facing in international markets were artificially high and that sooner or later there would be declines. Few public officials want to be the bearers of bad news. The third, and my final point, is that in the haste to recapture its lost export market share, the United States showed no concern about the effects of its actions on others. In terms of the public rhetoric, the European Community had been the villain primarily responsible for the United States' loss of market share through the continued expansion of production and the indiscriminate use of export subsidies. But who got bashed by the US actions? Not the EC farmer, at least not in the short run, though over time the lower international prices may lead to some reduction in EC intervention prices. The farmers who were hardest hit were in developing countries, such as the rice producers in Thailand and cotton producers in numerous low income countries. Only the United States and the USSR among the industrialised countries produce any significant amount of cotton; all the rest is produced in low income countries such as India, Egypt, China and Pakistan.

I have been quite critical of the policies of my own country and have said little about the worsening impact of the policies of the European Community and Japan on international markets for agricultural products. Protection levels in the European Community increased little, if at all, during the 1970s and actually declined during the early 1980s as the exchange value of the US dollar increased. But the Community deserves only condemnation for standing pat as it approached and surpassed self-sufficiency in grains, sugar, beef, poultry and cheese. In recent years the European Community has employed very large subsidies to produce oilseeds to add to its unneeded productive capacity. The Community has not shown the slightest concern for the effects of its farm price policies on any other country, except possibly some of its former colonies.

Japan has actually made some significant moves to open its markets, though most of these were instigated in the early 1960s when it was decided that the Japanese consumer should have access to more meat. But the intention was that it should be meat produced in Japan; to make that possible substantial imports of feedstuffs were required. Japanese policy makers have quite cynically played on the concerns of the Japanese consumer about food security to maintain support for a high cost and highly protected agriculture (Johnson 1986). Japanese policy makers played on food security fears that already existed as a result of the food shortages during and following the Second World War and, by forecasting world food stringencies and sharply higher real food prices, have attempted to keep these fears alive. The fact that current emphasis on self-sufficiency in rice does not provide a significant degree of food security for a country that imports all of the energy required to produce its fertiliser and other farm chemicals seems wholly beside the point to their policy makers.

Concluding Comments

When I wrote *World Agriculture in Disarray* I was rather optimistic that trade negotiations on agricultural policy interventions would have some success in reducing the barriers to trade in farm products. I concluded this, obviously incorrectly, because of the adverse consequences that I saw from a continuation of the farm and trade policies then prevailing. In the final chapter of *Disarray* I wrote the following:

> If the current agricultural and trade policies of the major industrial countries are continued throughout this decade, the following undesirable consequences are highly probable:
>
> 1. The level of costs of the farm policies in the industrial countries to taxpayers and/or consumers will continue to increase.
>
> 2. A substantial and probably increasing fraction of the world's agricultural output will be produced under high-cost conditions.
>
> 3. The percentage of the world's trade in agricultural products that is managed and manipulated through the use of export subsidies will increase from its current level.
>
> 4. The developing countries will face increasing difficulties in obtaining markets for any farm product that is directly competitive with farm products grown in temperate zones.
>
> 5. The degree of effective protection provided agriculture will gradually increase in several industrial countries (Johnson 1973, pp. 249–50).

Unfortunately each of the five projections has now been confirmed. Yet policy makers in the industrialised countries paid almost no attention.

Let me close with two quotations. In 1776 Adam Smith wrote of a change in Britain's Corn Laws in *The Wealth of Nations:*

> So far, therefore, this law seems to be inferior to the ancient system. With all its imperfections, however, we may perhaps say of it what was said of the laws of Solon, that, though not the best in itself, it is the best which the interests, prejudices, and temper of the times would admit of. It may perhaps in due time prepare the way for a better (Smith 1937, p. 510).

Unfortunately this great economist appears to have been unduly optimistic concerning the perfectibility of legislation dealing with agriculture and food. But being a realist, he probably would not have been wholly surprised in 1985 had he heard Senator Jesse Helms, then Chairman of the Agricultural Committee of the US Senate: 'I think we have an obligation to do something, even if it's wrong'.

Note

Plenary paper presented at the 31st Annual Conference of the Australian Agricultural Economics Society, University of Adelaide, Adelaide, 9–12 February 1987.

References

Bureau of Agricultural Economics (1985), *Agricultural Policies in the European Community: Their Origins, Nature and Effects on Production and Trade*, Policy Monograph No. 2, AGPS, Canberra.

Cloutier, P. and MacMillan, D. (1986), *Current Financial Difficulties of Canadian Agriculture*, Discussion Paper No. 310, Economic Council of Canada, Ottawa.

Commission of the European Communities (1984), *The Agricultural Situation in the Community*, 1983 Annual Report, Office for Official Publications of the European Communities, Luxembourg.

Johnson, D. G. (1973), *World Agriculture in Disarray*, Macmillan, London.

——— (1982), 'International trade and agricultural labor markets: farm policy as quasi-adjustment policy', *American Journal of Agricultural Economics* 64(2), 355–61.

——— (1986), 'Food security and Japanese agricultural policy', in U.S.-Japan Economic Agenda, *Issues in U.S.-Japan Agricultural Trade*, Carnegie Council on Ethics and International Affairs, New York.

OECD (1983), *Review of Agricultural Policies in OECD Member Countries 1980–1982*, OECD, Paris.

Schuh, G. E. (1974), 'The exchange rate and U.S. agriculture', *American Journal of Agricultural Economics* 56(1), 1–13.

Smith, A. (1937), *The Wealth of Nations*, Modern Library Edition, New York, first published in 1776.

Tyers, R. and Anderson, K. (1984), *Price Trade and Welfare Effects of Agricultural Protection: The Case of East Asia*, Pacific Economic Papers, No. 109, Australia-Japan Research Centre, Australian National University, Canberra.

17

Can There Be Too Much Human Capital?
Is There a World Population Problem?

Most of the emphasis in the study of human capital is on investment in people and the consequences of that investment—social and private returns, social and private costs, the effects of human capital upon productivity and national output growth, on-the-job training versus formal education, and the substitution between quality and numbers of children. I plan to discuss a rather different aspect of human capital, namely the value of a person—a human being—with little or no formal investment other than that common in the majority of the developing countries of the world.

I will approach this topic in an indirect manner since my interest in the value of a person whose primary human capital consists of a combination of physical capability, native intelligence and a limited education investment is in exploring whether population growth has a measurable negative effect upon real per capita incomes in developing countries. As I use the term, a *negative effect* exists if the total marginal product of an additional individual is less than the average product of the existing population. This definition of a negative effect does not require that the marginal product be negative, only that it be less the average product or per capita income level. This a very stringent test, much stronger than what is generally considered.

I believe if the majority of people were asked what was the relationship between population growth and per capita real incomes in developing countries they would answer that there was a negative relationship—population growth reduces real per capita incomes below what they would otherwise be. This was the common viewpoint expressed in the popular press prior to and during the 1992 world conference on the environment in

Reprinted by permission from *Human Capital and Economic Development*, edited by Sisay Asefa and Wei-Chiao Huang (Kalamazoo, Mich.: W. E. Upjohn Institute for Employment Research, 1994).

Rio de Janeiro. It is the intent of the Population Crisis Committee, certain agencies of the United Nations and the authors of *The Population Bomb* to create such an impression. It is certainly the answer given by officials in the People's Republic of China responsible for the country's population programs.

There have been relatively few voices putting forward a contrary view; Julian Simon has been the most prominent. A few years ago (1986) a Working Group was organized by the National Research Council of the National Academy of Sciences, and after many meetings and several studies commissioned, it issued a report (NRC 1986), *Population Growth and Economic Development: Policy Questions,* which has been ignored except for reviews in a few scholarly journals. I hope I will be forgiven if a significant part of this paper is devoted to presenting the major findings of that report which I believe have stood the test of time very well. But I may be biassed—I was the co-chairman of the Working Group.

Why Is Population Growth Bad?

It is useful to start with the arguments that support the view that population growth has adverse effects upon economic growth in developing countries. The case is quite simple and straightforward. The earth is considered to be finite, in terms of physical space and resources. The exploitation or use of its resources is subject to diminishing returns—additional inputs (including labor) applied to the land, water, minerals and forests will yield a diminished return. Therefore, other things equal, more people will mean a reduction in the marginal productivity of labor and, eventually, lower per capita incomes. These relationships seem so obvious that they are seldom spelled out so explicitly. The basic argument hasn't changed from that given by Malthus nearly two centuries ago.

The fact that only a small minority of the world's population eats less well today than did the majority of Europeans at the time Malthus wrote seems not to have dimmed the attractiveness of his model of human behavior (Fogel 1992). It is estimated that in 1781–90 that daily per capita calory consumption in France was 1,753. As of 1965 how many countries in the world had a smaller caloric supply? The answer is: Exactly two—Mozambique and Somalia (World Bank 1992). It was not until the second quarter of the nineteenth century that France's calory consumption equalled that of India today. At the end of the eighteenth century England's daily per capita production of calories was 10 percent below the Indian daily consumption in 1989. It is sad to note that in 1989 there were three countries, all torn by war and revolution, that had less than 1,753 calories per day—Mozambique, Ethiopia and Chad—and as of 1992 they have probably been joined by Somalia and Sudan. Yet the view that increased food supplies would soon be followed by increased mouths to feed should not be too surprising since this view was expressed more than two millennia ago in Ecclesiastes 5.11: "When goods are increased, they are increased that eat them."

The World Bank's *World Development Report 1984* dealt with the issue of population change and development. The discussion, which is both competent and balanced, gave rather little weight to the importance of diminishing returns to or scarcity of resources except for the possible adverse effects of population growth on the environment. Nor was the claim made that rapid population growth stopped development or caused a fall in real per capita income. Instead, the consequences were that rapid population growth, such as that occurring in most developing countries, slowed economic development (1984, p. 105). It attributes the negative effects of rapid population growth to two factors. The first was internal to the family—a large family reduces the investment in each child in terms of time and other resources—and since poor families have the most children, makes it more difficult to reduce poverty. The second was that "it weakens macroeconomic performance by making it more difficult to finance the investments in education and infrastructure that ensure sustained economic growth" (p. 105). This work represented an important shift in the scholarly discussion of population growth but seems not to have penetrated the popular discussion.

What Are the Facts?

What are the facts concerning the relationship between population growth and development as measured by the growth in real per capita incomes? This may seem like a simple question and, at one level, it is. But as will become clear it is far from a simple question. Or it may be a simple question with a complex answer or answers.

I shall present some empirical information concerning the relationships between population growth and economic development. Let me note in advance that the empirical relationships presented do not prove causality—they do not prove that population growth has either a positive or negative effect upon development and growth. For this reason the Working Group (NRC 1986) did not present any regressions relating population growth to various variables, such as per capita income growth. Spurious correlations abound. Why do I present such data then? I do so because those who believe that population growth is the source of most human ills either implicitly or explicitly claim that a negative relationship holds between population growth and economic and social development. A recent pamphlet of the Population Crisis Committee, *The International Human Suffering Index*, presents, in color no less, a comparison between an index which it calls the

Human Suffering Index and the annual rate of population increase. Two statements are made: "Most countries with high human suffering scores have very high rates of population growth," and "Virtually all the countries with low human suffering scores have low rates of population growth." But there appears to be a very weak statistical relationship between the two variables except between the 57 countries classified as having Minimal and Moderate Human Suffering versus the 83 others. Within the 83 other countries, whose fates include those classified as subject to Extreme Human Suffering, there was almost no relationship between population growth and the index; the square of the correlation coefficient is 0.06. This means that 94 percent of the variation in the suffering index was due to something other than differences in population growth rates. If one's only source of information on the status of the life of people in the developing countries were the discussion of the Human Suffering Index one would hardly guess that life expectancy at birth has increased from no more than 35 years in 1950 to 62 years in 1990 in the low income countries (less than $600 GNP per capita in 1990).

Why It Is So Hard To Know

Why is it so hard to determine how population growth affects economic development? As noted, the interrelationships between population and economic growth are very complex. First, if economic development is measured by changes in per capita real income, at certain income levels, increases in per capita income will result in an increase in population growth through influences that reduce mortality, especially infant mortality. For the developing economies, there is a strong negative relationship between income levels and infant mortality. Thus, if income and population growth are compared for the same time period it is likely that there will be a positive relationship between the two. Introducing lags as I have may not entirely eliminate the problems associated with the direction of causality. Second, the effects of population growth can differ in the short and long run, with the short run being perhaps as long as half a century. Other things equal, in the short run an increase in population growth would be expected to lower the rate of growth in real per capita incomes. An increase in population growth rates will be accompanied by an increase in the number of children and in the dependency ratio. The labor force declines as a percentage of the population. In the long run the effects of investment in human capital can be realized. Positive effects through invention and innovation and increasing returns to scale can result from population growth when there is time for adjustment and response to changing conditions. Slow or nil long-run increases in population also increases dependency as the population ages. Thus in the long run (after one or two generations) slow population growth can have similar effects upon per capita productivity as rapid population growth may have in the short run.

A possible reason for the belief that an increase in population in the developing countries will either lower the actual average income or the rate of growth of income is the implicit assumption that such individuals consume everything they produce. But this assumption is incorrect. The people of the developing countries do not consume everything they produce. According to World Bank data (1992, pp. 234–235), the domestic savings rates of the low income countries are equal to or greater than the rates for either middle or high income countries. Thus if resources are productively invested, over a lifetime an additional person has made possible an increase in the productive potential of a country. This conclusion assumes that savings rates are independent of population growth rates and this point is addressed below. It may be noted that the savings estimate excludes savings that result in human capital investment. If the succeeding generation has more education and more adequate health care than the current generation has, this increases the probability that a significant positive rate of population growth either will have a positive or nil effect on per capita real incomes in the long run. Consequently, in considering the effect of population growth on income growth we must consider the total effect and not solely the marginal product of labor. This is why I earlier used the term "total marginal product" of an additional person which includes not only the marginal product of labor but the increase in investment and any effects through increasing returns to scale.

Some Empirical Relationships

There are a number of empirical relationships or propositions that merit our attention. The first two are of a very general nature and the third involves regressions between population growth rates and real per capita incomes with the inclusion of other relevant variables. These empirical propositions raise some questions concerning the validity of the commonly held view that higher rates of population growth have adverse effects upon economic well-being. I do not claim that these empirical relationships imply causality but only that each is worth pondering and exploring.

The first of the empirical relationships is that the significant increases in the real incomes in the industrial countries of Europe and North America occurred in the eighteenth, nineteenth, and early twentieth centuries during a period of historically rapid population growth that followed a long period of slow population growth. From 1650 to 1750 population growth rates were very low in both the industrial and developing regions at 0.33 and 0.34 percent annually (see Table 1). At these rates population doubling

Table 1. Population Growth Rates and Distribution of World Population and Between Industrialized and Developing Regions, 1650–1986

| | Average Annual Rate | | | | |
Interval	Industrialized Regions (B)	Developing Regions (A)	Difference (B−A)	% of World Population in Developing Regions	% of Growth in Population in Developing Regions
1980–1986	0.66	1.98	−1.32	76.9	91.6
1970–1980	0.78	2.23	−1.45	74.4	89.6
1960–1970	1.04	2.41	−1.37	71.7	86.2
1950–1960	1.26	2.07	−0.81	69.9	79.8
1940–1950	0.35	1.44	−1.09	67.5	90.0
1930–1940	0.85	1.28	−0.43	66.4	77.3
1920–1930	0.91	1.11	−0.20	66.1	68.6
1900–1920	0.92	0.52	0.40	67.9	53.8
1850–1900	1.05	0.53	0.52	73.3	54.5
1800–1850	0.83	0.31	0.52	78.1	53.4
1750–1800	0.62	0.47	0.15	79.3	73.6
1650–1750	0.33	0.34	−0.01	79.3	79.4

Source: Donald J. Bogue, *Principles of Demography.* New York: John Wiley and Sons, Inc. 1969, p. 49. United Nations, *Demographic Yearbook 1986.* New York: United Nations, 1988.

Table 2. Expectation of Life at Birth for Six European Countries and Massachusetts in the United States: 1840 to 1955

Year	Expectation of life at birth	Average annual increase in e_o
1840	41.0	–
1850	41.5	0.05
1860	42.2	0.07
1870	43.5	0.13
1880	45.2	0.17
1890	47.1	0.20
1900	50.5	0.34
1910	54.3	0.38
1920	58.3	0.40
1930	61.7	0.34
1940	64.6	0.29
1955	71.0	0.43

Source: United Nations, *Population Bulletin No. 6,* Table IV.1, 1962.

would require two centuries. From 1750 to 1900 population growth rates were higher in the industrial than in the developing regions and this difference continued through the first two decades of this century.

As late as 1840 life expectancy at birth in six European countries and Massachusetts was 41 years (Table 2). It reached 50.5 years by 1900 and then increased rapidly reaching 71 years by 1955. While we have little knowledge of life expectancy at birth in the developing regions prior to 1950, it is unlikely that there was any significant improvement in the prior century. However, since 1950 the increase in life expectancy can only be described as spectacular, increasing from 35 years in 1950 to 62 years in 1990 (World Bank 1980 and 1992). Life expectancy at birth in the United States was approximately 42 years in 1880; it did not reach 60 years until 1930. Thus the lowest income countries in the world achieved a greater increase in life expectancy in 25 years than was achieved in the United States in twice that long. This improvement in the developing regions was achieved during a period of rapid population growth.

Similar rapid progress has been made in the developing world in reducing infant mortality and child death rates since 1950 (Table 3). The infant mortality rate declined from 165 in 1950 to 72 in 1985 while the child death rate fell from 27 to 11; both are rates per 1,000. The infant mortality rate in the United States in 1900 was 160, and it declined to approximately 80 over the next quarter century (U.S. Bureau of the Census 1971, p. 55).

The second of the empirical propositions is that the developing countries had rapid economic growth in the three decades 1950 to 1980 with population growth rates exceeding those ever realized in the industrial countries. Population growth rates were 2 percent or more while the per capita GDP grew at an annual rate of 2.6 percent during the three decades (Working Group 1986, p. 5). Prior to 1950 population growth rates had been much lower, generally 1 percent or less and there had been slow or nil increases in real per capita incomes.

Economic growth was relatively slow during the 1950s in the low income countries. If we look at a somewhat later period, namely 1965 to 1985, the rate of income growth was quite spectacular. The per capita gross national product for the low income countries grew 2.9 percent annually for 1965–85 and exceeded the 2.4 percent of the industrial economies (World Bank 1987). Since low income countries were defined as those with less than $400 per capita GNP in 1985, some rapidly growing countries that were poor in 1965 were excluded from the calculation because they grew out of the low income category by 1985; had these countries been included the growth of income would have been even higher.

The third of the empirical propositions is that the evidence does not support the view that for developing countries that the rate of population growth has a negative effect upon per capita income growth for the period since 1950. Some results are given in Table 4 for three decades. In the simple model it was assumed that per capita income growth in a decade was a function of population growth and per capita income growth in the prior decade. Enrollment ratios for primary schools and per capita gross domestic products were included. School enrollment is included to provide

Table 3. Life Expectancy at Birth, Infant Mortality Rate, and Child Death Rate

| | Life Expectancy at Birth | | Infant Mortality Rate[b] (under age 1) | | Child Death Rate[b] (ages 1–4) | | GNP per Capita (in $U.S.) |
	1960	1985[c]	1960	1985[c]	1960	1985[c]	1985[c]
Low-income economies[a]							
China	42	69	165	35	26	7	310
India	43	56	165	89	26	11	270
Other	43	52	163	112	31	19	200
Average	42	60	165	72	27	11	270
Africa							
Low-income							
Semi-arid	37	44	203	151	57	34	218
Other	39	49	158	112	37	22	254
Average	38	48	164	117	40	24	249
Middle-income[a]							
Oil importers	41	50	159	111	37	21	670
Oil exporters	39	50	191	113	51	21	889
Sub-Saharan	–	49	170	115	42	23	491
Middle-income economies[a]	51	62	126	68	23	10	1,290
Lower-middle income[a]	46	58	144	82	29	13	820
Upper-middle income[a]	56	66	101	52	–	–	1,850
Industrial Market Economies[a]	70	76	29	9	2	–	11,810

Source: World Bank, *World Development Report, 1987.*

[a] Income designations are based on per capita income (in 1985 U.S. dollars): low income, $390 or less; lower-middle income, $400–1,600; upper-middle income, $1,600–7,420.

[b] Rates are per 1,000.

[c] Data for Africa are for 1982.

Table 4. Cross-Country Regressions of Per Capita GDP Growth rates

Low income	1960–1970	1970–1980	1980–1988
Lagged Population Growth	0.053	0.115	0.319
Enrollment Primary School	−0.015	0.025	−0.015
Lagged GDP Growth	0.226	−0.221	−0.286
Lagged GDP Level	−0.001	−0.003^x	−0.00
Dummy: Africa	−0.931	−0.742	−3.371^x
$\bar{R}^2$	−0.004	0.163	0.059
η	12	32	33

Middle income			
Lagged Population Growth	−0.524^{xx}	0.319	−0.551
Enrollment Primary School	0.017	0.029	0.024
Lagged GDP Growth	0.236	0.163	0.320^x
Lagged GDP Level	−0.005	−0.000	−0.000^{xx}
Dummy: Africa	−0.711	−0.241	−0.792
Latin America	−0.859	−0.608	−2.115^{xx}
$\bar{R}^2$	0.163	−0.043	0.055
η	32	61	58

Notes: GDP data are from Robert Summers and Alan Heston, (1991), "The Penn World Table (Mark 5): An Expanded Set of International Comparisons, 1950–1988," *Quarterly Journal of Economics*, Vol CVII, No. 2; World Bank, *World Development Report*, Various years. Lagged population and lagged GDP growth rates are for the prior decade; Lagged GDP level is beginning year of the decade. Africa and Latin America dummies are 1 for country in region, 0 otherwise. Constant terms not presented.

^{x}Statistically significant at 5 percent level.

^{xx}Statistically significant at 10 percent level.

a rough indication of society's investment in human capital. Per capita GDP is included to determine if there is a convergence effect among the developing countries. Per capita GDP growth in the prior decade was included because of the possible causality between income growth and population growth and the possibility of continuity in per capita income growth from one decade to the next. The coefficient for population growth was positive but not significantly different from zero for each of the three decades for low income countries (Table 4). For middle income countries the coefficient was negative and significantly different from zero at the 10 percent level for the 1960s but was not significant for either of the other two decades. Thus, of the six coefficients for the developing countries, only one indicated that there was a (weak) negative relationship between past population growth and current growth of GDP per capita. The variables that were included had surprisingly little relationship to the real per capita growth rates. The correlation coefficients were very small and none of the coefficients were statistically significant at the conventional 5 percent level. Obviously other variables, including policy variables, are much more important in influencing income growth rates than population growth or the human capital measures that I have included.

Table 5 is from Levine and Renelt (1992) who included population growth as one of the variables in their analysis of the influence of policy factors on economic growth in 119 countries. In none of the regressions was there a statistically significant effect of population growth on the rate of economic growth. Several of the policy or policy related variables had a statistically significant coefficient, especially investment and government share. A study by Singh (1992) for twenty-nine developing economies obtained negative but insignificant coefficients for the population growth variable. In each of these studies the periods for income and population growth were contemporaneous.

Population Growth Only One Factor

Before turning to the summary of the results of the NRC study on population growth and economic development, I want to make a point that is all too often ignored in the popular discussions of the subject. Population growth is only one factor in determining the economic well-being of the citizens of a country. And, in my opinion, if we look around the world today at the observed differences in real per capita incomes, it is a rather minor factor in explaining such differences. Governmental policies are of far greater importance. Much of the human suffering that we have witnessed in recent years has not been caused by excessive population growth or too large populations; most of the real causes are to be found in civil war and strife and governmental mismanagement and failures. Bangladesh is a probable exception to this conclusion but I am not sure that there are any others.

China is a clear example of dominance of factors other than population growth in explaining real income growth over the past four decades. While there has been a reduction in population growth rates over the past three decades, the small decline in the population growth rate after the early 1970s can't account for any significant part of the sharp changes in the growth of income per capita. Using the Summers and Heston (1991) estimates, per capita GDP grew at an annual rate of 2.3 percent from 1960 to 1973, by 3.7 percent from 1973 to 1980 and, during major policy reforms, 7.8 percent from 1980 to 1988. Population growth rates in Taiwan closely parallel those of the mainland but from 1960 to 1980 the real income growth rate was double that of the mainland. This difference can only be explained by policy factors. From 1980 to 1988 the per capita income growth rate on the mainland exceeded that of Taiwan.

Most African countries unfortunately provide further evidence of the great importance of the national policy framework in dominating changes in economic welfare. From 1973 to 1980, 30 out of 46 African countries had negative real per capita income growth rates (Summers and Heston).

Table 5. Cross-Country Growth Regressions (Dependent Variable: Growth Rate of Real Per Capita GDP)

Independent Variable	Regression Period [Data Set]				
	(i) 1960–1989 [WB/IMF]	(ii) 1960–1985 [SH]	(iii) 1960–1989 [WB/IMF]	(iv) 1960–1985 [WB/IMF]	(v) 1960–1985 [SH]
Constant	−0.83	2.01	0.86	0.47	2.05
	(0.85)	(0.83)	(0.89)	(1.18)	(1.12)
Initial GDP Per Capita	−0.35*	−0.69*	−0.30*	−0.40*	−0.57*
(RGDP60)	(0.14)	(0.12)	(0.11)	(0.13)	(0.12)
Investment Share (INV)	17.49*	9.31*	16.77*	13.44*	10.15*
	(2.68)	(2.08)	(2.62)	(3.13)	(2.43)
Population Growth (GPO)	−0.38	0.08	−0.53	−0.15	−0.02
	(0.22)	(0.19)	(0.18)	(0.19)	(0.19)
Secondary-School	3.17*	1.21	–	0.63	0.33
Enrollment (SEC)	(1.29)	(1.17)	–	(1.26)	(1.23)
Primary-School	–	1.79*	–	0.91	1.07
Enrollment (PRI)	–	(0.58)	–	(0.73)	(0.70)
Government Share (GOV)	–	−6.37*	–	−0.59	−6.80*
	–	(2.03)	–	(3.73)	(2.30)
Growth of Government	–	−0.08	–	–	–
Share (GSG)	–	(0.06)	–	–	–
Socialist Economy (SOC)	–	−0.25	–	−0.21	−0.17
	–	(0.38)	–	(0.45)	(0.43)
Revolution/Coups	–	−1.76*	–	−0.86	−1.75*
(REVC)	–	(0.52)	–	(0.62)	(0.59)
Africa Dummy	–	−1.24*	–	−1.36*	−1.78*
(AFRICA)	–	(0.37)	–	(0.48)	(0.44)
Latin America Dummy	–	−1.18*	–	−1.34*	−1.27*
(LAAM)	–	(0.33)	–	(0.38)	(0.36)
Growth of Domestic	–	–	0.019*	0.013	0.008
Credit (GDC)	–	–	(0.009)	(0.008)	(0.007)
Standard Deviation of	–	–	−0.009*	−0.006*	−0.003
Domestic Credit (STDD)	–	–	(0.003)	(0.003)	(0.003)
Export-Share Growth	–	–	0.090	0.023	−0.03
(XSG)	–	–	(0.052)	(0.047)	(0.041)
Civil Liberties (CIVL)	–	–	−0.22	0.01	0.15
	–	–	(0.11)	(0.13)	(0.13)
Number of Observations	101	103	83	84	86
R^2	0.46	0.68	0.61	0.67	0.73

Source: Ross Levine and David Renelt. "A Sensitivity Analysis of Cross-Country Growth Regressions," *American Economic Review*, Vol. 82, No. 4 (September 1992), p. 950.

Notes: Regressions (i), (iii), and (iv) use primarily World Bank and IMF data, while regressions (ii) and (v) use Summers and Heston data. Coverage includes all countries with data given by the sources except major oil exporters.

*Statistically significant at the $P = 0.05$ level.

The same number had negative per capita growth rates during 1980–1988. During these two periods the African countries generally followed import substitution policies with heavy taxation of agriculture, especially of export commodities. The beginning of the shift to market-oriented policies did not begin generally until the mid-1980s and is just now beginning to influence the pattern of growth.

Propositions—Population and Growth

In the report, *Population Growth and Economic Development: Policy Questions*, nine questions were posed and the available evidence was used to answer them. I shall now paraphrase the nine questions and provide a brief summary of the report's conclusions.

1. Would slower population growth increase the per capita supply of exhaustible resources? In reflecting upon this question, it should be remembered that over the past century the real prices of exhaustible natural resources have declined, not increased (Simon 1981). The decline in the prices of such resources relative to wages or earnings has been striking; for oil and coal the time cost today is about a fifth of what it was a century ago. The same pattern of decline in relative scarcity has prevailed for nonfuel resources, such as copper.

An important point is that no exhaustible resource is essential or irreplaceable. As the easily available supplies of such a resource are extracted, the real cost of extraction and the price increases to encourage economizing on the use of that resource and stimulate the search for substitutes. One reason that real copper prices have declined is that aluminum, which is derived from a plentiful resource (bauxite) has been substituted for it in many uses, especially wire.

Assume that a resource is exhaustible. A little reflection will convince you that the number of people who will enjoy the use of that resource will be the same whether population growth is slow or fast. Exhaustion will occur sooner in time with rapid than with slow population growth, but the number of people who will have used the resource will be the same. If you reject that conclusion because you believe that with more time provided by slower population growth that it would be possible to learn how to economize on the use of the resource, you have already lost most of your argument that rapid population growth results in undue resource exhaustion. The application of more resources to find ways to economize on the resource is an effective substitute for more time. With a larger population there are more human resources to devote to the problem at an earlier time.

2. Would slower population growth increase the per capita availability of renewable resources and thus increase per capita income? Up to present the decline in the per capita availability of renewable resources, such as land, has not resulted in a reduction in per capita incomes. Instead, the long-run trend in the prices of farm products that depend upon land, especially the food crops, has been a declining one over the past century (Simon 1981).

We need to bear in mind that the use of renewable resources is influenced by existing institutional arrangements. Some institutional arrangements, such as common property or property rights that are limited and uncertain, may have disastrous effects upon the conservation of renewable resources. Consider the near extinction of certain kinds of whales or the depletion of certain fishing areas. These occurred where common property existed and, until recently, no effective mechanism existed for avoiding the "tragedy of the commons". The tragedy of the commons occurs because it is in the interest of the individual to use or harvest the commons as long as what he harvests has a value greater than his costs even though his activity reduces the total output. Consequently, if serious problems do occur due to the diminution of the productivity of renewable resources, it will probably be because appropriate institutional arrangements do not exist.

3. Will slower population growth lead to more capital per worker and higher per capita worker output and income? Does slower population growth increase the rate of saving? The Working Group, after reviewing various ways in which population growth could influence the saving rate concluded (p. 87): "We have found little evidence that the aggregate savings rate depends on growth rates or the age structure of the population." Subsequently A. C. Kelley (1988) presented the results of his research on the effect of population growth on savings and investments in developing countries. He considered three possible reasons for negative effects upon economic growth—age dependency, capital shallowing and investment diversion. The age dependency effect is due to the large proportion of children who do not work; the children may also be responsible for investment diversion to consumption. The capital shallowing is nothing more than a reduction in the amount of capital per worker if savings decline on a per capita basis. He concluded that the empirical research had not substantiated any of these effects (p. 459). He also noted that where the economic analysis was rather more sophisticated and included second order effects such as economizing on resources and supply effects, the "puzzle of reconciling the apparent divergencies between theory and fact disappears" (p. 460).

4. Do lower population densities cause lower per capita incomes by reduced stimulus to innovation and reduced economies of scale? One of Adam Smith's greatest contribution to economics was his analysis of the interrelationships between specialization, the extent of the market and the existence of economies of scale. The Study Group concluded that for manufacturing, the economies of scale can be achieved at an unspecified moderate sized city. The liberalization of international trade in agricultural products and the international flow of capital and technology probably does mean that low rates of population growth are without significant effect upon innovation in productivity in manufacturing. And the same answer seems to apply if the question is put differently: Does rapid population growth reduce the growth of labor productivity in manufacturing? The answer also seems to be in the negative.

The Study Group concluded that low density of population could and probably did have adverse effects upon productivity in agriculture. This effect resulted from the costs of the infrastructure, such as roads, communication, marketing services, agricultural research and extension. While Boserup (1981) showed how increased population density influenced the intensity of cultivation and encouraged technological change, it cannot be ruled out that after a certain density of agricultural population is reached that the productivity effects of increased density are either nil or negative. This still seems to be an open question.

5. Will slower population growth increase per capita levels of schooling and health? This question needs to be considered in two parts. The first is the effect of an increase in the average number of children in a family upon family expenditures on education and schooling. The evidence supports the conclusion that more children reduces the amount spent on each child.

The second is the response of the public sectors to an increase in the number of children. T. P. Schultz (1987) found that the percentage of children of school age that were enrolled in school was not associated with the size of the school age population relative to the total population. He found, however, that expenditures per child were a negative function of the relative size of the school age population. These results are quite striking in indicating that during recent years of rapidly growing populations in developing countries these countries were able to increase the places in school rapidly enough to keep pace; in fact, in almost all developing countries the percentage of school age children in school has increased. Increasing the number of places, unfortunately, was apparently associated with little increase in total expenditures on education and the percentage of the population in school increased.

The Study Group could not isolate any effects of population growth on health expenditures due to the poor quality of the available data.

6. Will slower population growth decrease the degree of income inequality? In the short run the effects of slower population growth on income inequality depends on the distribution of the fertility decline among income groups. If the fertility decline occurs first in the higher income groups, the short-run effects will be to increase the inequality of the income distribution. If the fertility decline is greater in urban than in rural areas, there will be a short-run increase in income inequality since urban incomes are significantly higher than rural incomes in developing countries.

In the long run the effect of a decline in fertility on the distribution of income is likely to work through increasing the amount of capital per worker as the number of entrants to the labor force declines. This conclusion depends on the earlier conclusion, namely, that savings rates are independent of population growth or age composition. In effect, if labor becomes more scarce it may command a higher share of the national product and in this way somewhat reduce the inequality in the distribution of income.

7. *Will slower population growth facilitate the transfer of workers into the modern sector and alleviate problems of urban growth?* It cannot be doubted that rapid urbanization has occurred in the developing countries since 1950, nor can one doubt that the rapid growth was a response to population growth rates. Approximately 60 percent of the increase in urban population has been due to natural population growth; the remainder has been due to migration from rural areas (NRC, p. 67).

But once one has noted these facts, it is important to recognize that part of the rapid growth of cities such as Mexico City and Cairo has been due to policies that subsidized living in cities, especially by providing large food subsidies that in most cases were not available to farmers. In addition, the income differences between urban and rural populations in developing countries have been very large, generally with urban per capita incomes ranging from 2.5 to 5 times rural incomes. Often the low rural incomes have reflected an urban bias and the taxation of agriculture through low prices for farm products and high prices for farm inputs.

Consequently, governmental policies have had a role in rapid urbanization in the developing countries. If agriculture and rural areas are permitted to share more fully in economic growth, there would be less concern about too rapid urbanization. But economic growth does result in a shift of population from rural to urban areas, and this shift can be avoided only at large cost. Except to protect the relatively high incomes of those now in the cities, it is not obvious why there should be objection to the growth of urban populations.

8. *Will slower population growth alleviate pollution and environmental degradation?* Environmental resources, such as air and water, were long considered to be common property. Unless appropriate institutional arrangements are made, these resources will be overused whether population is growing fast or slow or not at all. There is an optimum amount of pollution, namely where the marginal cost of reducing pollution equals the value of the marginal harm done by the pollutant. But where air and water are common property with no limitations on access, no enterprise or person has to bear more than a small part of the cost of pollution created. Public policy has the responsibility of internalizing the cost of pollution by adopting measures, such as regulations, fees, or incentives for pollution

abatement that force or induce enterprises (and consumers) to limit pollution to the socially optimum level. Please note that enterprise is used in a very broad sense and includes enterprises owned publicly, such as the Tennessee Valley Authority or the local school district, as well as those owned privately; no form of ownership has a monopoly on pollution.

The contributions of population growth, on the one hand, and higher incomes, on the other hand, to environmental change are complex in nature and differ significantly among the types of pollutants. In considering these effects, clarity demands that there be an understanding of what the environment is and what the sources of environmental disruption are. Most discussions of the environment leave the term undefined or it is evident that only a very particular aspect of the environment is relevant, such as particulate matter or sulfur dioxide emissions or global warning. I know of no better definition than from Webster's Unabridged: "Environment: the whole complex of climatic, edaphic, and biotic factors that act upon an organism or an ecological community and ultimately determines its form and survival."

The major types of pollution given greatest emphasis in the high income industrial countries are not those associated with rapidly growing populations but either with countries with high incomes or that had socialist economic systems. Feshback and Friendly (1992) chronicle the environmental degradation in the socialist economies where the resources "were owned by all the people." People with high incomes consume more than people with low incomes and their demand on production resources and the potential for certain kinds of environmental degradation is the greater though this tendency may be partially or fully offset by increased expenditures on pollution abatement. Higher incomes make it possible to spend more for the protection of the environment and, in fact, this is what occurs.

If population growth, as argued above, has little or no effect on income growth, more people do increase the potential for environmental disruption. But both the amount and type of environmental disruption or pollution is a function of many variables other than either per capita incomes or population.

When it is argued that higher per capita incomes result in increased pollution, reference is to particular kinds of pollution not to all kinds. Higher real per capita incomes have been associated with marked reductions of the kinds of pollution that have been responsible for the greatest loss of life over the centuries. Some forms of pollution associated with the burning of fossil fuels and the production and use of chemicals and radioactive materials that result in hazardous wastes may increase as real per capita incomes increase and high income countries produce more than their per capita share of such pollutants. But pollution sources with much greater and more immediate adverse effects on life expectancy and health are simply ignored when it is claimed that on a per capita basis rich countries contribute more to the world's pollution than poor countries do.

Unclean water and poor sanitation cause far more deaths and serious illnesses in the world today than all of the environmental hazards emphasized in high income countries. If we go back a century or less, unclean water and poor sanitation due largely to the way human and animal excrement was handled were major sources of pollution, causing high infant, child and adult death rates in North America and Europe. This and similar types of pollution are ignored in environmental discussions in the high income countries because these sources of pollution have been virtually eliminated through sewage systems and water treatment facilities. But for a large percentage of the world's population, unclean water and inadequate sewage disposal remain major carriers of disease and are important sources of premature deaths.[1]

Based on research done since the NAS report was published in 1986 it is now possible to say rather more about the effects of developing country economic growth on the state of the environment with environmental disruptions defined and weighted by the interests of all of the world's population rather than the special interests of the high income countries.

A major factor in economic growth is the openness of an economy to the world economy—the degree of trade liberalization (Levine and Renelt 1992). Because of the ongoing objections to the North American Free Trade Agreement by environmentalists, it is relevant to consider the probable effects of trade liberalization on the environment. This topic has been ably addressed by Kym Anderson (1992).

The probable direct effect of world trade liberalization would be to reduce several important sources of pollution, including some that are significant in high income countries. Agriculture is an important source of ground water pollution in high income countries. In these countries there is heavy use of chemicals in crop production and some types of livestock production are concentrated in large units. Trade liberalization for farm products would shift crop production from countries with high farm output prices and heavy chemical use to countries that have and would continue to have lower levels of use of fertilizers, insecticides and pesticides. If there were world-wide trade liberalization in agricultural products, grain production would shift from Western Europe to Argentina, Australia, Thailand and the United States. With lower crop prices in Western Europe, chemical use would be reduced significantly. While use of chemicals would increase in the countries expanding production, the use level would remain low due to the price relationships between crops and chemical inputs such as fertilizer.

A second effect of trade liberalization relates to the use of energy. Most developing countries hold energy prices below world market prices resulting in excessive use of energy per unit of output. In China and India coal prices have been kept at hardly more than half world market prices. Not only is an excessive amount of coal and other energy used, but energy is used inefficiently in ways that create a great deal of pollutants because it has not paid to invest in new and less polluting combustion technology. Inefficiency in combustion leads to a high level of pollution per unit of energy consumed. If developing countries liberalize trade, including in energy, energy prices will rise to world market levels and there will be incentives to conserve energy and reduce pollution (World Bank 1992, pp. 116–117).

Trade liberalization will result in an increased rate of growth in the developing countries. What if the people in the developing countries become rich? I am sure you have read what disasters would follow if they became rich, like we are.

One of the sources of environmental disruption in the low income countries is the harvesting of forests for firewood and the consequent erosion and silting of rivers. As labor becomes more valuable, wood becomes less competitive as fuel. The main cost of wood in developing countries is time since the institutional arrangements generally permit harvesting without paying for the trees. Deforestation occurs not only because labor is so cheap, but because the rights to utilize the forests are not well defined. Of course, deforestation occurs for many other reasons; the reduction of firewood collection is simply an example of the positive effects of increased value of time.

The available evidence indicates that pollution abatement is a normal good. This means that as real per capita incomes increase, people demand (and pay for) more pollution abatement. More effort and resources are devoted to pollution abatement and the acceptable standards for pollution become more stringent. Consequently, it can be expected that as per capita incomes increase in the developing countries, less and less pollution will be acceptable and the pollutants that are the greatest danger to health and life will continue to be reduced. True, some pollutants may increase in total while declining per unit of national output but overall the trend will be toward less pollution in total if reasonable weights are attached to the various types of pollutants.[2] Estimates of three forms of air pollution (World Bank 1992, ch. 4) indicate significantly lower levels in high than in low income countries. This is true for particulate matter and sulfur dioxide in cities and for indoor air pollution, especially in poor rural areas.

Experience strongly supports the conclusion that increasing real per capita incomes results in declining birth rates and, after a time, population growth rates. The immediate impact of higher real incomes on mortality is greater than on the birth rate and for a time higher real per capita incomes are associated with an increase in population growth as was generally true of the developing countries from 1950 to 1970. But it seems to me it is hard to argue against the view that this transition period is a cause for celebration rather than being viewed with alarm. Certainly you would celebrate if you were a mother who now expects that each of her children is likely to reach maturity instead of one out of five failing to survive the first year.

Those who oppose NAFTA or other trade liberalization efforts because of presumed environmental effects fail to consider the rational responses that do occur. Their's is a highly condescending and even insulting view of the people of developing countries. The progress developing countries have made in environmental improvement where it really counts, namely in saving lives, has been enormous over the past four decades. The sharp reductions in infant and child mortality have been due largely to environmental improvements—better handling of sewage and the provision of cleaner water. Mexico, for example, has had a decline in infant mortality from 91 per thousand births in 1960 to 40 in 1990. These improvements occurred because the people of Mexico had more income and allocated more resources to improving their environment. And the evidence is clear—the developing countries as a group have been highly successful in improving the environment where it really counted in saving lives and improving health. True, some forms of pollution, especially particulate matter and sulfur dioxide in the air, have not declined in the developing countries over the past two decades. But significant declines did occur in cities in middle and high income countries supporting the conclusion that higher incomes can and do result in reductions in important types of pollution.

9. Can a couple's fertility behavior impose costs on society at large? The apparent answer is in the affirmative. Where there are public goods such as parks and roads, another child increases congestion. Given that most schools are publicly financed, an additional child imposes costs upon taxpayers generally. Furthermore additional children will in due course increase the number of workers and thus reduce wages, resulting in a potential increase in income inequality. This is essentially the answer in the Study Group report.

It was an answer that I found unconvincing because it ignored the general thrust of the report, namely, that population growth had little or no effect on per capita income growth. True, some negative effects were noted, such as on the quality of education or at least upon per capita education expenditures. But there were also some positive effects, such as economies of scale. Thus on balance it seems to me that there is little evidence to support the position that a family imposes net negative externalities upon society when it chooses to have another child. The family does impose certain costs upon others, but the most important of these—the cost of schooling—is a cost that most societies have chosen to bear publicly for the general benefits that a society derives from having a well-educated population.

At least the Study Group came down against governmental coercion to control fertility decisions, although it was unwilling to use the word coercion and instead said that it preferred using changes in incentives rather than quantity rationing. Quantity rationing—limiting the number of children a couple may have—can only be imposed by the application of coercion. Finally, the following was stated: "It is important to note, however, that current data and theory are inadequate to quantify the size of the external effects; certainly there is no evidence to suggest that drastic financial or legal restrictions on child bearing are warranted" (p. 84).

What Is the Value of a Person?

The last of the questions, concerning whether an added birth imposes costs upon society generally, is not quite the question that I posed at the beginning of this paper. The question I posed was whether adding to the population lowered the level of per capita income. Put another way, the question was whether in a developing country the total marginal product of a person of average human capital was below the average product of that economy. I interpret the evidence, including the research and conclusions of the Study Group, to support the conclusion that the population growth rates that we have observed in recent decades have had little or no significant effect upon the level or the rate of growth of per capita income in either the long or the short run.

I wish to note two caveats to the conclusion that I have just stated. First, there are reasons to argue that moderate rates of population growth are more supportive of economic development than either low, including negative, or high rates. I define moderate rates of population growth in the approximate range of 1.25 to 2.5 percent. Population growth rates of 3 percent or more may reduce rates of per capita income growth primarily because of the stress placed upon institutions, such as education, health, and city governments, by the high rates of adjustment required. High rates of population growth may require rates of response that are beyond the capabilities of such institutions. Second, even if population growth rates have no significant effects upon economic development, governments should pursue many ordinary economic and social programs, such as universal elementary education, maternal and child health care, and institutions that assist individuals in providing personal financial security, that have the effect of reducing the rates of population growth although that is not the primary objective of such programs.

I conclude my paper with a brief discussion of what I call a positive population policy.

A Positive Population Policy

I strongly support a positive population policy that seeks the objective of assisting every family in a country to have the number of children that each family desires. By definition, such programs must be voluntary; coercion and compulsion are simply not consistent with each family achieving its objectives. Thus I strongly favor governments making both relevant information and contraceptive materials available to every family that desires such information and materials. As I argued above, a case has not been made that there is a significantly negative externality to the number of children a family has. But, because I believe that the welfare of families is enhanced if they are given the resources that they need to limit the number of children to the number each family desires, I believe that governments should accept the responsibility of assuring that such resources are available and their availability is highly publicized.

I would argue that governments should go well beyond what many would argue is a passive population policy such as I have described. While ruling out coercion, governments can and do influence the number of children a family desires as well as the number a family has. This can be done by influencing conditions that are recognized to have an effect upon the number of children desired. Education, especially of women, has major impacts on both the desired and the actual number of children. Creating conditions that reduce infant mortality leads to a reduction in the number of children born, though with a lag.

In many rural communities old age security is achieved primarily through having several children and, especially, male children. In rural China, for example, the rural reforms have not produced any alternative sources of old age security to having one or more sons. The ownership of land is one means for providing security in one's older years; however, private ownership of farm land does not exist in China and thus does not contribute to financial security in old age. Nor can rural families assume that the land use rights assigned to them can be readily marketed. There are other alternatives for making provision for the uncertainties of old age and death, such as life insurance and access to reliable savings institutions that are allowed to have positive real interest rates but these are generally not available in low income developing countries. The most immediate and direct approach for the provision of old age security in China would be the extension of the social security system to the entire population rather than restricting it to the employees of the government and state and some collective enterprises.

The cost of providing a pension to all rural residents 65 years and older at 80 percent of the average annual income of all rural residents would not be beyond the financial resources of the Chinese government. In 1989 there were 56 million persons living outside of cities who were 65 years of age or older. If the average pension were 50 yuan per month, the annual cost would be 34 billion yuan. This is less than the budgetary cost of the grain price subsidy in 1991! This subsidy went to urban residents who, on average, had far higher incomes than rural people. It appears that the grain price subsidy is being phased out. It would be hard to think of any alternative use for those funds that would create greater happiness and contribute more to the future viability and tranquility of the Chinese society than the creation of a universal social security system.

Such a system, especially if combined with providing the institutional and legal framework for individuals to make some provision for their financial security would achieve two major objectives. It would significantly alter the incentives to have large numbers of children and reduce the neglect and mistreatment of female babies. Put more directly, much of the pressure in rural China for violating the restraints of the coercive population policy would be eliminated if rural families had viable alternatives to a son or sons to provide security for one's old age.

Coercive population programs do great harm, not only to families but to a nation. Fortunately, policies and programs that will lead to reductions in population growth rates need not be coercive. But, let me add that the appropriate criteria for evaluating programs that directly influence the number of births is the degree to which families are assisted in having the number of children they desire. Lacking evidence that family decisions with respect to the number of children have adverse external effects on others, there is no basis for coercive behavior by governments. I must note that even if there were some adverse external effects, I would still hold that decisions with respect to the number of children should be made by the family. This is a fundamental right that should not be abrogated.

Notes

1. *The State of the World's Children 1985*, the annual report of the United Nations Children's Fund (1984) includes the following (p. 69): "It has long been known that three-quarters of all the illness in the developing world is associated, in one way or another, with inadequate water supply and sanitation. . . . For it remains a fact that most illness *is* related to unsafe excreta disposal, poor hygiene and water supplies which are inadequate in either quantity or quality. And it remains a fact that few changes can bring as many potential benefits to a community as ample quantities of clean water and safe and hygienic means of sanitation . . ."

The World Bank's *World Development Report 1992* was devoted to development and the environment. The report notes that in poor countries (1992, p. 44): "Diarrheal diseases that result from contaminated water kill about 2 million children and cause about 900 million episodes of illness each year. Indoor air pollution from burning wood, charcoal and dung endangers the health of 400 million to 700 million people. Dust and soot in city air cause between 300,000 and 700,000 premature deaths a year." In the same paragraph the adverse effects of soil erosion, salinization of irrigated land and the loss of tropical forests are noted as well as concerns over ozone depletion and loss of biodiversity and greenhouse warming are noted. It is the latter problems that get attention in the high income countries while the environmental factors associated with large annual losses of life in the developing countries are largely ignored.

2. The December 1992 issue of *American Journal of Economics* includes three papers and a discussion on trade and the environment that were given at the 1992 annual meeting of the American Agricultural Economic Association. Two of the papers (Lopez and Zilberman) dealt with issues closely related to the effects of trade and trade liberalization upon the environment in developing countries as did the discussion (Lutz). Lopez presents a relatively pessimistic picture of the possible negative global environmental impacts of opening up the developing countries to the outside world. Zilberman emphasizes the need to take into account environmental effects if the gains from trade are to be optimized and argues against using environmental regulations as trade barriers. In his discussion Lutz presents a general view that is more sympathetic to what I have expressed than did Lopez. The only point I wish to make is that the types of environmental disruption or destruction that Lopez considers are not those that cause large losses of life (unclean water and open sewers) but are natural resource depletions such as deforestation and soil erosion. In saying this I do not want to denigrate the importance of the thrust of his analysis but only emphasize that there are other forms of pollution reduction with significant short run benefits in reducing mortality and morbidity that will follow from higher real incomes in developing countries.

References

Anderson, Kym (1993, forthcoming), "Economic Growth, Environmental Issues and Trade," in M. Nolan, ed., *Pacific Dynamism and the International Economic System*. Washington, D.C.: Institute for International Economics.

Boserup, Esther (1981), *Population and Technological Change*. Chicago: University of Chicago Press.

Fogel, Robert (1992), "Egalitarianism: The Economic Revolution of the Twentieth Century," the 1992 Simon Kuznets Memorial Lectures, Yale University, April 22–24, 1992. Unpublished paper.

Feshbach, Murray and Alfred Friendly, Jr. (1992), *Ecocide in the USSR*. New York: Basic Books.

Johnson, D. Gale and Ronald D. Lee, editors (1987), *Population Growth & Economic Development: Issues and Evidence*. Madison: University of Wisconsin Press.

Kelley, Allen C. (1988), "Population Pressures, Saving, and Investment in the Third World: Some Puzzles," *Economic Development and Cultural Change*. Vol. 36, No. 3, pp. 449–64.

Levine, Russ and David Renelt (1992), "A Sensitivity Analysis of Cross-Country Growth Regressions," *American Economic Review*, Vol. 82, No. 4, pp. 942–63.

Lopez, Ramon (1992), "Environmental Degradation and Economic Openness in LDCs: The Poverty Linkage," *American Journal of Agricultural Economics*, Vol. 74, No. 5, pp. 1138–1143.

Lutz, Ernst (1992), "Trade and the Environment," *American Journal of Agricultural Economics*, Vol. 74, No. 5, pp. 1155–1156.

National Research Council (1986), *Population Growth and Economic Development: Policy Questions*, Working Group on Population and Economic Development. Washington: National Academy Press.

Population Crisis Committee (1992), *The International Human Suffering Index.*

Schultz, T. Paul (1987), "School Expenditures and Enrollments, 1960–80: The Effects of Income, Prices and Population Growth," in Johnson and Lee (1987), pp. 413–76.

Simon, Julian L. (1981), *The Ultimate Resource*. Princeton: Princeton University Press.

——— (1986), *Theory of Population and Economic Growth*. New York: Basil Blackwell.

Singh, Ram D. (1992), "Government induced price distortions and growth: Evidence From Twenty-nine Developing Countries," *Public Choice* 73, pp. 83–99.

Summers, Robert and Alan Heston (1991), "The Penn World Table (Mark 5): An Expanded Set of International Comparisons, 1950–1988," *Quarterly Journal of Economics*, Vol CVII, No. 2.

United Nations Children's Fund (UNICEF 1984), *The State of the World's Children.*

United States Bureau of the Census (1971), *Statistical Abstract of the United States.* Washington: Government Printing Office.

World Bank (1984, 1992), *World Development Report*. New York: Oxford University Press.

World Bank (1992), *World Development Report: Development and the Environment.* New York: Oxford University Press.

Zilberman, David (1992), "Environmental Aspects of Economic Relations Between Nations," *American Journal of Agricultural Economics*, Vol. 74, No. 5, pp. 1144–1149.

18

Role of Agriculture in Economic Development Revisited

Over the past two centuries there have been major changes in the role or roles attributed to agriculture in economic development. As the Nineteenth Century began agriculture was viewed as a major impediment to progress by the two greatest economists of the era—Malthus and Ricardo. The combination of a limited supply of land and diminishing (marginal) returns to the application of labor and capital to land were deemed to limit the improvement of welfare, even when productivity of labor in manufacturing increased over time. These were the conditions assumed to prevail in a developed country, such as England. At the same time, there were countries, such as America, with uncultivated land or land that was cultivated very extensively which could expand the output of food at constant or even declining prices. The strong support for the abolition of the English Corn Laws by many prominent economists was motivated in large part because free trade in grain was viewed as the only available means for freeing that country from the restraint that agriculture was imposing upon economic growth during the first half of the Nineteenth Century.

While the pessimism of Ricardo and Malthus was perhaps rather too great, we must remember the state of agricultural knowledge as of the early years of the Nineteenth Century. John Stuart Mill, writing at mid-century, saw what we would now call economic growth (rising real per-capita incomes) as complementary to improvements in agriculture that would hold in check rising real food prices (1920, p. 183). He wrote of the antagonism between "the law of diminishing return from land" and "the progress of civilization" (p. 183). His major points on how the progress of civilization acted to offset diminishing returns have a modern ring. He did not stop with noting that the progress of agricultural knowledge, skill and invention

Reprinted by permission of Elsevier Science Publishers from *Agricultural Economics* 8 (1993): 421–34.

permitted increased output from land or reduced the amount of labor per unit of output. He went on to note the positive effects of improved means of communication and transportation: "Good roads are equivalent to good tools." (p. 184). Reductions in the cost of bringing products to markets (or inputs to the farms) were equivalent to a reduction in the inputs required to produce agricultural products. Mill further recognized that if there were significant improvement in the productivity of labor in nonfarm pursuits an increase in the price of food need not prevent a rise in real consumption—the increased cheapness of clothing and lodging might more than make up for the increased cost of food. Finally, he noted that policy improvements, such as reductions in taxes, or the abolition of the corn laws "or of any other restrictions which prevent commodities from being produced where the cost of their production is lowest, amounts to a vast improvement in production." (p. 186). Unfortunately this important insight has been forgotten by policy makers far more often than it has been remembered.

In his *Principles of Economics*, Alfred Marshall had rather little to say about the relationship between agriculture and economic progress but two points are worth noting. In contrast to views that are now popular in certain circles, he argued that even when there exists diminishing returns in cultivation ". . . it may be possible for an increase in the population to cause a more than proportional increase in the means of subsistence." (1936, p. 166). How could population have such an effect? He stated: ". . . the pressure of population on the means of subsistence may be restrained for a long time by the opening up of new fields of supply, by the cheapening of railway and steamship communication, and by the growth of organization and knowledge." (p. 166). He did not argue that the effects of population growth could go unchecked indefinitely—". . . the evil day is only deferred; but it is deferred."

His second point was an empirical one, namely that even in agriculture ". . . the tendencies to increasing and diminishing return appear pretty well balanced, sometimes the one, sometimes the other being the stronger." (p. 670). There is little to argue with this conclusion a century later. One careful set of estimates found that for 1900–1982 that the real price of cereals declined by 0.8% annually, all food by 0.3% and all agricultural products by 0.8% (Diakosavvas and Scandizzo, 1991, pp. 244–245).

Hayami and Ruttan (1971, 1985, chapter 2) present excellent summaries of the Ricardian model and of the various stage theories of the relationships between agriculture and economic development. The growth-stage theories range from those of Friedrich List and Karl Marx to Eugene Rostow. The growth-stage theories call our attention to significant changes in variables and parameters as economic growth occurs. However, the assumption that there are clearly defined stages in the transition from primarily agricultural to primarily industrial economies cannot be supported empirically or, for that matter, theoretically. The process is one of continuous adjustment and change without artificial breaks or turning points. This is not to say that the adjustments occur without interruption or at a constant pace but rather that changes occur through product and factor markets that determine agriculture's relationship to the economy as a whole.

Dual Sector Models

In some ways the dual sector models of the 1950s and 1960s represented a retrogression from the neoclassical modifications of the Ricardian model of agriculture in economic development. The neoclassical modifications included the emphasis upon the role of productivity change in agriculture as an offset to the effects of diminishing returns and recognition of the low income elasticity of demand for food, following Engle, as major factors in explaining the declining relative importance of agriculture in the economy while holding in check any tendency toward long run increases in the real prices of food. In the development of the dual sector models neither Colin Clark's massive contribution to the understanding of economic growth or progress nor the important insights of G. B. Fisher on the structural changes that occur with economic growth appear to have been recognized. Had these contributions been recognized, some of the more unrealistic and unnecessary aspects of the earlier dual sector models might have been replaced by more appropriate assumptions with greater predictive power and which would have supported policies that were much more appropriate for agriculture and thus overall development. If there were ever a doubt about it, the experiences with respect to agricultural and development policies in the developing countries during the 1960s and 1970s prove that ideas count and are important. Unfortunately, all too often bad ideas based upon inadequate analysis dominate ideas that subsequently are proven to have been sounder.

The dual sector models appear to have dominated much of economic and policy thought during the 1950s and early 1960s, at least up to the publication of T. W. Schultz' *Transforming Traditional Agriculture* (1964). Unfortunately, whether or not intended by the major early contributors to the dual sector models (Lewis, 1954; Fei and Ranis, 1964), some implications of the models were interpreted as assigning to agriculture in the developing countries a negative or static role in economic growth. These implications were translated into policy frameworks that were disastrous in the role assigned to agriculture in the growth of developing countries and which adversely affected the welfare of farm people in most developing countries and, consequently, had adverse effects upon economic growth overall. The conclusion (or assumption, more accurately) that the value of the marginal product of labor in agriculture was zero over a wide range of employment gave intellectual support to the conclusion that agriculture's primary contribution to economic growth was to provide a costless supply of labor to support the growth of industry and the development of cities. The labor supply to industry was costless because agricultural output did not decline and the food surplus in the rural area created by transferring a worker from rural to urban areas could be appropriated to add to the capital of the urban areas. This conclusion was extended by accepting the assumption that the elasticity of supply of agricultural products was very low, approaching zero, and that agriculture could be continuously exploited to provide wage goods at low real prices to permit the maintenance of low labor costs in the industrial sector.

Another unfortunate development in economic writing was the position enunciated by Prebisch (1959) that the long run trends in real agricultural prices were adverse and investment in agriculture was not an appropriate use of the limited resources of developing countries. This view provided additional intellectual support for import substitution policies that stressed industrial development at the expense of agriculture. As noted above, during the Twentieth Century the overall trend in real prices of agricultural products in international markets has been a declining one. But declining real product prices does not mean that investment in agriculture would yield substandard returns or that real wages in agriculture must lag behind real wages elsewhere. Productivity improvements can and have more than offset the consequences of declining real prices for grains and other crop products in most countries. Where there have been increases in real per-capita incomes and factor markets have functioned reasonably well, the real returns to labor in agriculture have increased over time.

The conclusions derived from the dual sector models and the pessimism with respect to the trends in real prices of agricultural products combined to provide support for import substitution policies. Such policies were supposed to create a manufacturing sector that was to be a source of economic growth, something it was believed agriculture could not be. It is now clear that import substitution policies did not benefit the countries that adopted them. The evidence is overwhelming that countries that have been relatively open have had far superior economic growth compared to those that greatly restricted imports and, consequently, inhibited the development of exports (Krueger, 1980; Alam, 1991; Dollar, 1992; Levine and Renelt, 1992). These studies complement and support the many studies that have shown a positive relationship between export growth and GNP growth but the export-GNP relationship had been questioned because the direction of causality could have run either way, or both ways for that matter. What the supporters of import substitution policies ignored, and probably didn't understand, was that import duties are a tax on exports (Clements and Sjaastad, 1984). Consequently the import substitution policies resulted in slow growth, not only of imports, but also of exports. And since the import substitution policies were ineffective in generating rapid economic growth, it follows that a positive relationship between export (or total trade) and GNP growth was a relationship that should have been expected.

The model of economic growth and the role of agriculture presented by Colin Clark would have served policy makers and the farmers of the developing world far better than the inferences based on the dual sector models. In Clark's world the transformation of an economy with rising real incomes from one with most of the employment in agriculture to an economy emphasizing industry and service sectors was quite explicable—the increase in productivity in agriculture combined with income elasticities of demand for farm output that were both less than unity and declining as real per-capita incomes increase makes possible the transfer of labor from agriculture to the rest of the economy, where productivity is also increasing, while equating the supply and demand for farm products at constant or even declining real prices. The transfer of labor that occurs with economic progress was readily explained by the response of workers to differences in labor returns among the sectors (Clark, 1951, chapter X). Nor did Clark ignore the role of savings and capital accumulation in his analysis of economic progress (chapter XI). Clark's model, which follows from the neoclassical model of an enterprise economy, has stood the test of time and experience far better than the earlier dual sector models.

Agricultural Price and Income Policies

There is an enormous body of evidence to support the conclusion that policy makers in both the developing and developed countries have misunderstood the role of agriculture in economic development and how factor and product markets function. It has now been well established that during recent decades that there is an inverse relationship between real per-capita incomes and the degree of protection of agriculture (Miller, 1986; Binswanger and Scandizzo, 1983). This means, other things constant, that the lower a country's real per-capita income, the higher the level of taxation of agricultural output and the higher real per-capita incomes, the higher the level of subsidization.

The analyses undertaken by the World Bank under the direction of Anne O. Krueger, Maurice Schiff and Alberto Valdés have documented the degree of the discrimination against agriculture in 18 developing countries (Schiff and Valdés, 1992b). The period of the analyses was from 1960 to the mid-1980s. In addition to presenting measures of the extent of the negative protection of agriculture and whether due to direct or indirect mea-

Table 1. Direct and Indirect Taxation of Agriculture in 18 Countries, 1960–84 (period average in percent)

Country	Period	Indirect tax (negative protection)	Tax due to industrial protection	Direct tax	Total tax
Extreme taxers	1960–84	28.6	25.7	23.0	51.6
Ivory Coast	1960–82	23.3	23.2	25.7	49.0
Ghana	1958–76	32.6	32.4	26.9	59.5
Zambia	1966–84	29.9	21.4	16.4	46.3
Representative taxers	1960–86	24.2	32.8	12.0	36.4
Argentina	1960–84	21.3	39.5	17.8	39.1
Colombia	1960–83	25.2	37.8	4.8	30.0
Dominican Rep.	1966–85	21.3	20.8	18.6	39.9
Egypt	1964–84	19.6	27.5	24.8	44.4
Morocco	1963–84	17.4	13.4	15.0	32.4
Pakistan	1960–86	33.1	44.9	6.4	39.5
Philippines	1960–86	23.3	33.0	4.1	27.4
Sri Lanka	1960–85	31.1	40.1	9.0	40.1
Thailand	1962–84	15.0	13.9	25.1	40.1
Turkey	1961–83	37.1	57.4	−5.3	31.8
Mild taxers	1960–83	15.7	22.9	0.2	15.8
Brazil	1969–83	18.4	21.4	−10.1	8.3
Chile	1960–83	20.4	37.4	1.2	21.6
Malaysia	1960–83	8.2	9.9	9.4	17.6
Protectors	1960–84	13.6	13.9	−24.0	−10.4
Korea, Rep.	1960–84	25.8	26.7	−39.0	−13.2
Portugal	1960 84	1.3	1.0	−9.0	−7.7
Sample average		22.5	27.9	7.9	30.3

Source: Schiff and Valdés (1992a, p. 6).

sures, estimates of output effects are presented. Some of the more important results are presented in Table 1.

It should be obvious that under the policy conditions described by the measures of direct and indirect protection that agriculture's ability to contribute to economic progress was greatly circumscribed in all but two of the developing countries included in the study. In only two countries (Korea and Portugal) was there positive protection of agriculture. In the other 16 countries the protection was negative. For the three countries with the highest rates of negative protection it was estimated that the cumulative effect on agricultural output over the two-decade period was 23% of the output level in the final year. This means that if protection had been nil, at the end of the 20-year period agricultural output would have been 23% greater than it was. For the group of ten countries with an average negative nominal protection rate of 36%, the average output effect was 16%. For the group of three countries with a negative protection coefficient of 16% the output effect was estimated at 6%.

Table 2. Price Interventions and GDP Growth, by Country Group, 1960–85

Country group	Nominal rate of protection		Annual GDP growth
	Indirect	Total	
Extreme taxers	−28.6	−51.6	3.3
Representative taxers	−24.4	−36.4	5.1
Mild taxers	−15.7	−15.8	5.3
Protectors	−13.6	10.4	6.5

Source: Schiff and Valdés (1992a, p. 11).

   Presumably the negative protection of agriculture was designed to serve some objective or set of objectives. Certainly the objective was not that of making the distribution of income more equal (Schiff and Valdés, 1992a). While urban poor may have gained through lower food prices, both absolutely and relatively there are far more rural poor than urban poor. One of the puzzles of political economy is why the world aid community consistently ignored the huge transfers of income from low income rural areas to much higher income urban areas and in a number of ways through their aid actually abetted such transfers.

   If there had been any acceptable justification for the negative protection of agriculture it must have been that the transfers to governments and urban residents was a source of increased growth of the national economies. Without attempting to attribute causality, it is perhaps of interest to at least look at the experience of the 18 countries—Did discrimination against agriculture increase the rate of growth of GNP? Were the uses of the resources extracted from agriculture put to highly productive uses? The answers are clearly in the negative. A regression of per-capita real GNP growth on the average rates of nominal protection of agriculture for 1965–84 indicates that the higher the negative rate of protection, the lower the rate of national per-capita GNP growth. The regression coefficient between the protection rate and per-capita GDP growth was significant for the 1960–84 period at the one-percent level and the R-squared was at the relatively high level of 0.57. The regression coefficient indicates that the average rate of negative protection of 36% for the ten countries may have lowered their annual rate of per-capita GDP growth by as much as 2.5%. This result should not come as a total surprise; given the results of the studies of the relationships between trade orientation or openness of economies and the rates of economic growth referred to earlier. The countries with high negative rates of protection for agricultural products obviously had inward looking policies in all sectors of their economies. For a summary of the data used in the analysis, see Table 2.

   The industrial economies, as is well known (Johnson, 1973, 1991; Tyers and Anderson, 1992), have subsidized their agricultures, especially since the early 1960s. The policies followed by the European Economic Community, Japan and the United States have had a significant negative effect upon the international market prices for agricultural products. The rates of protection estimated in the World Bank studies did not reflect the distortions in world market prices due to the policies of the industrial countries. Consequently the discrimination against developing country agricultures was even greater than indicated; had the nominal rate of protection been zero, agriculture in the developing countries would have been adversely affected if they produced any of the temperate zone agricultural products.

Agriculture's Contribution to Economic Growth

I believe that it is now recognized, once again, that agriculture is capable of making several contributions to economic growth and does so if appropriate policies and conditions prevail. These contributions include (1) the release of labor for nonfarm employment; (2) the provision of an increased supply of food and fiber at constant or decreasing real prices; (3) production of an export surplus as an important source of foreign exchange to pay for capital goods and technical services not available domestically; and (4) savings to be invested in nonagricultural activities, either in rural or urban areas.

Each of these actual or potential contributions require a more or less continuous increase in output per worker. Without such productivity growth it is extremely difficult to significantly expand output in the non-agricultural sector of an economy in which agriculture accounts for most of the employment and output. It is through productivity change in agriculture that one of the major interconnections between agriculture and the rest of the economy becomes evident. It is through what John Stuart Mill called the "progress of civilization" that significant increases in resource productivity in agriculture becomes possible. The attributes of progress are of enormous variety—knowledge, research, roads, communication, markets, manufactured inputs, repair services, human skills.

Agriculture as a Declining Industry

It was noted above that one of agriculture's important contributions to economic growth was the transfer of labor to the nonfarm sector. Governmental policies universally fail to accept the transfer of labor from farms to nonagricultural jobs as essential for the economic health of the farm population. Instead it seems to be assumed that any decline in the farm population is an indication that policies have failed. If the labor transfer occurs slowly relative to the shifts in the demand for and supply of labor to agriculture, the incomes of farm families will grow more slowly than incomes in the rest of the economy. In economies with a large percentage of the labor force engaged in agriculture, say a quarter or more, rural per-capita incomes are significantly less than urban incomes. Consequently the labor transfer must be at such a rate as to not only to absorb the annually generated excess supply of labor in rural areas but to further reduce employment in agriculture to erase the differences in labor returns for individuals with comparable human capital.

There has been and is a reluctance of policy makers in industrial countries to accept the declining relative importance of agriculture and the transfer of labor out of agriculture that is inevitably associated with economic growth. Some farm price and income policies, such as those of the European Economic Community, attempt, albeit unsuccessfully, to limit the decline in farm employment and thus slow down the transfer of labor from the rural to urban areas. The evidence is very clear that such policies fail to achieve that result. Data for all the major industrial countries for the past three decades do not show a negative relationship between the level of protection and the rate of decline of farm employment (Johnson, 1991, chapter XI). In fact, countries with high rates of protection, such as Japan and members of the European Community, have had more rapid declines in farm employment since 1960 than the countries with the lowest rates of protection, such as Australia, Canada, New Zealand and the United States.

Governments have seldom adopted measures to facilitate the adjustment of rural areas to what economic growth requires. Such adjustments are inevitable, yet governments have not wanted to recognize this to be the case nor to accept responsibility for alleviating the costs that such adjustments impose upon farm and other rural people. To recognize that farm employment must decline does not mean that there must be a flood of migrants from the country to the city though such can be the outcome. In many cases the least costly way to assist the adjustment process is to make the countryside attractive for nonfarm activities that provide alternative employment opportunities for those who no longer find employment in agriculture an acceptable use of their human capital.

What is required to make rural areas more attractive for the creation of nonfarm employment? Basically it amounts to providing the necessary infrastructure in rural areas—roads, schools, communications, medical facilities, marketing structures, plentiful and reliable supplies of electricity. If these steps are taken, they make rural areas 'more attractive to rural people as well as those who make the investments required for the creation of nonfarm jobs in rural areas. Schools are key. It is a sad commentary upon rural policy making that it is only after agriculture has become relatively unimportant as a source of employment that schools in rural areas are of approximately the same quality and availability as urban schools.

One reason why the transition from a centrally planned to a market

economy is going to be so difficult in the former Soviet Union is the poor
state of the rural infrastructure in every regard—roads, schools, medical
facilities, communication. The transfer of labor from agriculture to the
rest of the economy, which will occur as economic growth occurs, will be
tragically difficult due to the decades of neglect of the rural infrastructure.
It will be a long time before the rural areas will be attractive for nonfarm
investment activities other than those directly related to agriculture.

A consequence of the myopic views of policy makers concerning the
agricultural adjustment process is that I know of no ministry of agriculture
that believes part-time farming merits its full support and encouragement.
This is evident in nearly the full range of activities that such ministries may
be involved in—research, extension, credit, adult education, agricultural
vocational education. Yet the experience of all of the market economies has
been that a large part of the adjustment of the farm population to economic
growth has been through part-time farming—the combining of farm and
nonfarm employment in the same household and often for the same person. As commercial as the agriculture of the United States is thought to
be, in most recent years 60% or more of the incomes of farm families have
come from nonfarm activities; hardly more than a quarter of all farms are
full-time farms in the sense that *more than half* of their family incomes
come from farm operations. In Japan and Taiwan as well as in Germany an
even larger fraction of farms are part-time. For millions of farm families
the opportunity to continue to live on their own farm, where homesteads
are dispersed, or in their village is a positive amenity. But this amenity is
available to large numbers only where nonfarm jobs have become acces-
sible to rural residents and it has been possible to combine these jobs with
farm work. Part-time farming has permitted the majority of people living
on farms in industrial countries to share in the increasing incomes created
by economic growth. If there had not been significant growth of nonfarm
jobs in rural areas permitting the development of part-time farming, there
would be far fewer farm households or there would be many more farm
households with low levels of income. The primary alternatives to part-
time farming were either a greater migration to the cities or the damming
up of more poor people in the countryside.

Alternative Organizations of Agriculture

The countries of Central and Eastern Europe are now trying to decide
how to organize their agricultures and, hopefully, will do so with the objectives of contributing the most to their economies and their farm people.
The experience of Poland during the socialist period makes it evident that
private ownership of land is not enough, by itself, to create a productive
agriculture. The major point that I wish to make in the next few paragraphs

is a simple one—it will take far more than deciding how farm units are organized to create efficient agricultures. The overall policy setting must be a congenial one in the sense that rural areas are not discriminated against in the provision of infrastructure, that farms are served by efficient input supply and output marketing systems, that there are adequate supplies of appropriate farm inputs such as fertilizers, machinery, petroleum products and electricity, and that farm organizations are permitted to make decisions that are in their best interests subject to appropriate restraints with respect to externalities. While it should not be necessary to state, farm output prices should not be manipulated for the benefit of urban consumers or for the benefit of government finances.

The organization of farm resources is an important aspect of the economic transition or transformation now taking place. The alternatives range from maintaining the large farm organizations, perhaps under the guise of joint stock companies, to creating much smaller farming cooperatives to family farms. The issue of the ownership of farm lands is as much a political as an economic issue. With appropriate policies, legal and institutional arrangements an efficient agriculture could exist under numerous different tenure arrangements—the terms under which a farmer or a farm organization obtains access to the use of land. It can be by ownership, by a use right or by a variety of rental arrangements (fixed rent in cash or kind or share rent).

In the United States where any tenure arrangement is possible, the most numerous form of organization is that of the family farm and of the family farms, part-ownership. In this case the farm operator family owns part of the land and rents the other part. It is important that the dark shadows of the past not prevail to either make land rental illegal or to encumber it with so many conditions that it is not a viable alternative. If land rental is to be consistent with efficient use of agricultural resources, there must be competition in the land rental market. Local authorities cannot be permitted to exercize monopoly powers in the rental of farm land, either in terms of establishing rental rates or in the determination of who the renters are on the basis of other than relevant criteria, such as past experience in paying rent. This point is made because it may be years before some of the republics of the former Soviet Union reach decisions concerning the private ownership of farm land. Until that time, family or even small scale cooperative farms can be viable only if the conditions under which farm land is available to farm operators is clearly defined and arbitrary and capricious decisions are minimized. If there are disputes between the farm operator and the owner of the land, there should be an independent agency for the resolution of the dispute. Such procedures generally do not now exist since the court systems in the republics are no where independent of the executive.

The success of the agricultural and rural reforms in China since 1978 show that not everything has to be perfect for rural people to make major contributions to economic development. What that example shows is that if the governmental restraints on the behavior of farm people are gradually relaxed and markets are permitted to develop and grow in significance, farm people will respond by working harder and more productively than under the old restrictive regime of the communes. But it was not only the form of farm organization that rural people found repressive. Very important was the many restraints on what farm people were permitted to do. At the beginning of the reform period, almost all forms of individual or private nonfarm activities were prohibited, such as selling directly to urban residents, engaging in production of handicrafts or simple manufacture, or buying and selling with the intent to make a profit. Individuals were not allowed to own the means of production, such as a tractor or truck. As these restraints were gradually lifted over the first half of the 1980s, all forms of rural productive activity expanded at rapid rates. Agricultural output grew at an unprecedented rate of 7% annually from 1979 to 1984 and rural industrial output has grown much more rapidly than the output of large and heavily subsidized urban state enterprises.

Farmers are now largely free to produce what they wish though there is still pressure on them to sell fixed quantities of a few products, such as grain, oilseeds and cotton, to the state. The remarkable agricultural production record was achieved even though they do not own the land they farm nor can they be sure that they will next year farm the land they farmed this year. True, as one would expect, there is evidence of underinvestment in maintaining or improving the productivity of land. Without security of tenure, long run investments are discouraged. In spite of the potential reforms that have not been carried out that would either increase output or improve resource productivity, the overall record in terms of increased real incomes of farm people, high rate of growth of agricultural output and the creation of nonfarm jobs for rural residents has been a remarkable one.

Concluding Comments

Revisiting the subject of the role of agriculture in economic development makes it clear that there was a great deal of wisdom in the thought of Mill and Marshall that appears to have been largely ignored in the formulation of economic policies affecting agriculture in the developing countries in the decades since World War II. But it was not only the policy makers that may have failed to understand the wisdom that history had provided, but many economists approached agriculture's role in development in a very different and, I would say, much less insightful manner. Nor did the massive empirical work of Colin Clark affect the models that were used, regardless of the intentions of their creators, to justify the exploitation of agriculture and rural people.

As one revisits the recent history of thought and policy related to the role of agriculture and economic development, it seems to me that all too often a very important fact has been ignored, namely "Farmers are as smart as the rest of us." It is simply wrong headed for any one to assume that farm people can be exploited over an extended period of time without there being negative effects, not only for the farm people, but for everyone.

References

Alam, M. S., 1991. Trade orientation and macroeconomic performance in LDCs: an empirical study. Econ. Dev. Cult. Change, 39: 839–848.

Binswanger, H. P. and Scandizzo, P. L., 1983. Patterns of agricultural protection. Rep. ARU 15, World Bank, Washington, DC.

Clark, C., 1951. The Conditions of Economic Progress. Macmillan, London.

Clements, K. W. and Sjaastad, L. A., 1983. How protection taxes exporters. Thames Essay 44, Trade Policy Centre, London.

Diakosavvas, D. and Scandizzo, P. L., 1991. Trends in the terms of trade of primary commodities, 1900–1982. Econ. Dev. Cult. Change, 39: 231–264.

Dollar, D., 1992. Outward-oriented developing economies really do grow more rapidly: Evidence from 95 LDCs. Econ. Dev. Cult. Change, 40: 523–544.

Fei, J. C. H. and Ranis, G., 1964. Development of a Labor Surplus Economy. Irwin, Homewood, IL.

Hayami, Y. and Ruttan, V. W., 1985. Agricultural Development: An International Perspective (Revised and Expanded Edition). Johns Hopkins Press, Baltimore, MD.

Johnson, D. G., 1973, 1991. World Agriculture in Disarray, 1st edition (1973), revised edition (1991). Macmillan, London.

Jorgenson, D. W., 1961. The development of a dual economy. Econ. J., 71: 309–334.

Krueger, A. O., 1980. Trade policy as an input to development. Am. Econ. Rev., 70: 288–292.

Levine, R. and Renelt, D., 1992. A sensitivity analysis of cross-country growth regressions. Am. Econ. Rev., 82: 942–963.

Lewis, W. A., 1954. Economic development with unlimited supplies of labor. Manchester School of Economics and Social Studies, 22: 139–161.

Marshall, A., 1936. Principles of Economics (8th Edition). Macmillan, London.

Mill, J. S., 1920. Principles of Political Economy. Longman S. Green, London.

Miller, T. C., 1986. Explaining agricultural price policy across countries and across commodities using a model of competition between interest groups. Ph.D. dissertation, Department of Economics, University of Chicago, IL.

Prebisch, P., 1959. Commercial policy in the underdeveloped countries. Am. Econ. Rev., 49: 251–273.

Schiff, M. and Valdés, A., 1992. The Plundering of Agriculture in Developing Countries. World Bank, Washington, DC.

Schiff, M. and Valdés, A., 1993. A Comparative Study of the Political Economy of Agricultural Pricing Policies. 4. Synthesis: The Economics of Agricultural Price Intervention in Developing Countries. World Bank, Washington, DC.

Schultz, T. W., 1964. Transforming Traditional Agriculture. Yale University Press, New Haven, CT. (Reprint: University of Chicago Press, 1973)

Tyers, R. and Anderson, K., 1992. Disarray in World Food Markets: A Quantitative Assessment. Cambridge University Press, Cambridge, UK.

BIBLIOGRAPHY OF D. GALE JOHNSON, 1942–1994

Articles reprinted in this volume are marked with an asterisk.

1942

1. "Existing Agricultural Price Policy." *Elements of a Price Policy for Agriculture.* Memo no. 1. Iowa State College. Mimeo.

2. "Current Price Relationships within the Livestock Feed and Wheat Complex." *Elements of a Price Policy for Agriculture.* Memo no. 2. Iowa State College. Mimeo.

3. "Current Prices and Price Relationships for Feed Grains, Including Wheat." *Elements of a Price Policy for Agriculture.* Memo no. 3. Iowa State College. Mimeo.

4. "Price Policy and the Fats and Oils Problem." *Elements of a Price Policy for Agriculture.* Memo no. 4. Iowa State College. Mimeo.

5. "Expansion of Livestock Production for Fat and Meat." *Elements of a Price Policy for Agriculture.* Memo no. 5. Iowa State College. Mimeo.

6. "Expanding Hog Production in Four Plains States." *Elements of a Price Policy for Agriculture.* Memo no. 6. Iowa State College. Mimeo.

1944

*7. "Contribution of Price Policy to Income and Resource Problems in Agriculture." *J. Farm Econ.* 26 (Nov.): 631–64.

8. With O. H. Brownlee. *Food Subsidies and Inflation Control.* Wartime Farm and Food Policy Series, pamphlet no. 10. Ames: Iowa State College Press.

1945

9. "A Price Policy for Agriculture, Consistent with Economic Progress that will Promote Adequate and More Stable Income from Farming." *J. Farm Econ.* 27 (Nov.): 761–72.

1946

10. With W. H. Nicholls. "The Farm Price Policy Awards: A Topical Digest of the Winning Essays." *J. Farm Econ.* 28 (Feb.): 267–73.

11. "Measurement of Marginal Productivities of Agricultural Resources." Washington, D.C.: Bureau of Economic Research. Mimeo.

1947

12. "Interrelationships between the Agricultural Policy and Foreign Economic Policy." Washington, D.C.: Department of State. Mimeo.
13. "Mobility and Agricultural Income." University of Chicago, Department of Economics. Mimeo.
14. *Forward Prices for Agriculture.* Chicago: University of Chicago Press. Repr. New York: Arno Press, 1976.

1948

15. "The Use of Econometric Models in the Study of Agricultural Policy." *J. Farm Econ.* 30 (Feb.): 117–30.
16. "The High Cost of Food: A Suggested Solution." *J. Pol. Econ.* 56 (Feb.): 54–57.
17. "Reconciling Agricultural and Foreign Trade Policies." *J. Pol. Econ.* 56 (Apr.): 567–71.
18. "Mobility as a Field of Economic Research." *So. Econ. J.* 15 (Oct.): 152–61.
*19. "Allocation of Agricultural Income." *J. Farm Econ.* 30 (Nov.): 724–49.
20. "Efficient Utilization of Agricultural Resources: A Project Outline." University of Chicago, Department of Economics. Mimeo.
21. With T. W. Schultz. "Price Policy of Germany, Food and Agriculture." B.C.O. Food and Agriculture. University of Chicago, Department of Economics. Mimeo.

1949

22. "High Level Support Prices and Corn Belt Agriculture." *J. Farm Econ.* 31 (Aug.): 509–19.
23. "Conflicts between Domestic Agricultural Policy and Foreign Economic Policy." Washington, D.C.: U.S. Department of State. Mimeo.

1950

24. "Domestic Agricultural Policy and Foreign Economic Policy." Washington, D.C.: U.S. Department of State, Foreign Service Institute. Jan.
*25. "Resource Allocation Under Share Contracts." *J. Pol. Econ.* 58 (Apr.): 111–23.
26. With O. H. Brownlee. "Reducing Price Variability Confronting Primary Producers." *J. Farm Econ.* 32 (May): 176–93.
*27. "The Nature of the Supply Function for Agricultural Products." *Amer. Econ. Rev.* 40 (Sept.): 539–64.
28. "Coming Developments in Agricultural Policy." *J. Farm Econ.* 32 (Nov.): 827–38.
29. *Agriculture and Trade: A Study of Inconsistent Policies.* New York: John Wiley and Son.

1951

*30. "Functioning of the Labor Market." *J. Farm Econ.* 33 (Feb.): 75–87.

31. "Rent Control and the Distribution of Income." *Amer. Econ. Rev.* 41 (May): 569–83.

32. With Marilyn Corn Nottenburg. "A Critical Appraisal of Farm Employment Estimates," *J. Amer. Statist. Assoc.* 46 (June): 191–205.

33. "Agriculture's Share of the National Income." Conference on the Church and Agricultural Policy. June. Mimeo.

34. "The Role of Support Prices in a Full Employment Economy." *Can. J. Econ. and Pol. Sci.* 17 (Aug.): 352–59.

35. "Policies and Procedures to Facilitate Desirable Shifts of Manpower." *J. Farm Econ.* 33 (Nov.): 722–29.

1952

36. "Are Farmers Getting Too Much? Comment." *Rev. Econ. and Statist.* 34 (Aug.): 255–58.

37. "U.S. Foreign Economic Assistance and the Demand for American Agricultural Products." *J. Farm Econ.* 34 (Dec.): 662–70.

38. "Economics of Agriculture." *A Survey of Contemporary Economics.* Ed. B. F. Haley. Homewood, Ill.: American Economic Association, pp. 223–56.

39. "Some Effects of Region, Community Size, Color, and Occupation on Family and Individual Income." *Studies in Income and Wealth.* Vol. 15. New York: National Bureau of Economic Research.

1953

*40. "Comparability of Labor Capacities of Farm and Non-farm Labor." *Amer. Econ. Rev.* 43 (June): 296–313.

41. "The Million American Families with Poor Opportunities—Possibilities for their Future." Farm Policy Forum, Ames, Iowa. June.

42. "Economic Problems of Proper World Milk Utilization." Washington, D.C.: First World Congress for Milk Utilization. 20 Nov.

43. "Report of the American Farm Economic Association Committee on Farm Employment Estimates." *J. Farm Econ.* 35 (Dec.): 976–87.

1954

44. "Milk's Impact on World Economy." Proceedings of the First Congress for Milk Utilization, Dairy Industries Society International, 1953, pp. 59–66; *Agr. Inst. Rev.* (Jan.–Feb.): 28ff; *Nordisk Mejeri—Tedsskrift.* pp. 76ff.

45. "The Income Problem in Agriculture." Kansas State College Bulletin. Vol. 38. 15 Jan.; Farm Policy Forum Proceedings. 5–6 Oct., pp. 5–13.

46. "Competition in Agriculture: Fact or Fiction?" *Amer. Econ. Rev.* 44 (May): 107–15.

47. "The Functional Distribution of Income in the United States, 1850–1952." *Rev. Econ. and Statist.* 36 (May): 175–82.

48. "Agricultural Price Policy and International Trade." Essays in International Finance. Department of Economics, Princeton University, June.

1955

49. "Analysis of Krushchev's Report in Livestock Product." American Committee for Liberation from Bolshevism, Inc. New York, 15 Mar. Mimeo.

50. "The Role of Farm Prices in Agricultural Production." In American Assembly, *United States Agriculture*, March 1955, pp. 45–53.

51. "Opening Foreign Markets." Farm Policy Forum, Ames, IA, Spring, pp. 22–30.

52. "A Study of the Growth Potential of Agriculture of the U.S.S.R." Santa Monica, Calif.: Rand Corporation. RM 1561, Oct.

53. "Notes on Visit of American Agricultural Delegation to U.S.S.R." Oct. Mimeo.

54. "Effects of Federal Taxation on Agriculture." Federal Tax Policy for Economic Growth and Stability. Joint Committee Print, 84th Congress, 1st session, 9 Nov.

55. "Corn and Commissars." *University of Chicago Magazine*. Nov.

56. Statement in Low-Income Families Hearings. Subcommittee on Low-Income Families, Joint Committee on Economics Report, 84th Congress, 1st session, Nov., pp. 715–19.

57. Statement in Foreign Economic Policy Hearings. Subcommittee on Foreign Economic Policy, Joint Committee on Economic Report, 84th Congress, 1st session, Nov., pp. 357–60.

1956

*58. "Eye-Witness Appraisal of Soviet Farming, 1955." *J. Farm Econ.* 38 (May): 287–95.

59. "Stabilization of International Commodity Prices." *Policies to Combat Depressions.* New York: National Bureau of Economic Research, pp. 357–70.

1957

60. "Comment on Edwin Mansfield's, City Size and Income, 1949." *Regional Income. Studies in Income and Wealth*, Vol. 21. National Bureau of Economic Research. Princeton, N.J.: Princeton University Press, pp. 307–14.

61. "Economics and the Educational System." *School Review* 16 (Autumn): 260–70. Chicago: University of Chicago Press.

62. "Farm Prices, Resource Use and Farm Income." *Policy for Commercial Agriculture, Its Relations to Economic Growth and Stability.* Joint Economic Committee, 22 Nov. 1957. Washington, D.C.: U.S. Government Printing Office, pp. 448–59.

1958

63. "The Farmer's Interest in Labor Unions." *Farm Policy Forum.* Spring, pp. 2–8. Ames: Iowa State College Press.

*64. "Government and Agriculture: Is Agriculture a Special Case?" *J. Law and Econ.* 1 (Oct.): 122–37.

65. "A Economia e o Sistema Educacional." *Educacao e Ciencias Sociais* 4: 123–37.

66. "An Appraisal of the Data for Farm Families." *An Appraisal of the 1950*

Census Income Data. Princeton, N.J.: Princeton University Press, National Bureau of Economic Research.

67. "Labor Mobility and Agricultural Adjustment." *Agricultural Adjustment Problems in a Growing Economy*. Ames: Iowa State University Press.

1959

68. "The United States Agricultural Price Policy and International Trade." *Malayan Econ. Rev.* 4 (Apr.): 80–94.

69. With Arcadius Kahan. "Comparisons of the United States and Soviet Economies." U.S. Congress, Joint Economic Committee, part 1, pp. 201–37. Washington, D.C.: U.S. Government Printing Office.

70. "The Dimensions of the Farm Problem." *Problems and Policies of American Agriculture*. Ames: Iowa State University Press, pp. 47–63.

1960

71. "Khrushchev's Farm Problem—How to Get More with Less." *Challenge* (May): 24–38, New York University Institute of Economic Affairs.

72. "Policies to Improve the Labor Transfer Process." *Amer. Econ. Rev.* 50 (May): 403–13.

73. "The Experience of the More Highly Developed Countries." Proceedings of the Tenth International Conference of Agricultural Economists, Mysore, India. London: Oxford University Press, pp. 34–42.

74. "Output and Income Effects of Reducing the Farm Labor Force." *J. Farm Econ.* 42 (Nov.): 779–96.

75. "Agricultural Economic Policy in the United States." Washington, D.C.: National Academy of Economics and Political Science. Special Publication Series, no. 16, pp. 7–11.

1961

76. "Let's Not Keep Them Down on the Farm." *Context* 1 (Spring): 23–25.

77. Testimony, Report of the Commission on Money and Credit Hearings. Joint Economic Committee, 87th Congress, 1st session, 14–18 Aug., pp. 335–42.

1962

78. With Arcadius Kahan. *The Soviet Agricultural Program: An Evaluation of the 1965 Goals*. Santa Monica, Calif.: Rand Corporation, RM-2248-PR, May.

79. "Soviet Agriculture: Structure and Performance." University of Chicago, Office of Agricultural Economics Research, paper no. 6208, 4 Oct.

80. "Welfare Economics and Agricultural Policy." University of Chicago, Office of Agricultural Economics Research, paper no. 6207, 4 Oct.

81. "Agricultural Employment Statistics." Paper prepared for President's Committee to Appraise Employment and Unemployment Statistics. Published in large part in *Measuring Employment and Unemployment by the President's Committee*. Washington, D.C., 96–102, 123–29.

82. "Prospects of Food Supply in Highly Developed Regions." *Food—One Tool in International Economic Development*. Iowa State Center for Agricultural and Economic Adjustment. Ames: Iowa State University Press, pp. 244–66.

83. With Robert L. Gustafson. *Grain Yields and the American Food Supply: An Analysis of Yield Changes and Possibilities.* Chicago: University of Chicago Press.

1963
84. "Efficiency and Welfare Implications of U.S. Agricultural Policy." *J. Farm Econ.* 45 (May): 331–42.

1964
85. "Agricultural Credit, Capital and Credit Policy in the United States." *Federal Credit Programs.* Commission on Money and Credit. Englewood Cliffs, N.J.: Prentice-Hall, Inc., pp. 355–424.
86. "The Credit Programs Supervised by the Farm Credit Administration." *Federal Credit Agencies.* Commission on Money and Credit. Englewood Cliffs, N.J.: Prentice-Hall, Inc., pp. 259–318.
87. "Agricultural Production." *Economic Trends in the Soviet Union.* Ed. A. Bergson and S. Kuznets. Cambridge, Mass.: Harvard University Press, pp. 203–34.
88. "The Role of Agriculture in Economic Development." *Natural Resources and International Development.* Ed. Marion Clawson. Washington, D.C.: Resources for the Future, Inc., pp. 3–26.
89. "Soviet Agriculture." *Bulletin for Atomic Scientists.* Jan.
*90. "Agriculture and Foreign Economic Policy." *J. Farm Econ.* 46 (Dec.): 915–29.

1965
91. "The U.S. Economy." *AMERIKA.* Hamburg: Hoffman and Campe Verlag.
92. "Econometric Analysis and Agricultural Development Plans." *Study Week on the Econometric Approach to Development Planning.* Vatican City. Part 2, pp. 1141–96.
93. "Alternative Trade Policies: Their Impact on U.S. and World Production and Trade." Proceedings of the Fifth Annual Farm Policy Review Conference: *Our Stake in Commercial Agriculture, Rural Poverty and World Trade.* CAED Report no. 22. Center for Agricultural and Economic Development, Iowa State University.
94. "Alternative Trade Policies." *Farm Policy Forum* 17: 6–13.
95. "Agriculture and Foreign Economic Policy." *Agr. Econ.* 7 (July).
96. "Prospects for U.S. Grain Exports." University of Chicago, Office of Agricultural Economics Research, paper no. 65:11, 26 May.
97. "Agriculture and Foreign Economic Policies: Implications to Producers." University of Chicago, Office of Agricultural Economics Research, paper no. 65:12, 1 June.
98. "Implications of the World Food Situation for American Agricultural Policy." *People and Food: Centennial Symposium, 1865–1965.* University of Kentucky, Oct., pp. 43–53.

1966

99. "The Environment for Technological Change in Soviet Agriculture." *Amer. Econ. Rev.* 56 (May): 145–53.

100. "Trade Policies and U.S. Agriculture." *J. Farm Econ.* 48 (May): 339–50.

101. "Agriculture as a Declining Industry: A Neglected Possibility." *Taylor-Hibbard Newsletter.* Madison: University of Wisconsin, pp. 2–6.

102. "Potential for Expanding Agricultural Exports in Increased East-West Trade." Proceedings of the Upper Midwest Conference on Agricultural Export Trade, University of Minnesota, May, pp. 57–65. Reprinted in part as "What Is the Potential for Expanded East-West Trade?" *Foreign Agriculture* 4 (13 June): 7–9.

103. "Economia do bemestar e political agraria." *Revista Brasileira de Economia* 20 (March): 1–18.

1967

104. "Changes in the Economy and Agriculture." *Journal of the American Society of Farm Managers and Rural Appraisers* 31 (April): 6–11. Reprinted in *Cargill Crop Bulletin* 42 (April): 9–12.

105. With others. *Food and Fiber for the Future.* Report of the National Advisory Commission on Food and Fiber. Washington, D.C.: U.S. Government Printing Office, 1967.

106. "Agricultural Trade and Foreign Economic Policy." *Foreign Trade and Agricultural Policy* 6. National Advisory Commission on Food and Fiber. Washington, D.C.: U.S. Government Printing Office, pp. 1–34.

107. "Sugar Programs: Costs and Benefits." *Foreign Trade and Agricultural Policy* 6. National Advisory Commission on Food and Fiber. Washington, D.C.: U.S. Government Printing Office, pp. 35–50.

108. "United States Food Problems and Policies: Pricing Developments." *National Farm Institute: Farmers and a Hungry World.* Ames: Iowa State University Press, pp. 85–94.

109. *The Struggle Against World Hunger.* Headline Series no. 184. New York: Foreign Policy Association.

110. "Sind die Weltmarkpreise fur Agrarprodukte Manipuliert?" *Wirtschaftsdienst* (Nov.): 585–90.

1968

111. "Food." *Toward the Year 2010.* Ed. Foreign Policy Association. New York: Cowles Education Corporation, pp. 126–35.

112. "Agriculture: Price and Income Policies." *International Encyclopedia of the Social Sciences.* Vol. 1. New York: Macmillan and Free Press, pp. 250–66.

113. Testimony on International Grains Arrangement of 1967. Hearings before a Subcommittee of the Committee on Foreign Relations, United States Senate, 90th Congress, 2d session, 26 Mar., 4–5 Apr., pp. 67–71.

1969

114. "Conflicts between Domestic Agricultural Policies and Trade Policy." *Agriculture and International Trade.* Ed. R. C. Brooks. Raleigh: North Carolina State University Press, pp. 107–26.

115. Editor, with Karl Fox. *Readings in the Economics of Agriculture* 13. American Economic Association Series of Republished Articles on Economics. Homewood, Ill.: Irwin.

116. "The New Agricultural Protectionism in the Industrial Countries." *Rivista Internazionale di Scienze Economiche e Commerciali* 16: 46–62.

1970

117. "Food Supply." *Encyclopaedia Britannica.*

118. "Famine." *Encyclopaedia Britannica.*

119. *The Yearbook of Agriculture.* Chaps. 41 and 63.

120. "Soviet Agriculture Revisited." *Amer. J. Agr. Econ.* 53 (May): 257–68.

121. "Perspectives on the International Grain Trade." University of Chicago, Office of Agricultural Economics Research, paper no. 70:8, 10 July.

122. "Economic Growth and Agriculture." University of Chicago, Office of Agricultural Economics Research, paper no. 70:9, 15 July.

1971

123. "Agricultural Economics." *Encyclopaedia Britannica.*

124. "Population Balance." *Toward Policies for Balanced Growth.* Ed. Donald Nelson. Washington, D.C.: Graduate School, U.S. Department of Agriculture, pp. 25–33.

125. "Agricultural Trade—A Look Ahead—Policy Recommendations." *United States International Economic Policy in an Interdependent World.* Vol. 1. Washington, D.C.: Commission on International Trade and Investment Policy, pp. 873–96.

1972

126. "Agricultural Price Policies and Effects on Trade: Some Examples from the United States and Western Europe." *Obstacles to Trade in the Pacific Area.* Ed. H. E. English and Keith A. J. Hay. Fourth Pacific Trade and Development Conference, Carleton University, Ottawa, pp. 81–94.

127. "Comparative Advantage and U.S. Exports and Imports of Farm Products." University of Chicago, Office of Agricultural Economics Research, paper no. 72:1. Prepared for the National Agricultural Outlook Conference, Washington, D.C., 23 Feb.

128. "New Directions for Agricultural Policy in the Industrial Countries." University of Chicago, Office of Agricultural Economics Research, paper no. 72:4, 22 Mar.

1973

129. "The Impact of Freer Trade on North American Agriculture." *Amer. J. Agr. Econ.* 55 (May): 294–300.

130. *World Agriculture in Disarray.* London: Macmillan; New York: St. Martin's Press and New Viewpoints.

131. *Farm Commodity Programs: An Opportunity for Change.* Washington, D.C.: American Enterprise Institute.

132. "Obstacles to Agricultural Trade." *Round Table,* no. 250 (Apr.): 161–74.

133. "Export Controls on Agricultural Products." Testimony, Senate Committee on Banking, Housing and Urban Affairs, June.

134. "Policies, Programs and Procedures to Maximize U.S. Farm Exports." University of Chicago, Office of Agricultural Economics Research, paper no. 73:20, Dec.

135. "What Difference Does Trade Make in World Community Welfare." *U.S. Trade Policy and Agricultural Exports.* Iowa State University Center for Agricultural and Rural Development. Ames: Iowa State University Press, pp. 49–68.

136. "Social Impacts of Increased Production Efficiency." Proceedings, Twenty-first Annual Meeting of the Agricultural Research Institute. Washington, D.C.: National Academy of Sciences, pp. 9–16.

137. "Soviet Economic Prospects for the Seventies: Some Comment and Extensions." Testimony, Joint Economic Committee, U.S. Congress, 18 July. 1973; University of Chicago, Office of Agricultural Economics Research, paper no. 73:17, 12 July.

*138. "Government and Agricultural Adjustment." *Amer. J. Agr. Econ.* 55 (Dec.): 860–67.

139. "Economics of Population: Discussion." *Amer. Econ. Rev.* 63 (May): 88–89.

1974

140. With John A. Schnittker. *U.S. Agriculture in a World Context: Policies and Approaches for the Next Decade.* New York: Praeger Press, for the Atlantic Council of the United States.

141. *The Sugar Program: Large Costs and Small Benefits.* Evaluative Studies 14. Washington, D.C.: American Enterprise Institute for Public Policy Research.

142. "Soviet Agriculture and World Trade in Farm Products." *Prospects for Agricultural Trade with the USSR.* U.S. Department of Agriculture, Economic Research Service, ERS-Foreign 356, Apr., pp. 43–50.

143. "World Food Problems." University of Chicago, Office of Agricultural Economics Research, paper no. 74:8, July.

144. "Population, Food and Economic Adjustment." *American Statistician* 28 (Aug.): 89–93.

145. "Are High Farm Prices Here to Stay?" *The Morgan Guaranty Survey* (Aug.), pp. 9–14.

146. "Project Independence and Agriculture." Testimony, hearings held by the Federal Energy Administration in Chicago, 21 Sept.

147. "Are Tight Food Supplies and High Farm Prices Here to Stay?" University of Chicago, Office of Agricultural Economics Research, paper no. 74:10, Sept.

148. "The Soviet Livestock Sector: Problems and Prospects." *Association for Comparative Economic Studies Bulletin* 16 (Fall): 41–62.

149. Testimony, National Nutrition Policy Study. Part 2: *Nutrition and the International Situation.* Hearings before the Select Committee on Nutrition and Human Needs of the U.S. Senate, 93rd Congress, 2d session, 19 June, pp. 279–312. Includes report of Panel on Nutrition and the International Situation. Also available as *National Nutrition Policy Study: Report and Recommendations* 6 (June).

150. "Politics and Hunger: U.S. Foreign Relations and Agricultural Trade."

Transcript of the proceedings: *Feeding the World Hungry—The Challenge to Business.* Continental Bank, Public Affairs Division, Chicago.

151. "Impact of Farm-Support Policies on International Trade." *In Search of a New World Economic Order.* Ed. Hugh Corbet and Robert Jackson. London: Croom Helm, pp. 211–28.

152. "Objectives of International Training in Agricultural Economics." *Amer. J. Agr. Econ.* 56 (Dec.): 1176–81.

153. "World Food Problems in Perspective." University of Chicago, Office of Agricultural Economics Research, paper no. 74:11, Oct.

154. "Determination of Optimal Grain Carryover." University of Chicago, Office of Agricultural Economics Research, paper no. 74:12, Oct. Revised Mar. 1975 with D. Sumner and Y. Danin.

155. "Implications of the World Food Conference." University of Chicago, Office of Agricultural Economics Research, paper no. 74:13, Dec.

1975

156. "Free Trade in Agricultural Products: Possible Effects on Total Output, Prices and the International Distribution of Output." University of Chicago, Office of Agricultural Economics Research, paper no. 71:9, July. Also in *Trade, Agriculture and Development.* Ed. George Tolley. New York: Ballinger Publishing Co., pp. 3–20.

157. *World Food Problems and Prospects.* Washington, D.C.: American Enterprise Institute for Public Policy Research, June.

158. "World Food Problems in Perspective." *Occasional Paper Series* 1 (Mar.). Raleigh: University of North Carolina, Institute of Nutrition. March 1975.

159. "The Soviet Grain Shortage: A Case of Rising Expectations." *Current History* 68 (June): 245–48.

160. *World Agriculture in Disarray.* Japanese translation, Tokyo, 1975.

161. With Roger Gray and Jimmye S. Hillman. "Food Reserve Policies for World Food Security: A Consultant Study on Alternative Approaches." Expert Consultation on Cereal Stock Policies Relating to World Food Security. Food and Agriculture Organization of the United Nations, ESC: CSP/75/2, Jan.

162. "The Nature of the Food Crisis and What Japan Should Do." *Toyo Keizai Shinposha* (Oriental Economist), no. 34 (22 Oct.): 34–40. In Japanese.

*163. "World Agriculture, Commodity Policy, and Price Variability." *Amer. J. Agr. Econ.* 57 (Dec.): 823–28.

164. "Theory and Practice of Soviet Agriculture." University of Chicago, Office of Agricultural Economics Research, paper no. 75:28, 15 Dec.

165. "The Role of Agriculture in Economic Growth." *Modernization of Agriculture: East and West.* Eleventh International Seminar, Urbino, 3–5 July, Milan, pp. 1–24.

1976

166. "Food for the Future: A Perspective," *Pop. and Devel. Rev.* 2 (Mar.): 1–20.

167. "World Food Problems: An Optimistic Appraisal." *Trends.* Washington, D.C.: U.S. Information Agency, Apr.

*168. With Daniel Sumner. "An Optimization Approach to Grain Reserves for Developing Countries." *Analyses of Grain Reserves, A Proceedings.* Comp. David

J. Eaton and W. Scott Steele. U.S. Department of Agriculture, Economic Research Service, in cooperation with National Science Foundation. Economic Research Service Report no. 634, Aug., pp. 56–76.

169. "American Food and Agricultural Policies: Responsibilities and Opportunities." The Anna Howard Shaw Lecture, Bryn Mawr College, 14 Sept.

170. "Food Problems in the World." *Kokusai Shigen* (Natural Resources of the World), Oct. In Japanese.

1977

171. "Eating Less Won't Provide More Food." *PHP*, Nov.: 2–11. Also in Japanese, Apr., pp. 45–48.

172. "The Soviet Grain Disaster of 1975." *1977 Britannica Book of the Year.* Chicago: Encyclopaedia Britannica, pp. 122–23.

173. "Grain Reserves in the World Context." Proceedings of the Eighth Annual Meeting of the Canada Grains Council, 5–6 Apr., pp. 50–54.

1978

174. "World Food Problems and U.S. Agriculture." *Lectures in Agricultural Economics.* Bicentennial Year Lectures Sponsored by the Economic Research Service. Washington, D.C.: U.S. Department of Agriculture, Economics Research Service, June, pp. 123–40.

175. As member of steering committee. *World Food and Nutrition Study: The Potential Contributions of Research.* Washington, D.C.: National Research Council, National Academy of Sciences.

176. "Increased Stability of Grain Supplies in Developing Countries: Optimal Carryovers and Insurance." *The New International Economic Order: The North-South Debate.* Ed. Jagdish Bhagwati. Cambridge, Mass.: MIT Press.

177. "Resource Adjustment in American Agriculture and Agricultural Policy." *Contemporary Economic Problems.* Ed. William Fellner. Washington, D.C.: American Enterprise Institute for Public Policy Research, pp. 203–38.

178. "Food Production and Marketing: Economic Developments in Agriculture." *Food and Agricultural Policy.* Proceedings of the American Enterprise Institute Conference, 10–11 Mar. Washington, D.C.: American Enterprise Institute, pp. 9–28.

179. "Soviet Agriculture and U.S.-Soviet Relations." *Current History* 73 (Oct.) 1977: 118–22.

180. "Post-War Policies Relating to Trade in Agricultural Economics." *A Survey of Agricultural Economics Literature.* Vol. 1. Ed. Lee R. Martin. Minneapolis: University of Minnesota Press, pp. 295–325.

181. *Food Production Potentials in Developing Countries: Will They be Realized.* Occasional Paper no. 1, Bureau of Economic Studies, Macalester College, St. Paul, Minnesota, Dec.

182. "International Food Security: Issues and Alternative." *International Food Policy Issues, A Proceedings.* Economics, Statistics, and Cooperative Services, USDA, Foreign Agricultural Economics no. 143, Jan., pp. 81–89.

183. "Estimating Appropriate Levels of Grain Reserves for the United States:

A Research Report." University of Chicago, Office of Agricultural Economics Research, paper no. 77:26; revised 10 Feb.

184. "The Outlook for Food Production in Southeast Asia and the World by Year 2000." *Food and Agriculture Malaysia 2000.* Ed. H. F. Chin, I. C. Enoch, and Wan Mohamad. Malaysia: Faculty of Agriculture, Universiti Pertanian Malaysia, pp. 459–70.

185. "Limitations of Grain Reserves in the Quest for Stable Prices." *World Economy* 1 (June): 289–300.

186. "International Prices and Trade in Reducing the Distortions of Incentives." *Distortions of Agricultural Incentives.* Ed. Theodore W. Schultz. Bloomington: Indiana University Press, pp. 195–215.

187. "Climate Changes and World Food Production." The University of Chicago, Office of Agricultural Economics Research, paper no. 78:20, 12 Oct.

188. "The Food and Agriculture Act of 1977: Implications for Farmers, Consumers, and Taxpayers." *Contemporary Economic Problems 1978.* Ed. William Fellner. Washington, D.C.: American Enterprise Institute, pp. 167–209.

189. "World Food Institutions: A 'Liberal' View." *International Organization* 32 (Summer): 837–54; *The Global Political Economy of Food.* Ed. Raymond F. Hopkins and Donald J. Puchala. Madison: University of Wisconsin Press, pp. 265–82.

190. "Potential Role of Humanitarian Efforts." *World Agricultural Trade: The Potential for Growth.* Proceedings of a symposium sponsored by the Federal Reserve Bank of Kansas City, 18–19 May, pp. 94–108.

191. "National Agricultural Policies and Market Relations." *Amer. J. Agr. Econ.* 60 (Dec.): 789–92.

192. With Umberto Colombo and Toshio Shishido. "Reducing Malnutrition in Developing Countries: Increasing Rice Production in South and Southeast Asia." *Triangle Papers* 16. New York: The Trilateral Commission.

193. "Grain Insurance, Reserves and Trade: Contributions to Food Security for LDCs." Paper presented at International Maize and Wheat Improvement Center and International Food Policy Research Institute Conference, in Mexico, November.

194. "Agriculture in the International Economy." *Tariffs, Quotas and Trade: The Politics of Protectionism.* San Francisco: Institute for Contemporary Studies, pp. 269–80.

1979

195. "World Agricultural and Trade Policies: Impact upon U.S. Agriculture." *Contemporary Economic Problems 1979.* Ed. William Fellner. Washington, D.C.: American Enterprise Institute, pp. 293–324.

196. "Soviet Agriculture in the Late 1970s." University of Chicago, Office of Agricultural Economics Research, paper no. 79:4, 29 Jan. Prepared for presentation at a seminar on "Economic Results in 1978: Is There a Slowdown in the Soviet and East European Economy?" Russian Research Center, Harvard University, 5 Feb.

197. *The World Grain Economy and Climate Change to the Year 2000: Implications for Policy.* Apr. Published by National Defense University Press, Washington, D.C., 1983.

198. "Strategies for Rapid Agricultural Development." University of Chicago, Office of Agricultural Economics Research, paper no. 79:20, 25 June. Prepared for Congressional Roundtable on World Food and Hunger, 24–25 July. Reprinted in *Future Dimensions of World Food and Population*. Ed. Richard G. Woods. Boulder, Colo.: Westview Press, 1981, pp. 281–98.

199. "Policies for Achieving Farm Income Stability." Paper presented at Seventeenth Annual Meeting of the Brazilian Agricultural Economics Association, 1 Aug., Brasilia.

200. "Research and Technological Transfer in Agriculture." University of Chicago, Office of Agricultural Economics Research, paper no. 79:22, 14 Aug. Published in Spanish in *Agricultura Tecnica* 39 (July-Sept.): 69–75.

201. "Food Reserves and International Trade Policy." *International Trade and Agriculture Theory and Policy*. Ed. Jimmye S. Hillman and Andrew Schmitz. Boulder, Colo.: Westview Press, pp. 239–52.

1980

202. "Agricultural Policy for the 1980s." University of Chicago, Office of Agricultural Economics Research, paper no. 80:13, 7 Apr. Paper prepared for presentation at Chase Econometrics, Apr., Bala Cynwyd, Penna.

203. Editor. *The Politics of Food: Producing and Distributing the World's Food Supply*. Chicago: The Chicago Council on Foreign Relations.

204. "Increasing Food Security of Low Income Countries." *The Politics of Food: Producing and Distributing the World's Food Supply*. Ed. D. Gale Johnson. Chicago: The Chicago Council on Foreign Relations, pp. 183–206.

205. "U.S. Agriculture and the World Economy." Laurence J. Norton Distinguished Lecture, 25 Sept. University of Illinois, Urbana; *The Role of Government in a Market Economy*. Ed. Lowell D. Hill. Ames: Iowa State University Press, pp. 75–90.

206. "The Food Gap: Catastrophe or Illusion." *Transactions/Society* 17 (Sept.–Oct.): 26–28.

207. "The Place of the American Farmer in the World Food System." *1980 National Farm Credit Directors' Conference Report*. 12–16 Oct., Denver, pp. 98–106.

208. "Inflation, Agricultural Output and Productivity." *Amer. J. Agr. Econ.* 62 (Dec.): 917–23.

1981

209. Editor. *Food and Agriculture Policy for the 1980s*. Washington, D.C.: American Enterprise Institute.

210. "Agricultural Policy Alternatives for the 1980s." *Food and Agriculture Policy for the 1980s*. Ed. D. Gale Johnson. Washington, D.C.: American Enterprise Institute, pp. 183–209.

211. "Food and Agriculture of the Centrally Planned Economies: Implications for the World Food System." *Essays in Contemporary Economic Problems: Demand, Productivity and Population*. Ed. William Fellner. Washington, D.C.: American Enterprise Institute, pp. 171–214.

212. "The World's Poor: Can They Hope for a Better Future?" *Social Service Review* 55 (Dec.): 544–56.

213. "Agricultural Productivity in the United States: Some Sources of a Remarkable Achievement." University of Chicago, Office of Agricultural Economics Research, paper no. 81:38. Presented as the 1981 Seaman Knapp Memorial Lecture in Washington, D.C., 9 Nov. at the meeting of the National Association of State Universities and Land Grant Colleges.

1982

214. "Prospects for Soviet Agriculture in the 1980s." *Soviet Economy in the 1980s: Problems and Prospects.* Joint Economic Committee, U.S. Congress, 97th Congress, 2d session, 31 Dec., pp. 7–22.

215. "Agricultural Organization and Management in the Soviet Union: Change and Constancy." *The Soviet Union Toward the Year 2000.* Ed. Abram Bergson and Herbert S. Levine. London: Allyn and Unwin, pp. 112–42.

216. "The Economics Major: What It Is and What It Should Be: Panel Discussion." *Amer. Econ. Rev.* 64 (May): 138–40.

217. "The World Food Situation: Developments During the 1970s and Prospects for the 1980s." *Contemporary Economic Problems 1980.* Washington, D.C.: American Enterprise Institute, pp. 301–31. A slightly revised version published in *U.S.-Japanese Agricultural Trade Relations.* Ed. Emery N. Castle and Kenzo Hemmi. Washington, D.C.: Resources for the Future, Inc., pp. 15–57.

218. "International Trade and Agricultural Labor Markets: Farm Policy as Quasi-Adjustment Policy." *Amer. J. Agr. Econ.* 64 (May): 355–61.

219. *Progress of Economic Reform in the People's Republic of China.* Washington, D.C.: American Enterprise Institute.

220. "Trends in the Development of International Trade in Agricultural Products and their Influence on Sino-American Bilateral Relations." University of Chicago, Office of Agricultural Economics Research, paper no. 82:6. Prepared for a conference sponsored jointly by the Brookings Institution and the Chinese Academy of Social Sciences in Beijing, 31 May–4 June.

*221. "Agriculture in the Centrally Planned Economies." *Amer. J. Agr. Econ.* 64 (Dec.): 845–53.

222. "Agricultural Research Policy in Small Developing Countries." *Managing Renewable Natural Resources in Developing Countries.* Ed. Charles W. Howe. Boulder, Colo.: Westview Press.

223. "International Capital Markets, Exchange Rates and Agricultural Trade." University of Chicago, Office of Agricultural Economics Research, paper no. 82:17, 25 Aug.

224. "The United States, the Soviet Union and the World Grain Economy." University of Chicago, Office of Agricultural Economics Research, paper no. 82:19, 31 Aug. Prepared for presentation at the Conference on Agriculture and Food Trade in the 1980s: The United States and Soviet Union, Grinnell College, 27 Sept.

225. "Food and Agriculture in the USSR." University of Chicago, Office of Agricultural Economics Research, paper no. 82: 25, 30 Nov. Presented at a joint session of the American Farm Economic Association and American Economic Association, New York, Dec.

1983

226. "International Trade and the Third World Situation." *Issues and Perspectives in Third World Development.* Ed. Thomas R. Ford and Carolyn Holmes. Center for Developmental Change, University of Kentucky, CDC paper no. 20, June, pp. 102–07.

227. With Karen Brooks. *Prospects for Soviet Agriculture in the 1980s.* Bloomington: Indiana University Press.

228. "The Current World Food Situation." Prepared for presentation at the Conference on The Role of Markets in the World Food Economy, 14–16 Oct. Minneapolis; *The Role of Markets in the World Food Economy.* Ed. D. Gale Johnson and G. Edward Schuh. Boulder, Colo.: Westview Press.

229. "Agriculture and U.S. Trade Policy." *Agriculture in the Twenty-First Century.* Ed. John W. Rosenblum. New York: John Wiley and Sons, pp. 297–309.

230. "Agriculture—Management and Performance." *Bulletin of the Atomic Scientists* 39 (Feb.): 16–22.

231. "Warren Nutter and the Study of Economic Growth in the USSR." Prepared for delivery at the Nineteenth International Seminar, CESES, Florence, 8–10 Sept.

232. "Trade Liberalization and Resource Adjustments in American Agriculture." University of Chicago, Office of Agricultural Economics Research, paper no. 83:19, 24 Aug. Prepared for presentation in the Faculty and Graduate-Student Seminar Series, Brigham Young University, Provo, Utah, 7 Oct.

233. "International Trade and the World Food Situation." University of Chicago, Office of Agricultural Economics Research, paper no. 83:20, 25 Aug.

234. "Rural Infrastructure and Agricultural Productivity in the Soviet Union." University of Chicago, Office of Agricultural Economics Research, paper no. 83:22, 28 Sept. Paper prepared for presentation at a session on the rural community in Russian economic history held in honor of Arcadius Kahan at the annual meeting of the American Association for the Advancement of Slavic Studies, 22 Oct., Kansas City.

235. "Future Food Imports of the Soviet Union and Other Centrally Planned Economies." University of Chicago, Office of Agricultural Economics Research, paper no. 83:23, 29 Sept. Paper prepared for presentation at the Sixtieth Annual Agricultural Outlook Conference, U.S. Department of Agriculture, 31 Oct.– 3 Nov., Washington, D.C.

236. "The Dilemma of Free Versus Regulated Markets: The Economic Dimensions of the Problems." University of Chicago, Office of Agricultural Economics Research, paper no. 83:29, 31 Oct. For presentation at London and Continental Bankers, Ltd., Agribusiness Symposium, 16–17 Nov., in London.

237. *The World Grain Economy and Climate Change to the Year 2000: Implications for Policy.* Washington, D.C.: National Defense University Press.

1984

238. "World Food and Agriculture," *The Resourceful Earth: A Response to Global 2000.* Ed. Julian L. Simon and Herman Kahn. Oxford: Basil Blackwell, pp. 67–112.

239. With Barbara Huddleston, Shlomo Reutlinger, and Alberto Valdés. *Fea-*

sibility of Financial Arrangements for Food Security. Baltimore: Johns Hopkins University Press.

240. "United States Agricultural Policy—National and International Contexts." University of Chicago, Office of Agricultural Economics Research, paper no. 84:3, 27 Feb.

241. "Cracks in the Iron Rice Bowl." *Across the Board* (Feb.)

242. "A World Food System: Actuality or Promise." Prepared for presentation at the Seventy-fifth Anniversary Colloquium on World Food Policy, Harvard Business School, 8–11 Apr.

243. "World Grain Trade Beyond 2000." University of Chicago, Office of Agricultural Economics Research, paper no. 84:17, 11 June. Prepared for presentation at the Centennial Forum, A Symposium to Celebrate the 100th Session of the International Wheat Council, 28–29 June, Ottawa, Canada.

244. "Domestic Agricultural Policy in an International Environment: Effects of Other Countries' Policies on the United States." *Amer. J. Agr. Econ.* 66 (Dec.): 735–44.

1985

245. "Alternatives to International Food Reserves." *World Food Security: Selected Themes and Issues.* FAO Economic and Social Development Paper, no. 53. Rome: FAO.

246. "The Performance of Past Policies: A Critique." *Alternative Agricultural and Food Policies and the 1985 Farm Bill.* Ed. Gordon C. Rausser and Kenneth C. Farrell. Berkeley: Giannini Foundation of Agricultural Economics, pp. 11–36.

247. "World Commodity Market Situation and Outlook." *U.S. Agricultural Policy: The 1985 Farm Legislation.* Ed. Bruce Gardner. Washington, D.C.: American Enterprise Institute, pp. 19–55.

248. "The World's Poor: Any Good News Lately?" Prepared for delivery at the 395th Convocation of the University of Chicago, Mar.

249. With Kenzo Hemmi and Pierre Lardinois. *Agricultural Policy and Trade: Adjusting Domestic Programs in an International Framework.* Task Force Report. New York: The Trilateral Commission.

250. "Rural Infrastructure and Agricultural Productivity in the Soviet Union." *Current History* 84 (Oct.).

1986

251. "A World Food System: Actuality or Promise?" *Perspectives in Biology and Medicine* 29 (Winter): 108–98.

252. "Agricultural Policies and World Trade: The U.S. and the European Community at Bay: Comment." *America and the World Economy.* Ed. Loukas Tsoukalis. New York: Blackwell, for the College of Europe, pp. 77–80.

253. "Agricultural Reforms in China: Achievements and Some Unfinished Business." The 1986 John Nuveen Lecture, presented 9 Apr. at the Divinity School of the University of Chicago; University of Chicago, Office of Agricultural Economics, paper no. 86:9.

254. "The Current Setting of U.S. Agriculture." Paper prepared for presenta-

tion at the Federal Reserve Bank, Chicago, 25 June; University of Chicago, Office of Agricultural Economics Research, paper no. 86:11, 20 May.

255. "The Economic and Policy Frameworks for Agriculture in the USSR and People's Republic of China." Paper presented at the International Farm Managers Association, Minneapolis, 30 June.

256. "Recent Developments in the Food Situation in Developing Countries: A Mixed Picture." Paper prepared for the Chicago Council on Foreign Relations for the Conference on Food Policy in Latin America, Chicago, 24 Oct.

257. "Domestic Agricultural Policies and World Grain Markets." Paper prepared for Conference on Grain Marketing, sponsored by Fundacion Cargill, Buenos Aires, 3 Nov.

258. "Policy Issues in Rainfed Agriculture." In *Development of Rainfed Agriculture Under Arid and Semiarid Conditions: Proceedings of the Sixth Agriculture Sector Symposium.* Ed. Ted J. Davis. Washington, D.C.: The World Bank.

259. "Reducing World Hunger: An Economist's View." *Global Hunger: A Look at the Problem and Potential Solution.* Ed. Malcolm H. Forbes and Lois J. Merrill. Evansville, Ind.: University of Evansville Press, pp. 105–38.

260. With Ronald D. Lee, as cochairman of Working Group on Population Growth and Economic Development. *Population Growth and Economic Development: Policy Questions.* Washington, D.C.: National Academy Press.

1987

261. "Constraints to Price Adjustments: Structural, Institutional and Financial Rigidities." University of Chicago, Office of Agricultural Economics Research, paper no. 87:7, 31 Mar. 1987. Presented at Nomisma International Conference on the Agro-Technological System Towards 2000, Bologna, 18–20 Sept.

262. "Agricultural Structural Policies." University of Chicago, Office of Agricultural Economics Research, paper no. 87:2, 7 Jan.

*263. "World Agriculture in Disarray Revisited." *Aus. J. Agr. Econ.* 31 (Aug.): 142–53.

264. "IMF Conditionality and Agriculture in the Developing Countries." *The Political Morality of the International Monetary Fund, Ethics and Foreign Policy.* Vol. 2. Ed. Robert J. Myers. New Brunswick, Conn.: Transaction Books, pp. 127–40.

265. "Is Population Growth the Dominant Force in Development." *Cato Journal* 7 (Spring-Summer): 187–93.

266. "Excess Capacity: The Evil of Modern Agricultural Policy." University of Chicago, Office of Agricultural Economics, paper no. 87:19, 28 July. Prepared for presentation at annual meeting of the American Agricultural Economics Association, East Lansing, Mich., 4 Aug.

267. "Making Efficient Use of the World's Agricultural Resources." Proceedings of the Sixteenth Pacific Science Congress, Seoul. Nov., pp. 154–63.

268. "Agricultural Policy Reforms in the USSR and the People's Republic of China." University of Chicago, Office of Agricultural Economics Research, paper no. 87:28, 2 Dec.

269. Editor, with Ronald D. Lee. *Population Growth and Economic Development: Issues and Evidence.* Madison: University of Wisconsin Press.

270. Editor. *Agricultural Reform Efforts in the United States and Japan.* New York: New York University Press, pp. viii, 1–4, 90.

271. "Policy Options and Liberalizing Trade in Agricultural Products: Addressing the Interests of Developing Countries." *Development with Trade: LDCs and the International Economy.* Ed. Anne O. Krueger. San Francisco: International Center for Economic Growth.

272. "The Soviet Union Today: Agriculture." *The Soviet Union Today: An Interpretive Guide.* 2d ed. Ed. James Cracraft. Chicago: University of Chicago Press, pp. 198–209.

273. "Policy Implications." *Soy Protein and National Food Policy.* Ed. F. H. Schwartz. Boulder, Colo.: Westview Press, pp. 1–10.

274. "U.S.-Canadian Agricultural Trade Challenges—Developing Common Approaches—Summing Up the U.S. Perspective." *U.S.-Canadian Agricultural Trade Challenges: Developing Common Approaches.* Proceedings of a Symposium. Ed. Kristen Allen. Washington, D.C.: Resources for the Future, pp. 189–92.

*275. "Economic Reforms in the People's Republic of China." *Econ. Devel. and Cult. Ch.* 30 supp. (Apr.): S225–46.

276. "Constraints on Price Adjustments: Structural, Institutional and Financial Rigidities." *The Agri-Technical System Towards 2000.* Ed. G. Antonelli and A. Quadrio-Curzio. North-Holland: Elsevier Science Publishers B.V., pp. 81–102.

277. "Paradoxes in World Agriculture." *Thinking About America: The United States in the 1990s.* Ed. Annelise Anderson and Dennis L. Bark. Stanford, Calif.: The Hoover Institution Press, 1988.

278. "Target Prices in the United States: A Reform That Failed." *Agriculture and Governments in an Interdependent World.* Proceedings of the Twentieth International Conference of Agricultural Economists, Buenos Aires, 24–31 Aug. Aldershot, England: Dartmouth Publishing Company, pp. 466–83.

279. "Trade Liberalization and Other Desirable Agricultural Policies." University of Chicago, Office of Agricultural Economics Research, paper no. 88:17, 28 Sept. Paper prepared for the International Seminar on Agricultural Policy, 26–28 Oct., São Paulo.

1989

280. "Guidelines for Agricultural Price Interventions by Governments." 25 Apr. University of Chicago, Office of Agricultural Economic Research.

281. "New Developments in USSR Agricultural Price Policy: Report of USDA Team on Agricultural Prices." University of Chicago, Office of Agricultural Economics Research, paper no. 89:12, 16 June.

282. "Soviet and Chinese Agriculture and Trade." University of Chicago, Office of Agricultural Economics Research, paper no. 89:14, 31 Aug. Paper prepared for presentation to the U.S. Congress, Joint Economic Committee, 7 Sept.

283. "Continuity and Change in the Beef Industry." 25 Sept. Paper prepared for presentation at the National Broiler Council Thirty-fifth Annual Conference, Washington, D.C., 13 Oct.

284. With John M. Connor, Timothy Josling, Andrew Schmitz, and G. Edward Schuh. *Competitive Issues in the Beef Sector: Can Beef Compete in the 1990s?* Hubert Humphrey Institute Report no. 1, Minneapolis, University of Minnesota, Oct.

285. "An Overview: Agricultural Productivity Growth." University of Chicago, Office of Agricultural Economics Research, paper no. 89:16, 16 Oct. Paper prepared for presentation at ERS Conference on Public Policy, Emerging Technologies in Agriculture and Agricultural Productivity Growth, Washington, D.C., 17–18 Oct.

1990
286. "Possible Impacts of Agricultural Trade Liberalization on the USSR." *Comp. Econ. Studies* 32 (Summer): pp. 144–54.

287. "Comparisons of Political, Economic and Institutional Conditions in Centrally Planned Economies." University of Chicago, Office of Agricultural Economics Research, paper no. 90:11, 31 May. Paper prepared for presentation at the Annual Meeting of the American Agricultural Economics Association, Vancouver, British Columbia, 6 Aug.

288. "Population, Food and Well-Being." University of Chicago, Office of Agricultural Economics Research, paper no. 90:13, 9 July.

289. "The People's Republic of China: 1978–1990." *International Center for Economic Growth, Country Studies Number* 8. San Francisco: Institute for Contemporary Studies Press.

1991
290. "European Community Integration and East European Agriculture," *Economic Systems* 15 (Oct.): 295–307.

291. *World Agriculture in Disarray.* Rev. ed. London: MacMillan.

292. "Agriculture in the Overall Liberalization Process." *Liberalization in the Process of Economic Development.* Ed. Lawrence Krause and Kim Kihwan. Berkeley: University of California Press.

1992
293. "The Relevance of the Wealth of Nations to the Transitions to Market Economies." University of Chicago, Office of Agricultural Economics Research, paper no. 92.2, 22 Apr. Presented at the Adam Smith Conference, Adam Smith Institute, Paris, 11–12 May.

294. "Economic vs. Noneconomic Factors in Chinese Rural Development." *Rural Development in Taiwan and Mainland China.* Ed. Peter Calkins, Wen S. Chern, and Francis C. Tuan. Boulder, Colo.: Westview Press, pp. 25–42.

1993
295. "Historical Experience of Eastern and Central European and Soviet Agriculture." *The Agricultural Transition in Central and Eastern Europe and the Former U.S.S.R.* Ed. Avishay Braverman, Karen M. Brooks, and Csabi Csaki. Washington, D.C.: The World Bank, pp. 11–26.

296. With Chi-Ming Hou. *Agricultural Policy and U.S.-Taiwan Trade.* Washington, D.C.: AEI Press.

297. "U.S. Agricultural Programs as Industrial Policy." *Industrial Policy for Agriculture in the Global Economy.* Ed. S. R. Johnson and S. A. Martin. Ames: Iowa State University Press, pp. 307–16.

*298. "Role of Agriculture in Economic Development Revisited." *Agr. Econ.* 8 (1993): 421–34.

299. "Trade Effects of Dismantling the Socialized Agriculture of the Former Soviet Union." *Comp. Econ. Studies* 35 (Winter): 21–31. Edited from "Il settore agro-alimenatare della ex Unione Sovietica: problemi del passaggio ad una economia di mercato." *Innovazione E Materie Prime*, no. 3 (1992): 14–29.

1994

300. "Does China Have A Grain Problem." *China Economic Review* 4.

301. "Policies and Institutions Affecting Rural Population Growth in China." *Population and Development Review* 20: 503–31. Earlier version published (in Chinese) in *Population Science of China* 5, no. 3: 19–33.

302. "The Limited But Essential Role of Government in Agriculture and Rural Life." University of Chicago, Office of Agricultural Economics, paper no. 94:04. The Elmhirst Lecture of the International Agricultural Economics Association, Harare, Zimbabwe, 22 Aug.

303. "Institutional and Policy Changes Affecting Chinese Agriculture." University of Chicago, Office of Agricultural Economic Research, paper no. 94:5, 18 July. Presented at Annual Meeting of American Agricultural Economics Association, San Diego, 8 Aug.

304. "Can There Be Too Much Human Capital? Is There a World Population Problem?" *Human Capital and Economic Development*. Ed. Sisay Asefa and Wei-Chiao Huang. Kalamazoo, Mich.: W. E. Upjohn Institute for Employment Research.